Electromagnetic Pulse Simulations Using Finite-Difference Time-Domain Method

Electromagnetic Pulse Simulations Using Finite-Difference Time-Domain Method

Shahid Ahmed
Senior Application Engineer,
ANSYS, Inc.

Home Address: 26611 Boulder Cove Ct,
Katy, Texas 77494,
USA

WILEY

Registered Office
John Wiley & Sons, Inc., 111 River Street, Hoboken, NJ 07030, USA

Editorial Office
John Wiley & Sons, Inc., 111 River Street, Hoboken, NJ 07030, USA

For details of our global editorial offices, customer services, and more information about Wiley products visit us at www.wiley.com.

Wiley also publishes its books in a variety of electronic formats and by print-on-demand. Some content that appears in standard print versions of this book may not be available in other formats.

Library of Congress Cataloging-in-Publication Data applied for

ISBN: 9781119526179

Cover design by Wiley
Cover image: © Anna Bliokh/Getty images

MIX
Paper from responsible sources
FSC
www.fsc.org
FSC® C013604

Dedicated to My
Wife Nigar Naushi
&
Daughters Aishah and Fatimah

Contents

Acknowledgments

I sincerely thank my wife Nigar Naushi for her encouragement and support in writing this book. Without the inspiration and sacrifice of my wife, this book would not have been completed. My daughters Aishah Shahid and Fatimah Shahid demonstrated enormous patience, love, and support by giving ample time for this accomplishment.

Preface

Various natural and artificial processes, such as lightning discharges and nuclear explosions, can produce a strong pulse of broad-band electromagnetic radiation, called an electromagnetic pulse (EMP). The large electric fields in such a pulse can cause damage to electronic and control equipment. A number of laboratories around the world have developed EMP simulators that can produce EMPs of different types, with the objective of testing the susceptibility of systems exposed to EMP. These are generally driven by a high-voltage pulsed-power source. "Bounded-wave" type simulators, with which the study in this book is concerned, concentrate energy within the workspace of the system itself.

This book presents a self-consistent, three-dimensional, finite-difference time-domain (FDTD) simulations for a bounded-wave EMP simulator, including a realistic geometry and test object. The simulator consists of a constant-impedance transverse electromagnetic (TEM) structure driven by a charged capacitor, discharging through a fast closing switch. These simulations yield the detailed 3-D electromagnetic fields within the TEM structure, in its immediate vicinity, as well as the radiated far-field. To the best of author's knowledge, this is the first time such a detailed study has been performed.

The prepulse seen in these simulations can be explained quantitatively in terms of capacitive coupling across the switch and the known charging waveform across the capacitor. To the best of author's knowledge, this is the first quantitative explanation of prepulse through 3D simulations.

Placement of a test object within the simulator significantly modifies the electric fields within the test volume, in terms of field strength as well as the frequency spectrum. This means that, for a given simulator, larger objects would be subjected to somewhat lower frequencies. Also, regardless of the size of the test object, the waveforms match only up to the first peak in the electric field. Multiple scattering off the simulator structure and test object produces features in the simulated waveform that are markedly different from the free-space case.

Radiation leakage out of the bounded volume can result in severe electromagnetic interference with surrounding equipment. Shorter closing times for the switch, which allow access to higher frequencies, have a concomitant cost of higher radiation leakage. The use of a matched sheet termination shows significant reduction of leakage, as compared to the use of two parallel rods. For a given termination, larger test objects marginally reduce leakage. These conclusions have also been explained in terms of the distribution of Poynting flux, and in terms of the current induced in the test object. Increasing the test volume by the use of longer parallel plates offers the advantage of reduced backflow of power toward the pulser. However, this comes at the cost of increased energy leakage. The methodology used in this analysis can be extended to study leakage from essentially arbitrary combinations of bounded-wave simulators and test objects.

We have analyzed the electromagnetic mode structure inside the simulator, by the application of the singular value decomposition (SVD) method to time-domain FDTD data. This combination of two powerful techniques yields a wealth of information about the internal mode structure which cannot otherwise be obtained. The TEM mode is dominant throughout the simulator length. The TM_1 mode dominates over TM_2 over most of the length. Close to the termination, the TM_2 mode becomes stronger. This has been explained in terms of the induced current distribution in the resistive sheet and the resulting magnetic field in the vicinity of the termination.

Placement of a perfectly conducting test object in the parallel-plate section produces major changes in the mode structure near the object. Higher order TM modes become stronger, at the cost of the $m = 0$ mode. This finding has considerable practical significance, since it implies that the object would be subjected to electromagnetic fields that deviate significantly from the desired $m = 0$ form. We have provided a physical explanation for the deviation in terms of induced currents on the object and the resulting electromagnetic fields in its vicinity.

The TEM-cell radiates mainly in the forward direction, with much weaker side lobes. We have calculated the temporal cross-correlation between the modes near the termination and the forward-radiated far-field. The principal TEM mode turns out to be poorly correlated with the far-field. The dominant TM modes, viz. TM_1 and TM_2, exhibit fairly strong correlation with the far-field. TM_1 exhibits long-time correlation with the far-field, while TM_2 only shows short-time correlation. This shows that, for a given mode amplitude, higherorder TM modes are more effective in producing radiation leakage as compared to the TEM mode. We have also studied the effect on mode structure of the angle between the simulator plates, and of the switching time. The use of an optimized spatial mode filter reduces the TM_1 mode, without significantly modifying the desired TEM mode. However, the TM_2 mode remains largely unchanged. These observations could suggest ways to improve simulator design from the point of view of electromagnetic interference with surrounding equipment and the purity of the TEM mode seen by the test object.

FDTD modeling of pulsed experiments is normally done with only the load set up inside the FDTD domain, the evolution of the pulsed energy source being modeled by some other method. In order to permit a truly self-consistent study, we have also set up the capacitor and switch within the FDTD domain. Apart from self-consistency, this would also allow inclusion of distributed parameter effects in energy storage capacitors. Such effects are important when capacitors are used to drive fast-risetime experiments. The study has identified critical numerical issues that must be taken into account in such work.

Katy, Tx *Shahid Ahmed*
29 August 2019

1

Electromagnetic Pulse

Various natural and artificial processes, such as lightning discharges and nuclear explosions, can produce a strong pulse of broad-band electromagnetic radiation called an electromagnetic pulse (EMP). EMP has been the subject of research since World War II, as Fermi anticipated the electromagnetic effects resulting from a nuclear explosion [1]. The large electric fields in such a pulse can cause damage to electronic and control equipment. The generation of EMP during nuclear tests was first observed in the 1950s, where it sometimes resulted in instrumentation failure [2]. EMP occurring in lightning discharges, and during fast switching of high-voltage circuits, is also known to cause damage to electrical and electronics systems. The experimental and theoretical study of different sources of EMP, and their effects on systems, is an active field of study around the world [2].

1.1 Sources of EMP

There are various natural and artificial sources of EMP. A common natural source is lightning. Artificial sources include high-voltage fast switches, power stations and distribution systems, nuclear explosions, ultra-wideband radar, etc. EMP generated by lightning is called lightning electromagnetic pulse (LEMP), while that due to nuclear explosions is called nuclear electromagnetic pulse (NEMP). More details are available in [3].

Figure 1.1 illustrates the basic mechanism of NEMP generation. A nuclear detonation releases a stream of energetic gamma-ray photons. This primary gamma, γ_p, produces Compton electrons e_c following a collision with free electrons available in the atmosphere. The current channel formed by the Compton electrons gives rise to a large $d\vec{J}/dt$, producing NEMP [1].

Figure 1.2 shows the temporal as well as spectral waveforms of LEMP and NEMP. This figure is adapted from [4]. The electromagnetic fields in a NEMP

Electromagnetic Pulse Simulations Using Finite-Difference Time-Domain Method, First Edition.
Shahid Ahmed.

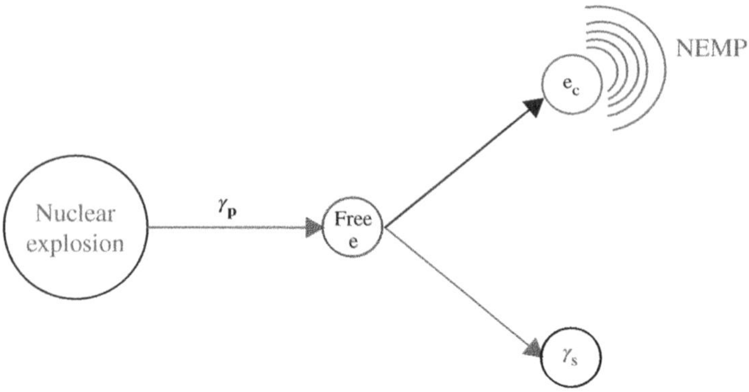

Figure 1.1 Schematic of basic mechanism for NEMP generation.

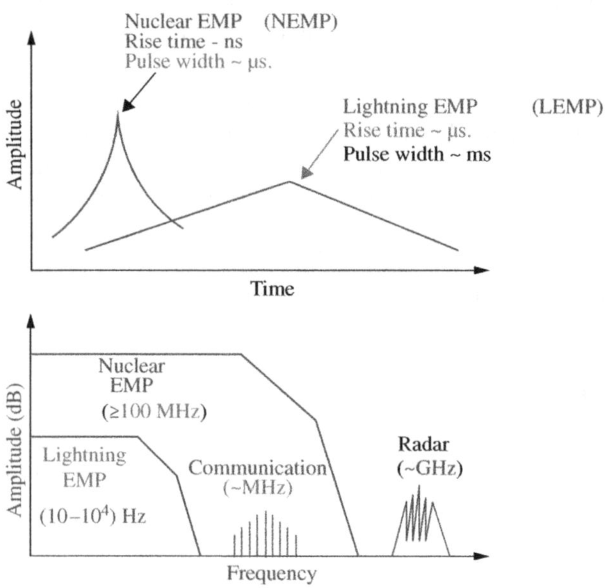

Figure 1.2 Temporal and spectral waveform of different kinds of EMP [4].

follow a double-exponential temporal waveform given by [5]:

$$E(t) = A(e^{-\beta t} - e^{-\alpha t})$$

where A, α and β are constants that govern the amplitude, inverse of rise and fall times, respectively. The rise-time and pulse-width of NEMP are of the order of nanoseconds and microseconds. For LEMP, these parameters are typically microsecond and ~millisecond, respectively. Both have an ultra-wideband nature.

1.2 EMP Coupling and its Effects

We have seen that EMP from different sources covers a broad range of the electromagnetic spectrum, with frequencies ranging from a few hertz to hundreds of megahertz. This corresponds to a wide range of free-space wavelengths. The longer wavelengths can couple to large objects such as overhead transmission lines, while small wavelengths couple to small objects such as control equipment and semiconductor devices. The coupling mechanism can be divided into two broad types, viz. "front door" and "back door" coupling. Front door coupling refers to energy that enters through the antennas of systems containing a receiver or transmitter. Back door coupling denotes energy that leaks into systems through apertures and seams in their enclosures [6].

The amount of front-door coupling depends upon the design frequency of the antenna and is maximum around its bandwidth. Back-door coupling through apertures and vents is maximum for wavelengths of the order of the aperture size and falls off steadily with increase in the wavelength. Figure 1.3 shows a schematic of front- and back- door coupling of EMP generated following a nuclear detonation, to electronic and electrical equipment [3]. EMP can enter the enclosure through overhead and underground transmission lines, telephone lines, windows, as well as utility ducts.

The high-intensity transient voltages and currents induced in electrical/electronic appliances can cause damage. The damage can be either temporary or permanent, depending upon the intensity of the incident pulse and the hardness of the exposed system [3].

1.3 EMP Simulators

A number of laboratories around the world have developed EMP simulators that can produce pulses of different types, with the objective of testing the susceptibility of systems exposed to EMP [7]. These are generally driven by a high-voltage pulsed-power source, e.g. a Marx capacitor bank. These simulators are used in two ways. The first is for assessing the effects of EMP on systems. The second is for testing the effectiveness of shielding ("hardening") of these systems.

Simulators can be divided into two broad categories: bounded-wave (closed) and radiate-wave (open). In a radiate-wave simulator, an ultra-wideband antenna, e.g. transverse electro-magnetic (TEM) horn, is used to radiate the electromagnetic field. This type of EMP simulator is used when systems to be tested are spread over a wide area [8]. Bounded-wave type simulators, with which this study is concerned, concentrate energy within the workspace of the system itself [9].

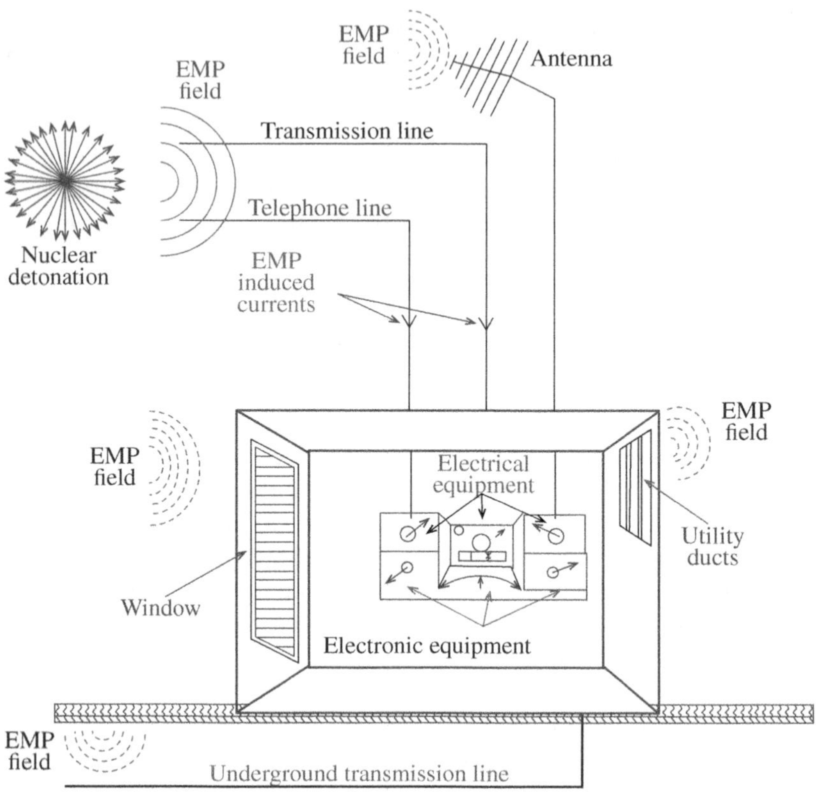

Figure 1.3 Schematic showing EMP coupling to electrical and electronics. This schematic is taken from [3]. (Source: Ghose [3]. 1984, Don White Consultants.)

A bounded-wave EMP simulator, in its simplest form, consists of two electrically conducting triangular plates, making up a TEM structure, separated by a parallel plate region [10]. This is illustrated schematically in Figure 1.4.

The front plate, which displays a near-constant impedance over a wide frequency range, plays a significant role in determining the EMP waveform, while the middle and rear plates serve to guide the signal [10]. The object to be tested is mounted in the bounded volume of the parallel-plate region.

There are several variants of the geometry shown in Figure 1.4. Some simulators do not have the rear plate, while others dispense with the parallel-plate section as well. The tapered section could also have some other shape, e.g. conical.

Figure 1.5 shows the setup of a bounded-wave EMP simulator, details of which have been reported in Ref. [9]. Only the tapered section of the simulator is shown here – the test section consists of a parallel-plate section, several meters in length.

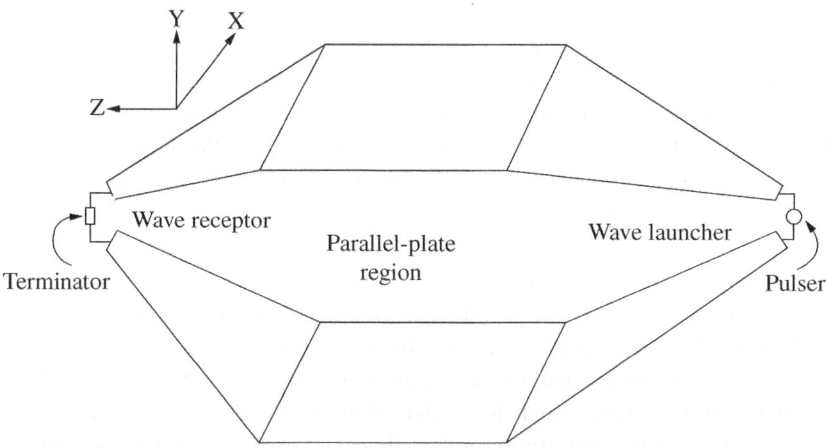

Figure 1.4 Schematic of parallel-plate transmission-line type of EMP simulator. (Source: Adapted from Giri et al. [11] .)

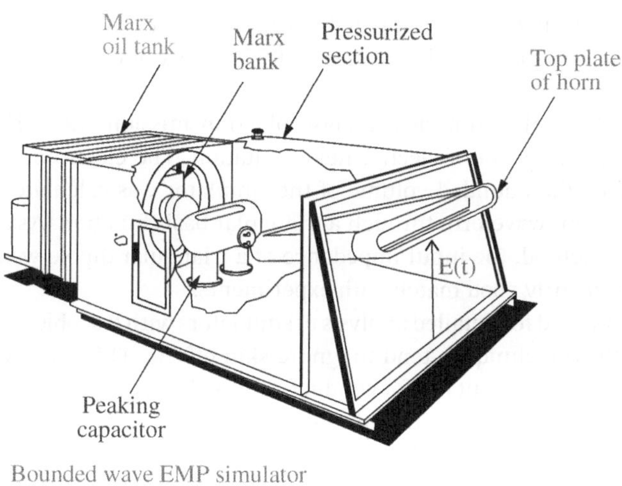

Figure 1.5 Experimental setup of a bounded-wave EMP simulator. (Source: Adapted from Schilling et al. [9].)

1.4 Review of Earlier Work

In this section, we examine earlier work in different areas relevant to modeling of EMP simulators.

We first consider earlier work involving overall analysis of bounded-wave simulator performance. Several time- and frequency-domain models have been reported. However, these analyses are based on several simplifying assumptions. For example, the conducting plates of a simulator have been approximated by wire grids or meshes. The current induced on the wires has been solved in the time-domain using a space–time-domain technique [12]. It has also been solved in the frequency domain using the method-of-moments (MOM) [13].

The transient electromagnetic field distribution inside a simulator has been studied through a space–time-domain technique [12]. The problem was formulated in terms of the radiation of a transient waveform from perfectly conducting wires, which involves the computation of the induced currents on the wires. This has been solved using the space–time-integral equation [14]. King and others [10, 15] have theoretically analyzed the transient behavior of a rhombic EMP simulator. In their simplified model, simulator plates are approximated by a hexagonal wire structure which is located along the edge of the metal plates. This approximation is based upon the fact that the largest current density in the parallel-plate simulator is found along the edge of the metal plates. Klaasen [16] has numerically analyzed the transient behavior of a bounded-wave simulator using a space–time-domain technique. The basic waveform of the current induced in the wire has been taken from [10]. As compared to King and workers [15], that study more accurately models the plates by increasing the number of wires in the calculation.

Hoo [13] has used MOM for the numerical analysis of a transmission line EMP simulator using a known waveform of electric field excitation. The electromagnetic field structure inside the bounded volume of the simulator was calculated by approximating the current waveform through a triangular basis function. As a check on the numerical method, the input impedance of a triangular dipole was calculated, which shows a fairly good match with experiments.

These methods are not suited for detailed analysis of simulators with test objects, for two reasons. First, these techniques tend to ignore skin effects. This may be acceptable for the simulator plates, but not for conducting test objects, particularly over the low-frequency portion of EMP. Second, MOM is not suited for handling objects with small apertures and internal cavities, where it becomes extremely demanding computationally. It also becomes computationally expensive for analyzing structures having dimensions considerably larger than a wavelength corresponding to the applied frequency.

King and others [17] have experimentally measured the electric field in the working volume of the simulator. However, their experimental measurements were only made at high frequencies, where the simulator no longer behaves like a terminated TEM mode transmission line.

In certain studies of EMP simulators, the excitation is assumed to follow a known form, e.g. a double exponential. Other workers have reported three-dimensional (3-D) finite-difference time-domain (FDTD) studies of horn antennas driven by specified excitation pulses [18, 19]. The excitation pulse is usually taken either as an idealized form like a Gaussian [18, 19], or an experimentally measured waveform [19]. To the best of author's knowledge, none of these studies have self-consistently evolved the pulser (e.g. capacitor bank) voltage and current along with the electromagnetic fields in the structure. Furthermore, since switching was not included in these simulations, none of the studies reports on the prepulse within the bounded volume.

We next consider earlier studies on the issue of radiation leakage from a bounded-wave simulator. Given the high electromagnetic (EM) power levels, it is important to minimize radiation leakage out of the bounded volume, which could cause electromagnetic interference (EMI) with surrounding equipment. Yang and Lee have reported that the parallel-plate extension plays an important role in the confinement of EM energy inside the bounded volume [20]. They have used the conformal mapping transformation method for the calculation of charge distribution over the plate and the field distribution in and around the plate region. Baum [21] has qualitatively shown that the EM energy leakage from a bounded-wave simulator can be reduced by using a smaller angle β between the tapered plates. Sancer and Varvatsis [22] have numerically determined the surface current density in the parallel-plate section and used that for estimating the EM energy radiated out of the working volume of the simulator. Their results support those of Baum [21]. King and Blejer [17] showed a safe frequency limit of operation for a bounded-wave simulator, which can efficiently guide only those waves for which the lowest wavelength is at least of the order of aperture height. Wavelengths shorter than the aperture height are radiated into the surrounding space. Their work is, however, limited only to single frequencies of excitation.

Matched resistive terminations are commonly used for reducing reflections in the case of transmission lines. Several alternatives to this option have also been examined in the literature some of which are summarized below.

Robertson and Morgan [23] have stated that the parallel-plate extension to a TEM horn structure plays an important role in reducing reflections, especially for low-frequency signals. Kanda [24] have resistively loaded a TEM horn from the apex to the aperture, such that the resistance varies continuously from a low value at the apex to a high value at the aperture. The functional form of this variation was similar to that proposed by Wu and King for reducing reflections from the open ends of a dipole antenna [25]. The small impedance at the feed was used for matching to the feeding transmission line and the high impedance at the aperture for matching to free-space. Baum [26] has proposed the use of an admittance sheet for terminating broadband signals. The resistive part effectively terminates lower frequencies, while the reactive part is useful at high frequencies. Hence, the termination is effective over a wide frequency range. The physical shape of the resistive termination is also important, as it significantly affects low-frequency performance [27].

The purpose of an EMP simulator is to provide an electromagnetic field that approximates the plane traveling wavefront produced at some distance by a nuclear explosion. However, the field in a transmission line simulator is quite different from that of a plane traveling wave, hence, it is very important to determine the differences between these two fields [12]. Most authors have treated this problem theoretically via the frequency domain, generally in terms of the TEM and higher parallel-plate modes [10]. Rushdi et al. [28] have reported that the field inside the simulator is a superposition of the TEM mode and higher-order transverse-electric (TE) and transverse-magnetic (TM) modes. Their work was an extension of earlier works, was confined only to the parallel-plate region, and based on a single frequency analysis. Giri et al. [11] have used the conformal mapping transformation method for the modal analysis of a bounded-wave EMP simulator. However, their analysis was limited to the TEM mode and did not address higher-order TE and TM modes. All these analyses are limited in their scope, either due to restrictions on the allowable geometry, or to single-frequency excitation, or only apply to a particular portion of the simulator. Most important, none of the analyses examines the modifications in simulator performance in the presence of the test object, which is crucial to a real-life simulator.

Rushdi et al. [28] and Giri et al. [29] have reported that higher-order TE and TM modes tend to cause energy leakage. However, these studies have not accurately quantified the relative contribution of different modes to radiation leakage. It is also desirable to find some way of suppressing these modes. Giri et al. [29] have suggested the use of electrical conductors called spatial mode filters (SMF) for suppressing certain modes. However, that analysis has been done under several simplifying assumptions. For example, the calculation of the angle of inclination of the SMF with respect to the plates of the horn is based upon the high-frequency assumption, i.e. frequencies for which the wavelengths are small as compared to the aperture height of the TEM horn.

Finally, let us consider the issue of setting up the entire pulsed-power driver for an EMP simulator within the FDTD domain. Different techniques have been reported in the literature for depositing charge within the FDTD domain. For example, Wagner and Schneider [30] have used a single-element filamentary current source at a particular grid location for studying grid capacitance effects. However, to the best of author's knowledge, none of these studies has systematically examined the conditions that must be satisfied by the numerical scheme for obtaining the proper capacitance.

The literature includes a large number of papers dealing with EMP-related experiments and modeling, as discussed in Section 1.4. However, we have found that the papers on modeling generally make simplifying assumptions that are not quite valid in a real-life EMP simulator. Also, there are some important issues that have not been addressed quantitatively. The major issues where work is required are summarized below.

There are several numerical and analytical analyses of TEM horn structures driven by specified voltage pulses. The input voltage waveform to the horn structure is normally taken either from experimental measurements, or as some analytical form like a Gaussian. None of these takes into account the self-consistent evolution of the pulser (e.g. capacitor bank) voltage and current along with the electromagnetic fields in the simulator structure. Such a self-consistent analysis is absolutely essential, if we are to take into account the effect of capacitor and switch design on simulator performance.

We recall that the simulator is supposed to subject a test object to a desired EMP environment. Now, the placement of a finite-sized test object inside the simulator itself modifies the electromagnetic fields to which it is exposed. This modification can be significant in terms of both the field strength and the frequency spectrum. It depends on the material composition as well as the size of the object. This must also be determined self-consistently. Such a treatment is not available in the existing literature.

An EMP simulator is meant to subject test objects to fields similar to those that would be encountered when the pulse comes through free-space [8]. This means that the wavefront incident on the test object should be planar, i.e. a TEM wave. In practice, however, wavefronts propagating in the test volume deviate from planarity due to the excitation of higher-order modes [29]. This deviation increases further with the placement of a test object. It is necessary to determine the extent to which the desired TEM mode is "contaminated" by higher-order modes, since it would affect the interpretation of data from the simulator. This requires an analysis of the mode structure inside the bounded volume. In the literature, techniques such as conformal mapping and Fourier spectral analysis have been employed for modal analysis of the parallel-plate section of a bounded-wave EMP simulator for discrete frequencies. However, these analyses do not discuss the mode structure for ultra-wideband excitation, which a bounded-wave simulator actually requires. Moreover, these analyses have been performed only in the limiting cases of plates that are very wide or very narrow in relation to the interplate separation [11]. Hence, these analyses fail when the plate widths and heights are comparable, which is often the case. A more general treatment, allowing arbitrary dimensions and shapes, is necessary.

Even the above modal analyses have been done only for the parallel-plate region, the tapered section generally being ignored. The structures of the transverse electric (TE) and transverse magnetic (TM) modes in the tapered section of the simulator have not been addressed [11]. Since the tapered section is the region where the mode structure evolves, it plays a key role.

The confinement of electromagnetic (EM) energy inside the test volume of a bounded wave simulator is an important issue. Leakage of EM energy out of the bounded volume is a waste of energy and leads to a reduction in the energy available for coupling with the test object. Furthermore, due to the high-power levels, it can produce severe EMI with surrounding equipment. Hence, it is necessary to accurately assess the amount of leakage, its angular distribution and its frequency spectrum. Earlier work in this area, discussed in Section 1.4, has focused on simple estimates of the leakage, or imposed limits on the operating frequency range, based on waveguide or antenna theory. However, to the best of author's knowledge, none of these involves a detailed assessment of the issues mentioned above.

Finally, only simple analyses are available of the link between the mode structure inside the simulator and radiation leakage. Some researchers have proposed the use of devices for suppressing the generation of higher-order modes. It is necessary to examine the effectiveness of such mode filters in realistic geometries. The studies presented in this book aim to address all these issues.

1.5 Overview of this Book

In this book, several issues that are important for the design and optimization of a bounded-wave EMP simulator have been examined. A brief summary of the work presented in different chapters of the book is given below.

As generation and propagation of an EMP in real time can be visualized in time domain. On the other hand, the coupling of an EMP with different systems can be assessed by analyzing the frequency spectrum. It is important to learn the time and frequency domain aspects of an EMP, which has been discussed in Chapter 2. The important details of the numerical scheme "FDTD" method from the perspectives of EMP simulations have been presented in Chapter 3. In Chapter 4, we develop our understanding about the behavior of an EMP propagation in free-space and material media using the FDTD computational method.

The self-consistent analysis of bounded wave simulators requires that we set up not just the TEM structure but also the capacitor and switch, within the FDTD domain. In Chapter 5, we report on a three-dimensional FDTD analysis of the charging and discharging of a parallel-plate capacitor. This study has helped identify the critical numerical issues that must be taken into account in any such study.

The self-consistent analysis of a bounded-wave EMP simulator involves first the charging of a capacitor, and then its discharge into a TEM structure through a fast closing switch. FDTD modeling of pulsed experiments is normally done with only the load, e.g. an antenna, set up inside the FDTD domain, the evolution of the pulsed energy source being modeled by some other method. The present study has shown one way to set up the pulse-power driver within the same domain.

In Chapter 6, we report on self-consistent, 3-D, time-domain calculations for a simulator, including a realistic geometry and test object. The simulations yields the detailed 3-D evolution of electromagnetic fields within the structure, within the test object and in its immediate vicinity. The modification of the field structure in the presence of the test object is also determined and explained. We also study the prepulse produced by capacitive coupling across the fast closing switch.

In an "ideal" NEMP simulator, the pulse incident on the test object should be a "pure" TEM wave with a near-planar wavefront, as in free-space propagation. In a bounded-wave simulator, however, scattering from different parts of the simulator structure, reflection from the termination, and the finite size of the structure, can lead to the excitation of higher-order transverse electric (TE) and transverse magnetic (TM) modes. Furthermore, the presence of the test object itself modifies the fields to which it is subjected. Interpretation of the test results from an EMP simulator should thus take into account the strength of higher-order modes. The degree to which the TEM mode is "contaminated" by high-order modes depends strongly upon the design of the simulator, the frequency spectrum of interest, and the geometry and material composition of the test object. Given the number of variables, no simple estimate is possible. Hence, a detailed analysis becomes necessary. Chapter 7 involves the study of the electromagnetic mode structure inside a bounded-wave EMP simulator, both with and without a test object. This has been done by the application of the singular value decomposition (SVD) method to time-domain data generated by FDTD simulations. This combination of two powerful techniques yields a wealth of information about the internal mode structure which cannot be otherwise obtained.

Given the high electromagnetic (EM) power levels in a simulator, it is important to minimize radiation leakage out of the bounded volume. Leakage is undesirable for two reasons. First, it results in EMI with surrounding equipment. Second, it reduces the amount of energy available for coupling to the test object. A self-consistent, 3-D, time-domain study for calculation of radiation leakage from such a simulator is performed in Chapter 8. We have examined the sensitivity of leakage to design parameters such as the simulator geometry, the type of termination, and the closure time of the switch. We have also determined the sensitivity to the size of the test object. Physical explanations for some of the observations are also reported in that chapter.

Apart from affecting the purity of the TEM mode, higher-order modes can also enhance radiation leakage from the simulator. It is important to determine the effectiveness of different modes in radiating electromagnetic energy. The modal perspective of radiation leakage from a bounded wave simulator has been discussed in Chapter 9. It is also desirable to find some way of suppressing these modes. Certain mode suppression devices have been suggested in the literature, and their utility needs to be examined in a self-consistent analysis involving real-life geometry. The effectiveness of a mode suppression technique using the full-wave 3-D FDTD method has been analyzed in Chapter 10. An application of a bounded-wave EMP simulator in studying the EMP interaction with a human holding a cell-phone where the impact on biological tissues is an important concern has been illustrated in Chapter 11. Finally, Chapter 12 illustrates a full-wave 3-D FDTD computer program for simulating the EMP environment within the volume of the structure as well as in the far-field.

1.6 Summary

This chapter provides an overview of the subject area of this book. The basic mechanism of EMP generation by natural and artificial processes have been introduced. The inherent nature of EMP being broadband makes it easy to couple with different systems, and strong field associated may cause temporary or permanent failure. An overview of different kinds of EMP simulators and the research progress around the world have been surveyed. Important contributions of this book have been summarized and the contents of the different chapters have been highlighted.

2

Time and Frequency Domain Analysis

2.1 Introduction

The generation and propagation of an electromagnetic pulse (EMP) in real time, can be visualized in time domain because of its transient nature. A time-domain waveform is characterized by its peak value, the rise-time, and the duration of the pulse. However, the coupling of an EMP with different systems can be assessed by analyzing the frequency spectrum and, hence, can be very well understood in the frequency-domain through the lowest frequency, the highest frequency, and the energy content. This is why it is important to learn the time and frequency domain aspects of an EMP.

The simulation of electromagnetic environment generated by natural and artificial phenomena such as lightening, electrostatic discharges, nuclear explosion, etc., have been the subject of interest worldwide. The high strength of electromagnetic field associated with such phenomena can cause a temporary shutdown to our routinely used systems (computers, tablets, phones, and TV) or can be responsible for a national-level threat. Therefore, designing a system to simulate the electromagnetic environment posed by such phenomena is important, however, this is complex and challenging because it requires the knowledge of characteristic features of such events. For many years of significant research in these areas around the world has led the scientists and engineers to choose well-defined characteristic waveforms validated by measurements that can simulate the expected environment developed without initiating the undesirable events such as nuclear explosion.

In this chapter, we will learn the mathematical waveforms representing the nuclear electromagnetic pulse (NEMP) for simulating the electromagnetic environment in the event of nuclear explosion.

Electromagnetic Pulse Simulations Using Finite-Difference Time-Domain Method, First Edition.
Shahid Ahmed.

2.2 Nuclear Electromagnetic Pulse

An electromagnetic environment generated in the event of high-altitude nuclear explosion is called NEMP. Here we will discuss the two most commonly used waveforms of a high-altitude electromagnetic pulse (HEMP).

2.2.1 Differences of Two Exponentials Times in a Unit Step Function

The mathematical waveform of a transient electric field is represented [31]:

$$E(t) = E_0 \left[e^{-\beta t} - e^{-\alpha t}\right] u(t) \tag{2.1}$$

where,

E_0 = field intensity constant (V/m)
α = risetime constant (rad/s)
β = decay constant (rad/s)
$u(t)$ = unit step function.

A typical time-domain waveform of the kind shown in Eq. (2.1) having parameters $E_0 = 50$ kV/m, $\alpha = 5.0 \times 10^8$ rad/s, $\beta = 4.0 \times 10^6$ rad/s is depicted in Figure 2.1.

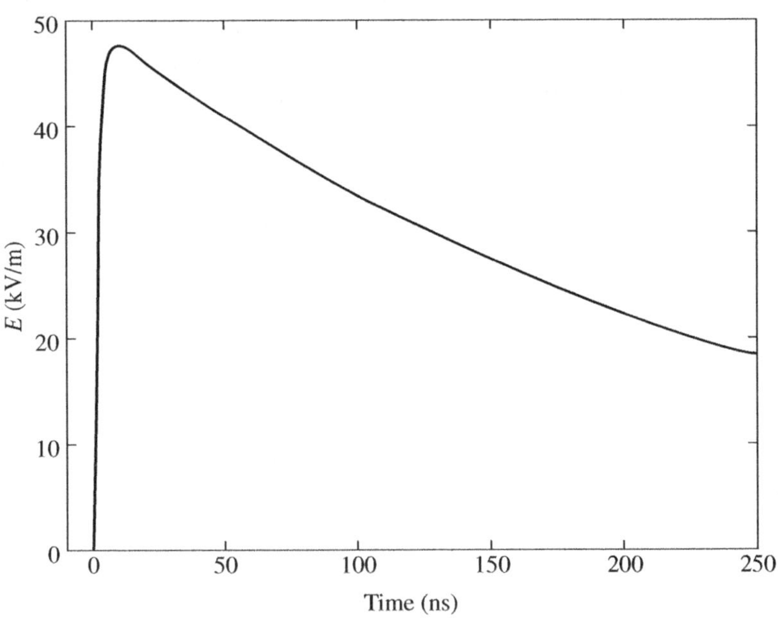

Figure 2.1 The graph shows waveform function (2.1) for the parameter set $E_0 = 50$ kV/m, $\alpha = 5.0 \times 10^8$ rad/s and $\beta = 4.0 \times 10^6$ rad/s.

As we see the pulse starts at zero with a finite slope at $t = 0$, rises to a peak value of 47.7 kV/m (slightly less than the chosen field intensity $E_0 = 50$ kV/m), and smoothly decays. This type of waveform is observed in a parallel-plate EMP simulator driven by a capacitive pulse generator and terminated by a perfectly matched frequency independent resistive load. A detailed study of such an EMP simulator together with the capacitor bank, opening/closing switches, and resistive termination can be learned through the details discussed in Chapter 6.

2.2.1.1 Time-Domain

The characteristics of a NEMP waveform in time-domain can be understood through the following parameters:

- *Rise time*: The risetime is a measure how fast an EMP rises from its low state (typically 10% of peak value) to the high state (90% of the peak value). This is called (10–90)% risetime, $t_{r(10-90)}$, which is most commonly used as a measure of the risetime of a pulse unless stated explicitly. The characteristic time during the rise of the waveform can be defined as the ratio of the maximum value to the maximum time derivative:

$$t_r = \frac{E(t)|_{max}}{\frac{d}{dt}E(t)|_{max}} = \frac{1}{\alpha - \beta} \tag{2.2}$$

  For the case of a risetime, much faster than the delay time ($\alpha \gg \beta \geq 0$), the maximum risetime $t_{mr} = \frac{1}{\alpha}$, which is e-folding time of the waveform (2.1). This maximum risetime provides the fastest time event, however, the commonly used risetime definition is $t_{r(10-90)} \simeq 2.2/\alpha$. For example, $\alpha = 5.0 \times 10^8$ rad/s, the maximum risetime $t_{mr} = 2$ ns and $t_{r(10-90)} = 4.4$ ns.
- *Pulse duration*: The pulse duration provides an estimate of how long an EMP lasts, which can be estimated by computing the complete time integral. For the waveform (2.1), the pulse duration (τ) can be obtained as

$$\tau = \frac{1}{E_0}\int_{-\infty}^{\infty} E(t)dt = \frac{\alpha - \beta}{\alpha\beta} = \frac{1}{\beta} \text{if } \alpha \gg \beta \geq 0 \tag{2.3}$$

  Therefore, for the case where $\alpha = 5.0 \times 10^8$ rad/s and $\beta = 4.0 \times 10^6$ rad/s, $\tau \simeq$ 250 ns.
- *Peak characteristics*: The time derivative of the waveform (2.1) becomes zero at the peak. Therefore, the time at which the waveform peaks, can be calculated by equating the first derivative to zero, that gives $t_{max} = \frac{1}{\alpha-\beta} \ln \frac{\alpha}{\beta} = \simeq 9.7$ ns.

2.2.1.2 Frequency-Domain

The spectral content of the pulsed waveform illustrated in (2.1) can be obtained by performing the Laplace transform, which is represented in the frequency domain as follows:

$$\tilde{E}(\omega) = E_0 \left[\frac{1}{j\omega + \beta} - \frac{1}{j\omega + \alpha} \right] \tag{2.4}$$

The magnitude and phase of the spectrum can be obtained from the following:

$$|\tilde{E}(\omega)| = E_0 \frac{\alpha - \beta}{[(\alpha^2 + \omega^2)(\beta^2 + \omega^2)]^{1/2}} \tag{2.5}$$

$$\varphi = \arg(\tilde{E}(\omega)) = -\arctan\left(\frac{\omega}{\alpha}\right) - \arctan\left(\frac{\omega}{\beta}\right) \tag{2.6}$$

The graphical waveform of the magnitude of the electric field in the log–log scale is depicted below.

Let us discuss the characteristic of different frequency regime.

1) *Low frequency*: In the low-frequency limit $\omega \simeq 0$, the magnitude and phase are defined as $|\tilde{E}(0)| = E_0 \frac{\alpha - \beta}{\alpha\beta}$ and phase $\varphi = 0$. So, in the low frequency limit, $|\tilde{E}(\omega)|$ is almost constant and hence can be considered as independent of the frequency. This frequency range can be considered dc to 1 MHz, which is very clear from Figure 2.2. The corresponding phase change 0° to −50° is shown in Figure 2.3.
2) *Intermediate frequency characteristics*: The intermediate frequency range can be attributed as $\alpha \gg \omega \gg \beta$, which transforms the magnitude and phase as

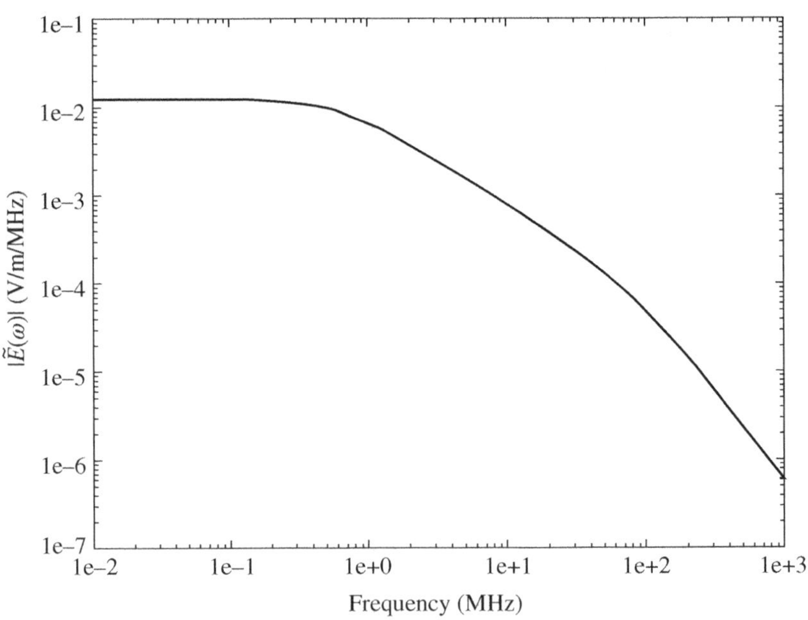

Figure 2.2 Log–log plot: frequency spectrum of the waveform (2.1).

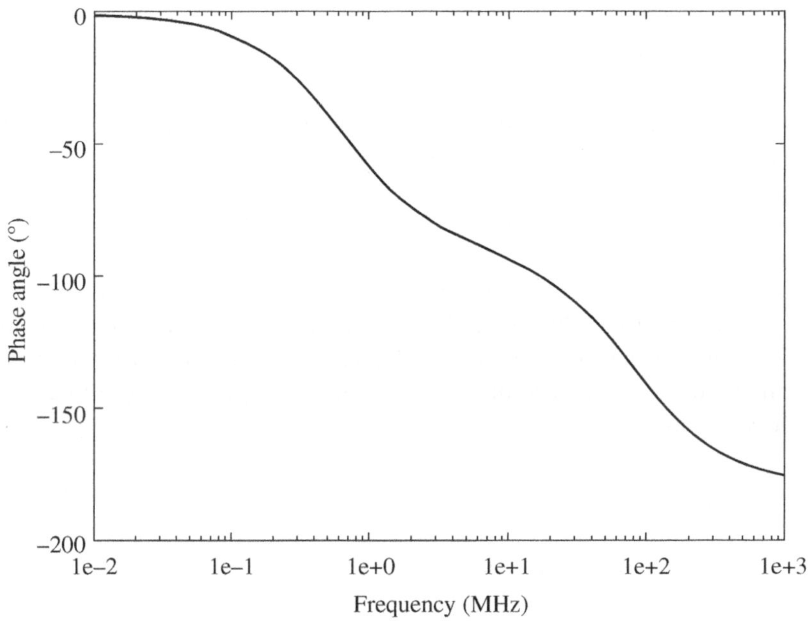

Figure 2.3 Semilog plot (x-axis is in log-scale): variation of phase angle with frequency.

$|\tilde{E}(\omega)| = E_0/\omega$, $\varphi = -\arctan(\frac{\omega}{\alpha})$. The transition between the low and intermediate frequency occurs at $\omega = \beta$, that is indeed seen in Figure 2.2 by the linear roll-off of $|\tilde{E}(\omega)|$ as the frequency increases. This transition can also be seen by the kink around 1.2 MHz ($\beta/2\pi$); the corresponding phase angle $-90°$. We do see another kink around 100 MHz, which attributes the high-frequency transition as discussed below.

3) *High frequency characteristics*: In the high-frequency range $\omega \gg \alpha \gg \beta$, the magnitude and phase can be written as $|\tilde{E}(\omega)| = E_0\alpha/\omega^2$, $\varphi = -\pi$. In this high-frequency regime, we see that the magnitude is inversely proportional to the ω^2, $|\tilde{E}(\omega)|$ falls more steeply as compared to the intermediate frequency characteristic. We do see the high-frequency transition kink around 100 MHz, which is pretty close to the inverse of risetime constant $\alpha/2\pi$. This suggests that the high- frequency content of an EMP can be controlled by controlling the pulse risetime.

2.2.2 Reciprocal of the Sum of Two Exponentials

The time domain waveform of an EMP represented by the reciprocal of the sum of two exponentials is illustrated in the following:

$$E(t) = \frac{E_0}{\left[e^{-\alpha(t-\tau)} + e^{\beta(t-\tau)}\right]} \quad \text{V/m} \tag{2.7}$$

where τ is the reference time or the time-shift and the other symbols have their usual meaning. It is important to note that $\alpha \gg \beta$, and hence α dictates the rise-time and β determines the decay time of the pulse. The pulsed waveform represented in Eq. (2.7) corrects the discontinuity that is inherent with the earlier waveform shown in Eq. (2.1). The risetime represented by the waveform (2.7) is consistent with the rising nature of the electromagnetic radiation in the process of nuclear effects. The graph depicted in Figure 2.4 shows the temporal evolution of the waveform represented in Eq. (2.7). Clearly, the waveform rises exponentially in the early stage of evolution and develops smoothly, therefore, avoids discontinuities.

2.2.2.1 Time-Domain Characteristics

Here we will learn the features of time domain characteristics through the rise-time, decay time, and the magnitude.

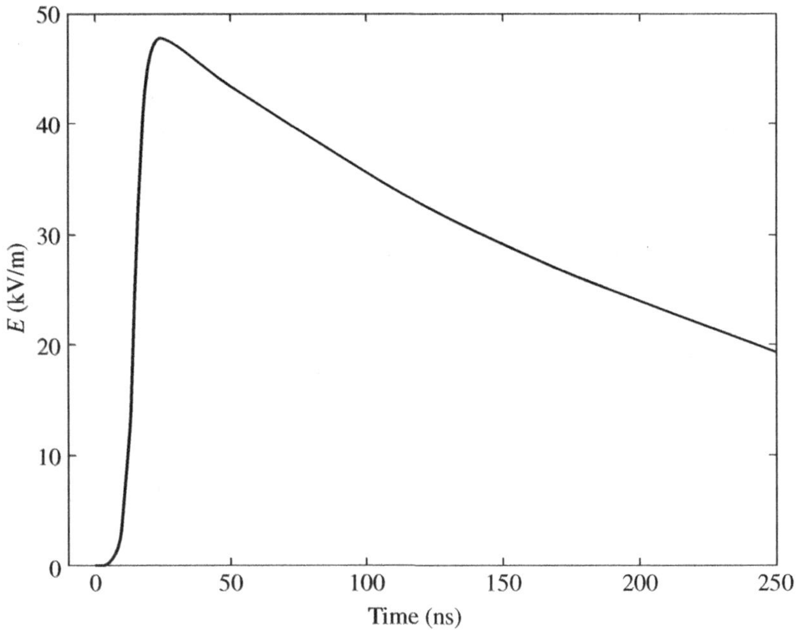

Figure 2.4 The graph shows time-domain waveform (2.7) for the parameter set $E_0 = 50$ kV/m, $\alpha = 5.0 \times 10^8$ rad/s and $\beta = 4.0 \times 10^6$ rad/s, $\tau = 15$ ns.

- *Risetime characteristics*: For determining the dependence of early time characteristic, we consider $\alpha \gg \beta$ and $\beta \to 0$, Eq. (2.7) turns out

$$E(t) = \frac{E_0}{e^{-\alpha(t-\tau)} + 1} \tag{2.8}$$

The characteristic risetime that measures the e-folding time is $t_r = 1/\alpha$, and the 10–90% risetime is given by $t_{r(10-90)} \simeq 4.4/\alpha$. At $t = \tau$, $E(t) = E_0/2$ which is considered 50%-point because the amplitude falls to half of the peak value E_0. The maximum risetime t_{mr} at $t = \tau$ is given by

$$t_{mr} = \frac{E_0}{\frac{d}{dt}E(t)} = \frac{4}{\alpha} \tag{2.9}$$

- *Peak characteristics*: The EMP waveform (2.7) peaks at the time when its time derivative becomes zero. The time t_{max} at which the peak occurs can be obtained by equating the time derivative of Eq. (2.7) to zero, which gives

$$t_{max} - \tau = \frac{1}{\alpha + \beta} \ln\left(\frac{\alpha}{\beta}\right) \tag{2.10}$$

Substituting the value of $t_{max} - \tau$ in Eq. (2.7) gives the characteristic peak value E_{pk}

$$E_{pk} = \frac{E_0}{\left[e^{-\alpha(t_{max}-\tau)} + e^{\beta(t_{max}-\tau)}\right]} \quad \text{V/m} \tag{2.11}$$

To have an estimate for the parameter set of our interest $E_0 = 50$ kV/m, $\alpha = 5.0 \times 10^8$ rad/s and $\beta = 4.0 \times 10^6$ rad/s, the characteristic peak value $E_{pk} =$ 47.738 kV/m, which is, slightly less than E_0 (see Figure 2.4).

- *Pulse duration*: The pulse duration (τ_{pd}) provides an estimate how long an EMP lasts, which can be determined by the complete time integral of the waveform function (2.7)

$$\tau_{pd} = \frac{1}{E_0} \int_{-\infty}^{\infty} E(t)dt = \frac{\pi}{\alpha + \beta}\left[\sin\left(\frac{\pi\beta}{\alpha + \beta}\right)\right]^{-1} \quad \text{if } \alpha \gg \beta \geq 0 \tag{2.12}$$

Therefore, for the case when $\alpha = 5.0 \times 10^8$ rad/s and $\beta = 4.0 \times 10^6$ rad/s, $\tau \simeq$ 250 ns.

2.2.2.2 Frequency-Domain

The spectral content of the EMP waveform illustrated in (2.7) can be obtained by the Laplace transform as

$$\tilde{E}(\omega) = \frac{E_0 \pi}{\alpha + \beta} \csc\left(\pi \frac{j\omega + \beta}{\alpha + \beta}\right) e^{-j\omega\tau} \tag{2.13}$$

The magnitude and phase of the spectrum can be obtained from the following:

$$|\tilde{E}(\omega)| = E_0 \frac{\pi}{\alpha + \beta}\left[\sinh^2\left(\frac{\pi\omega}{\alpha + \beta}\right) + \sin^2\left(\frac{\pi\beta}{\alpha + \beta}\right)\right]^{-1/2} \tag{2.14}$$

$$\varphi = \arg(\tilde{E}(\omega)) = -\omega\tau - \arctan\left[\tanh\left(\frac{\pi\omega}{\alpha + \beta}\right)\cot\left(\frac{\pi\beta}{\alpha + \beta}\right)\right] \tag{2.15}$$

- *Low-frequency characteristics*: In the low-frequency limit $\omega \to 0$, Eq. (2.13) turns to

$$\tilde{E}(\omega) = \frac{E_0\pi}{\alpha + \beta}\csc\left(\frac{\pi\beta}{\alpha + \beta}\right) \tag{2.16}$$

- *Intermediate frequency characteristics*: In this frequency regime $\alpha \gg \omega \gg \beta$, which, in turn simplifies Eq. (2.13) as

$$\tilde{E}(\omega) = E_0 \frac{e^{-j\omega\tau}}{j\omega} \tag{2.17}$$

The magnitude $|\tilde{E}(\omega)| = E_0/\omega$ falls linearly with the frequency. The transition between the low frequency and intermediate frequency occurs at $\omega_{tr} = \beta$. The transition frequency (ω_{tr}) is the reciprocal of the decay time constant β.

- *High-frequency characteristics*: In order to analyze the high-frequency response of an EMP, we can express (2.13) as

$$\tilde{E}(\omega) = E_0 e^{-j\omega\tau}\frac{2\pi j}{\alpha + \beta}\left[e^{j\pi\frac{j\omega+\beta}{\alpha+\beta}} - e^{-j\pi\frac{j\omega+\beta}{\alpha+\beta}}\right]^{-1} \tag{2.18}$$

In the high-frequency limit $\omega \to \infty$, we can write

$$|\tilde{E}(\omega)| = E_0 \frac{2\pi}{\alpha + \beta} e^{-\frac{\pi\omega}{\alpha+\beta}} \tag{2.19}$$

It is important to note that the spectral content of the EMP waveform falls off exponentially with the frequency, which is correlated with the smooth rise nature of the early risetime of EMP waveform (2.7). In contrast, the high-frequency spectral content of the EMP waveform (2.1) falls off $1/\omega^2$ thus presents discontinuities through the generation of higher frequencies in its early risetime. The transition between the intermediate and high frequencies occurs at a corner frequency $\omega_H \propto \alpha$, where the proportionality constant varies how we define the corner frequency. The important point to note here is that the higher-corner frequency is dictated by the risetime constant.

A graph showing the comparative analysis of frequency response of the two waveforms (2.1) and (2.7), respectively, representing the exponentials times a unit step function (DEXP) and the reciprocal of the sum of two exponentials (QDEXP) is depicted in Figure 2.5. It is very clear that the "QDEXP" waveform falls off smoothly in contrast to its counterpart "DEXP" as we go to high frequency, which, in turn, represents the smooth nature of the early rise in the time domain waveforms (see time-domain comparison in Figure 2.6).

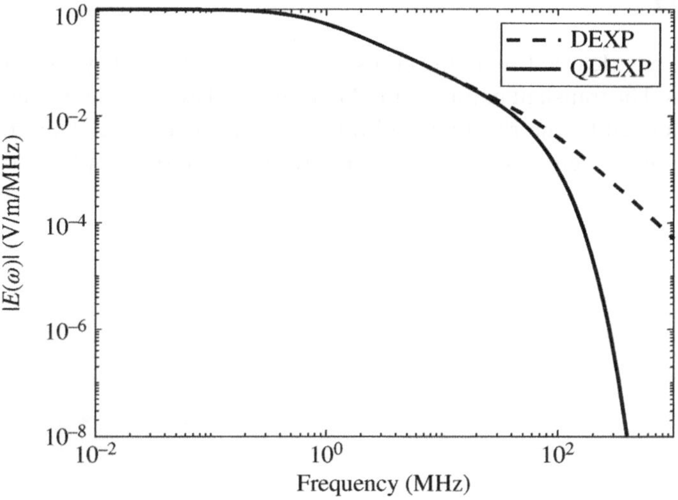

Figure 2.5 Log–log plot: comparison between the differences of two exponentials times a unit step function (DEXP) and the reciprocal of the sum of two exponentials (QDEXP).

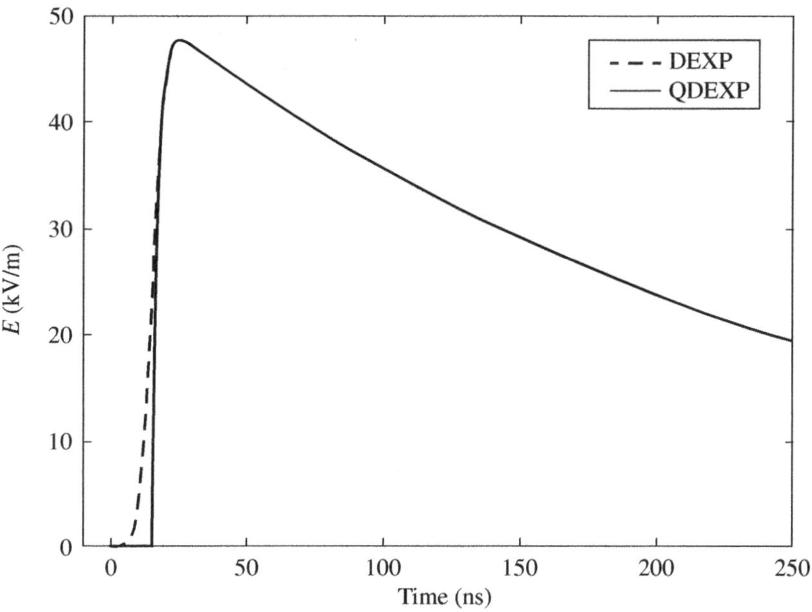

Figure 2.6 Comparison between the differences of two exponentials times a unit step function (DEXP) and the reciprocal of the sum of two exponentials (QDEXP).

2.3 Summary

In this chapter, we have learned about the necessity of time and frequency domain in an EMP analysis. The transient nature of an EMP is revealed by the time-domain analysis – characterized by the rise-time, fall-time, and the peak value. The frequency domain analysis helps us to understand how does an EMP couples with different systems.

3

Simulations Using FDTD Method

3.1 Introduction

This chapter will provide an overview of the finite-difference time-domain (FDTD) method adapted in this book. A more detailed discussion can be found in [32] and [33]. The FDTD method is a versatile and robust numerical technique. It has been most commonly used in the field of electrical engineering technologies, ranging from defense applications to communication systems. All these problems require the numerical solution of electromagnetic fields, governed by the Maxwell's equations because of the complexity of the problem. The FDTD method is a direct time-domain approach for solving Maxwell's differential (curl) equations on spatial grids or lattices, based on a technique introduced by Yee [34]. The rapid growth in computer hardware (disk space, RAM, and CPU) technologies over the past decade has led to explosive growth in the number of problems to which FDTD is being applied [33].

FDTD offers a variety of advantages. The first is its ability to handle complex, multimaterial geometries. The second is a natural treatment of ultra-wideband (e.g. impulse) problems. The third is the ability to handle nonlinearities in materials. The fourth is the explicit computational scheme, which makes the computer codes relatively simple. The explicit scheme also makes parallelization easier, an important consideration for handling large computational domains. It is important to note that the explicit scheme restricts the time-step size, which means the scheme is not efficient for handling long time-scale problems. However, variants of the scheme that allow larger time-steps are available, but a discussion of their features lies beyond the scope of this book.

Electromagnetic Pulse Simulations Using Finite-Difference Time-Domain Method, First Edition.
Shahid Ahmed.

3.2 Need for FDTD Analysis of an EMP Simulator

The high cost of setting up electromagnetic pulse (EMP) simulators means that fabrication must be preceded by numerical optimization. Also, it sometimes becomes necessary to use an existing transverse electro-magnetic (TEM) structure with a different pulse-power driver, e.g. one having a shorter risetime than the original driver for which the structure was designed. In such cases, numerical simulation is necessary to assess overall simulator performance.

The general behavior of a simulator is complex due to the existence of a wide-band pulse and complex geometry of the simulator and test object. It can be further complicated by the presence of different materials, both dielectrics and conductors in the test object. The combination of the TEM structure, capacitor, switch, and test object is a complex three-dimensional (3-D) object. The input impedance $Z(\omega)$ of the TEM structure is a function of frequency. $\vec{E}(\omega)$ inside the structure depends upon the input current waveform $I(t)$, which in turn depends upon $Z(\omega)$. This means that the electric field distribution within the structure and $I(t)$, must be determined self-consistently. Hence, there is a need for a self-consistent, 3-D analysis.

An important output required of such an analysis is the electromagnetic field variation, $\vec{E}(\vec{r}, t)$ and $\vec{B}(\vec{r}, t)$, in the vicinity of the test object. Now, the object could itself be complex, consisting of various materials, having internal cavities, small apertures, and so on. Placement of the object within the test volume would, therefore, modify the fields to which it is subjected. The fields are further affected by scattering of waves off the object as well as the simulator structure. This means that details of the simulator, such as launcher angles, finite plate widths, earth conductivity, etc., would also play a role [8].

Several time- and frequency-domain models have been reported for the analysis of EMP simulators. These analyses are based on several simplifying assumptions. For example, the conducting plates of a simulator have been approximated by wire grids or meshes. The current induced on the wires is solved in the time- or frequency-domain using a space–time-domain technique [12] or the method-of-moments (MOM) [13]. The transient electromagnetic field distribution inside a simulator has been studied through a space–time-domain technique [12]. MOM has been used to analyze a high-frequency band of the EMP spectrum [13]. These methods are not suited for detailed analysis of simulators with test objects, for two reasons. First, these techniques tend to ignore skin effects. This may be acceptable for the simulator plates, but not for conducting test objects, particularly over the low-frequency portion of EMP. Second, MOM is not suited for handling objects with small apertures and internal cavities.

The FDTD method is a powerful tool for analyzing problems involving 3-D objects with complex geometries and multiple materials [32]. Hence, it is ideally suited for analysis of EMP simulators and is used in this book.

3.2.1 Choice of Method for Self-consistent Analysis

In certain studies of EMP simulators, the excitation is assumed to follow a known form, e.g. a double exponential. Other workers have reported 3-D FDTD studies of horn antennas driven by specified excitation pulses [18, 19]. The excitation pulse is usually taken either as an idealized form like a Gaussian [18, 19], or an experimentally measured waveform [19]. To author's knowledge, none of these studies have self-consistently evolved the pulser (e.g. capacitor bank) voltage along with the electromagnetic fields in the structure.

A self-consistent analysis could be done in two ways. The first would be to model the simulator structure through the 3-D FDTD equations [32] and treat the capacitor and switch as lumped elements whose evolution is represented by ordinary differential equations. The two sets of equations would be coupled at the feed point of the TEM structure by matching the voltage and current. This would require an iterative solution at each FDTD time-step, which is computationally expensive. The second method would be to set up the capacitor and switch, along with the simulator structure, in the FDTD grid itself. This would, naturally, require idealizations to be made in the geometry of the capacitor and switch. However, assuming that the geometry is reasonably well represented, this method offers the advantage of taking into account distributed parameter effects within the capacitor and switch, which could be significant for short time-scale phenomena. The addition of the switch and capacitor marginally increases the size of the FDTD domain, with a resulting increase in the computational effort per time-step. However, this is more than compensated by avoiding the iterations, which are inherent in the first method. We have, therefore, opted for the second approach.

3.3 Maxwell's Equations and the Yee Algorithm

The electromagnetic fields ($\vec{E}(\vec{r}, t)$ and $\vec{B}(\vec{r}, t)$) inside an EMP simulator can be obtained by the self-consistent solution of the Maxwell's equations:

$$\nabla \times E = -\frac{\partial B}{\partial t} \tag{3.1}$$

$$\nabla \times H = J + \frac{\partial D}{\partial t} \tag{3.2}$$

$$\nabla \cdot B = 0 \tag{3.3}$$

$$\nabla \cdot D = \rho \tag{3.4}$$

$$D = \epsilon E \tag{3.5}$$

$$B = \mu H \tag{3.6}$$

where $E, D, H, B, J, \rho, \epsilon$, and μ represent the electric field intensity, electric displacement, magnetic field intensity, magnetic induction, conduction current density, charge density, permittivity, and permeability of the medium, respectively. The Maxwell's curl equations (3.1) and (3.2) can be represented in the form of FDTD formulation as follows:

$$\frac{\partial H}{\partial t} = -\frac{1}{\mu}(\nabla \times E) - \frac{\sigma_{\mathrm{M}}}{\mu}H \tag{3.7}$$

$$\frac{\partial E}{\partial t} = \frac{1}{\epsilon}(\nabla \times H) - \frac{\sigma_{\mathrm{E}}}{\epsilon}E \tag{3.8}$$

where the relation $J = \sigma_{\mathrm{E}}E$ incorporates the electrical losses through the conductivity σ_{E}, however, the magnetic conductivity σ_{M} accounts for the losses in the magnetic materials. In the Cartesian coordinate system, the above Eqs. (3.7) and (3.8) can be expressed in terms of coupled partial differential equations as follows:

$$\frac{\partial H_x}{\partial t} = -\frac{1}{\mu}\left(\frac{\partial E_z}{\partial y} - \frac{\partial E_y}{\partial z}\right) - \frac{\sigma_{\mathrm{M}}}{\mu}H_x \tag{3.9}$$

$$\frac{\partial H_y}{\partial t} = -\frac{1}{\mu}\left(\frac{\partial E_x}{\partial z} - \frac{\partial E_z}{\partial x}\right) - \frac{\sigma_{\mathrm{M}}}{\mu}H_y \tag{3.10}$$

$$\frac{\partial H_z}{\partial t} = -\frac{1}{\mu}\left(\frac{\partial E_y}{\partial x} - \frac{\partial E_x}{\partial y}\right) - \frac{\sigma_{\mathrm{M}}}{\mu}H_z \tag{3.11}$$

$$\frac{\partial E_x}{\partial t} = \frac{1}{\epsilon}\left(\frac{\partial H_z}{\partial y} - \frac{\partial H_y}{\partial z}\right) - \frac{\sigma_{\mathrm{E}}}{\epsilon}E_x \tag{3.12}$$

$$\frac{\partial E_y}{\partial t} = \frac{1}{\epsilon}\left(\frac{\partial H_x}{\partial z} - \frac{\partial H_z}{\partial x}\right) - \frac{\sigma_{\mathrm{E}}}{\epsilon}E_y \tag{3.13}$$

$$\frac{\partial E_z}{\partial t} = \frac{1}{\epsilon}\left(\frac{\partial H_y}{\partial x} - \frac{\partial H_x}{\partial y}\right) - \frac{\sigma_{\mathrm{E}}}{\epsilon}E_z \tag{3.14}$$

In 1966, Yee [34] originated a set of finite-difference equations for the time-dependent curl equations, viz., (3.1) and (3.2), for lossless materials, using second-order accurate, centered finite differences [33]. This algorithm updates the electric and magnetic field components in a fully explicit, time-stepping procedure. The Yee algorithm centers its $\vec{E}$ and $\vec{H}$ components in 3-D space so that every $\vec{E}$ component is surrounded by four circulating $\vec{H}$ components and vice versa. This is illustrated in Figure 3.1.

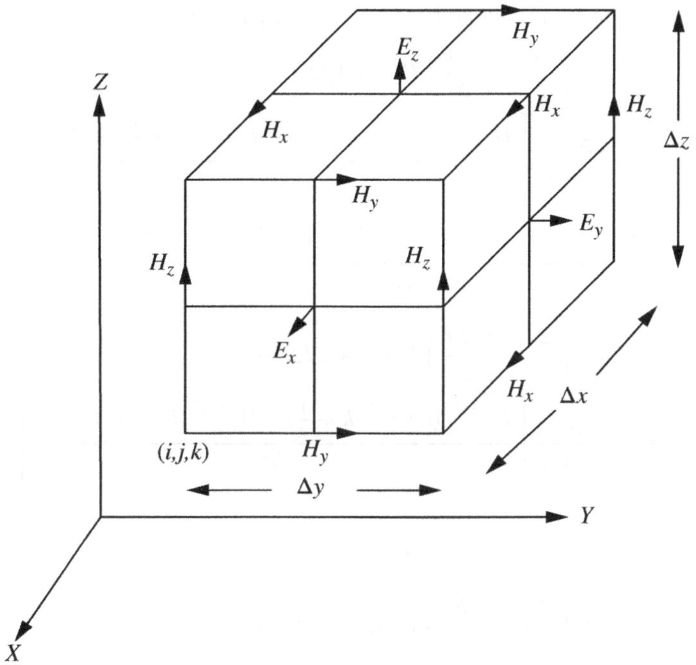

Figure 3.1 Position of electric and magnetic field vector components about a cubic cell of the Yee space lattice. This figure is based upon the schematic of a Yee cell given in [31]. (Source: Labratories [31]. © 1975, Bell Laboratories.)

3.4 FDTD Implementation

To model an electromagnetic problem, we define a spatial domain, consisting of n_x, n_y, and n_z numbers of discrete cells with cell sizes Δx, Δy and Δz in the x, y, and z-directions respectively. In Cartesian coordinates, the x-component of the electric field E_x is evaluated at a position represented in the index notation as $(i + 1/2, j, k)$, the y-component E_y at $(i, j + 1/2, k)$ and the z-component E_z at $(i, j, k + 1/2)$. Here, the Cartesian locations of an index point (i, j, k) are $x = i\Delta x, y = j\Delta y$ and $z = k\Delta z$. Similarly, the x-component of the magnetic field H_x is evaluated at a position $(i, j + 1/2, k + 1/2)$, the y-component H_y at $(i + 1/2, j, k + 1/2)$, and the z-component H_z at $(i + 1/2, j + 1/2, k)$. Furthermore, the electric and magnetic field components are displaced from each other in time by a half time step. Symbolically, the above field components evaluated at a spatial position and measured at an integer time point "n" are represented by $E^n_{x,i+1/2,\ j,k}$, $E^n_{y,i,j+1/2,\ k}$, $E^n_{z,i,j,k+1/2}$, $H^{n+1/2}_{x,i,j+1/2,\ k+1/2}$, $H^{n+1/2}_{y,i+1/2,\ j,k+1/2}$, and $H^{n+1/2}_{z,i+1/2,\ j+1/2,\ k}$. The Maxwell's curl equations expressed in terms of coupled partial differential equations (3.9)–(3.14) can now be implemented using the FDTD explicit scheme [32, 33] as follows:

$$E_x^{n+1}\ (i+1/2,j,k)=\left(\frac{2\epsilon-\sigma_{\mathrm{E}}\Delta t}{2\epsilon+\sigma_{\mathrm{E}}\Delta t}\right)E_x^n(i+1/2,j,k)$$
$$+\frac{2\Delta t}{2\epsilon+\Delta t\sigma_{\mathrm{E}}}\left[\frac{H_z^{n+1/2}(i+\frac{1}{2},j+\frac{1}{2},k)-H_z^{n+1/2}(i+\frac{1}{2},j-\frac{1}{2},k)}{\Delta y}\right.$$
$$\left.-\frac{H_y^{n+1/2}(i+\frac{1}{2},j,k+\frac{1}{2})-H_y^{n+1/2}(i+\frac{1}{2},j,k-\frac{1}{2})}{\Delta z}\right] \tag{3.15}$$

$$E_y^{n+1}\ (i,j+1/2,k)=\left(\frac{2\epsilon-\sigma_{\mathrm{E}}\Delta t}{2\epsilon+\sigma_{\mathrm{E}}\Delta t}\right)E_y^n(i,j+1/2,k)$$
$$+\frac{2\Delta t}{2\epsilon+\Delta t\sigma_{\mathrm{E}}}\left[\frac{H_x^{n+1/2}(i,j+\frac{1}{2},k+\frac{1}{2})-H_x^{n+1/2}(i,j+\frac{1}{2},k-\frac{1}{2})}{\Delta z}\right]$$
$$-\frac{H_z^{n+1/2}(i+\frac{1}{2},j+\frac{1}{2},k)-H_z^{n+1/2}(i-\frac{1}{2},j+\frac{1}{2},k)}{\Delta x} \tag{3.16}$$

$$E_z^{n+1}\ (i,j,k+1/2)=\left(\frac{2\epsilon-\sigma_{\mathrm{E}}\Delta t}{2\epsilon+\sigma_{\mathrm{E}}\Delta t}\right)E_z^n(i,j,k+1/2)$$
$$+\frac{2\Delta t}{2\epsilon+\Delta t\sigma_{\mathrm{E}}}\left[\frac{H_y^{n+1/2}(i+\frac{1}{2},j,k+\frac{1}{2})-H_y^{n+1/2}(i-\frac{1}{2},j,k+\frac{1}{2})}{\Delta x}\right]$$
$$-\frac{H_x^{n+1/2}(i,j+\frac{1}{2},k+\frac{1}{2})-H_x^{n+1/2}(i,j-\frac{1}{2},k+\frac{1}{2})}{\Delta y} \tag{3.17}$$

$$H_x^{n+\frac{1}{2}}\ (i,j+1/2,k+1/2)=\left(\frac{2\mu-\sigma_{\mathrm{M}}\Delta t}{2\mu+\sigma_{\mathrm{M}}\Delta t}\right)H_x^{n-1/2}(i,j+1/2,k+1/2)$$
$$+\frac{2\Delta t}{2\mu+\Delta t\sigma_{\mathrm{M}}}\left[\frac{E_y^n(i,j+\frac{1}{2},k+1)-E_y^n(i,j+\frac{1}{2},k)}{\Delta z}\right]$$
$$-\frac{E_z^n(i,j+1,k+\frac{1}{2})-E_z^n(i,j,k+\frac{1}{2})}{\Delta y} \tag{3.18}$$

$$H_y^{n+\frac{1}{2}}\ (i+1/2,j,k+1/2)=\left(\frac{2\mu-\sigma_{\mathrm{M}}\Delta t}{2\mu+\sigma_{\mathrm{M}}\Delta t}\right)H_y^{n-1/2}(i+1/2,j,k+1/2)$$
$$+\frac{2\Delta t}{2\mu+\Delta t\sigma_{\mathrm{M}}}\left[\frac{E_z^n(i+1,j,k+\frac{1}{2})-E_z^n(i,j,k+\frac{1}{2})}{\Delta x}\right]$$
$$-\frac{E_x^n(i+\frac{1}{2},j,k+1)-E_x^n(i+\frac{1}{2},j,k)}{\Delta z} \tag{3.19}$$

$$H_z^{n+\frac{1}{2}}(i+1/2,j+1/2,k) = \left(\frac{2\mu - \sigma_{\mathrm{M}}\Delta t}{2\mu + \sigma_{\mathrm{M}}\Delta t}\right) H_z^{n-1/2}(i+1/2,j+1/2,k)$$
$$+\frac{2\Delta t}{2\mu + \Delta t\sigma_{\mathrm{M}}}\left[\frac{E_x^n(i+\frac{1}{2},j+1,k) - E_x^n(i+\frac{1}{2},j,k)}{\Delta y}\right.$$
$$\left.-\frac{E_y^n(i+1,j+\frac{1}{2},k) - E_y^n(i,j+\frac{1}{2},k)}{\Delta x}\right] \quad (3.20)$$

The basic Yee algorithm in rectilinear Cartesian coordinates is described next. First, an electric field component, E_x, E_y or E_z is updated at the present time level, $n+\frac{1}{2}$, based on its initial value at the previous time level, $n-\frac{1}{2}$, and the values of the four surrounding magnetic field components at time level n. Once all the electric field components in the computational domain are updated, then the magnetic fields, H_x, H_y, and H_z in the computational domain are updated. A magnetic field component is updated now at the present level $n+1$, based on its value at the previous time level, n, and the values of the four surrounding electric field components at time level $n+\frac{1}{2}$. The process of updating the electric fields and then the magnetic fields is repeated. A simplified flow-chart of the FDTD code based on Yee algorithm is shown in Figure 3.2.

The appropriate boundary conditions are then placed along the boundaries of the object and also along the boundaries of the computational domain. The boundary condition along the boundary of the object is specified by the material of the object to be modeled. Similarly, the boundaries of the computational domain are truncated by a absorbing boundary condition (ABC), whose purpose is to absorb outgoing waves from the object. A variety of boundary conditions have been used by different workers [35]. For example, Shlager et al. [18] have used a second-order Liao radiation boundary condition for modeling the transient response of a TEM horn. A second-order outer radiation boundary condition (ORBC) assumes that electromagnetic fields radiated by the object assume the form of plane waves traveling normal to the boundary [32]. However, the plane waves of arbitrary incidence, polarization, and frequency can be matched using the perfectly matched layer (PML) boundary condition [36].

3.5 Numerical Issues

Given the mesh sizes, numerical stability of the explicit scheme requires that the time step Δt be determined from the **Courant stability criterion**, given by [32]

$$c\Delta t \leq 1 / \sqrt{\frac{1}{\Delta x^2} + \frac{1}{\Delta y^2} + \frac{1}{\Delta z^2}} \quad (3.21)$$

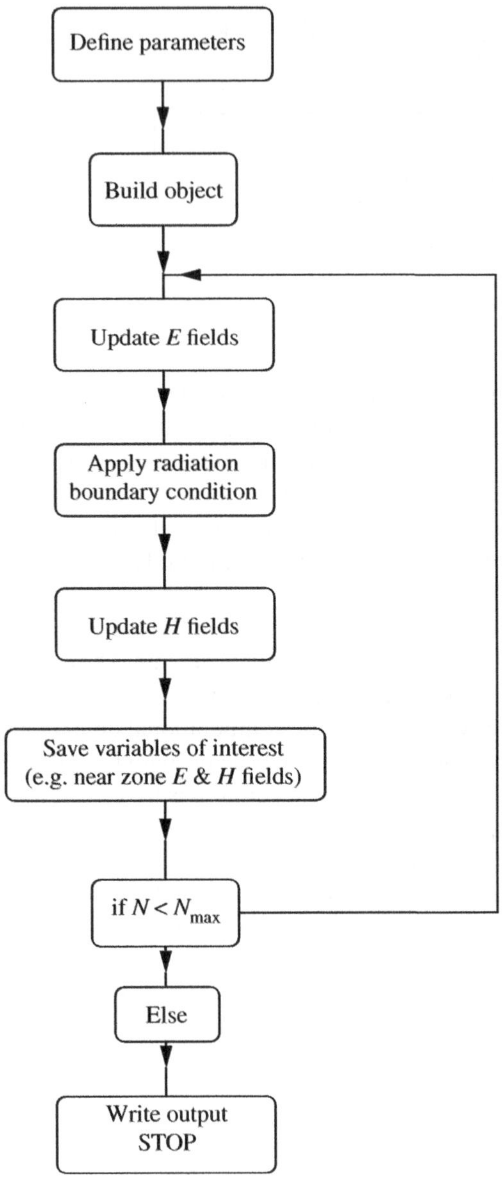

Figure 3.2 Flow chart of the FDTD algorithm used in the simulation. (Source: Kunz and Luebbers [32]. © 1993, Taylor & Francis Group.)

where c is the velocity of electromagnetic waves through the medium of propagation. Physically, this means that in one time step, any point of a plane wave propagating through the FDTD domain should not pass through more than one computational cell.

To obtain accurate results with the FDTD method, it is important to choose a grid spacing that is small compared to the wavelength corresponding to the highest significant frequency contained in the excitation pulse. A small enough grid spacing will minimize numerical dispersion errors, modeling errors and also "staircasing" errors.

Numerical dispersion errors in the FDTD method result from the inability of numerical waves to propagate at exactly the speed of light in the computational domain and results in phase shifts for different frequency components of the pulse. Typically, allowing a grid resolution on the order of 20 cells per wavelength will allow these errors to be at a sufficiently low level to be acceptable [32]. Here, the relevant wavelength is one corresponding to the highest frequency of interest.

Modeling errors are geometry dependent; they result from not accurately sampling the fields in the complex region, for example the feed of a TEM cell. Using a finer mesh in regions with high-field gradients can help to minimize this type of error.

"Staircasing" errors arise because a curved surface cannot be exactly modeled in a Cartesian mesh. These can also be minimized by using a finer mesh. Finally, the use of proper boundary conditions can help reduce spurious "numerical" reflections from the domain boundary.

3.6 Summary

In this chapter, we learned the necessity of a 3-D FDTD method for the self-consistent analysis of an EMP simulator. The FDTD method provides flexibility for setting up the complex 3D structure (capacitor bank, switch, TEM structure, the test volume occupying the 3-D device-under-test) in the FDTD grid space. Moreover, it provides a wealth of electromagnetic field information in a single run, which is ideally suited for the transient nature of EMP analysis. The necessary steps for EMP simulations using the FDTD method have been discussed for the benefit of researchers interested to extend their ideas for designing an EMP simulator.

4

Electromagnetic Pulse in Free Space and Material Media

4.1 Introduction

In this chapter, we will develop our understanding about the behavior of an electromagnetic pulse (EMP) propagation in free-space and material media using the finite-difference time-domain (FDTD) computational method. The exercises illustrated here will be helpful to write computer programs for the applications of our interest and visualize the simulated data. We will use analytical expressions to further strengthen our understanding.

4.2 Input Waveform

We can choose any waveform for pulse excitation but for the sake of simplicity and clarity of our illustrations a commonly used Gaussian waveform has been used in this chapter. A time-dependent Gaussian waveform for representing an electric field $E(t)$ can be expressed as follows:

$$E(t) = Ae^{-\alpha(t-\tau)^2} \tag{4.1}$$

where t = time variable ($0 \leq t \leq 2\tau$). For numerical calculation, the Gaussian pulse has been truncated at $t = 0$ and $t = 2\tau$ through the parameter $\alpha = (4/\tau)^2$ such that the signal magnitude falls to exp(−16) at the truncation. This avoids the generation of high-frequency spurious signals due to an abrupt truncation of the Gaussian waveform, resulting in an erroneous result. Therefore, satisfying these conditions, we may reasonably select the truncation (highest) frequency $f_{\max} = 5/\tau$. For illustration, let us consider the highest frequency $f_{\max} = 1$ GHz; $E(t)$ is shown in Figure 4.1. The function $E(t)$ can be a representative of an electric field excitation in a preferred x, y, or z-direction, which in this case can be E_x, E_y, or E_z.

Electromagnetic Pulse Simulations Using Finite-Difference Time-Domain Method, First Edition.
Shahid Ahmed.

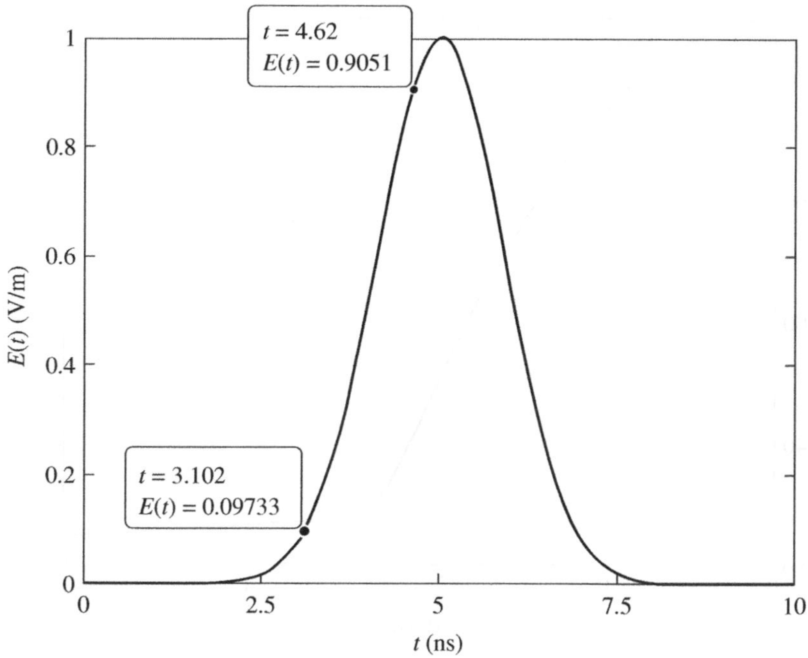

Figure 4.1 Gaussian waveform for the highest frequency 1 GHz.

From Figure 4.1, we see that $E(t)$ reaches its peak at $t = 5$ ns, which is exactly equal to τ. The rise-time of the pulse can be obtained from 10% to 90% of the peak value $A = 1$. Two boxes depicted in Figure 4.1, one in the bottom (10% of peak), and other in the top (90% of peak) give an estimate of rise-time $t_r = (4.6 - 3.1)$ ns = 1.5 ns. The frequency spectrum of the time-domain Gaussian waveform can be assessed by the fast Fourier transform (FFT), which is depicted in Figure 4.2.

4.2.1 MATLAB® Script for Visualization: Listing #1

A MATLAB [37] script for reproducing the graphs shown in Figures 4.1 and 4.2 is listed below.

```
%
% Purpose: Generate a Gaussian Pulse
clear all;
delt = 3.3e-11; % time stepsize (s)
fmax = 1E+9; % Maximum frequency (Hz)
tau = 5.0/fmax;
beta = round(tau/delt);
```

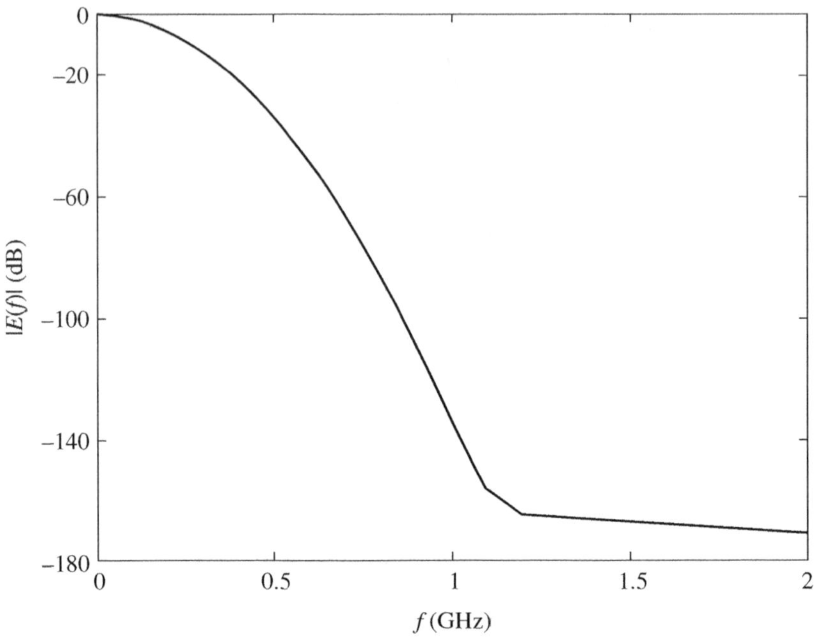

Figure 4.2 Fourier transform of the waveform shown in Figure 4.1.

```
betadt=beta*delt;
alpha = (4/betadt)^2;
t = [0:delt:2*betadt];
Len=length(t);
A = 1.0;% amplitude (V/m)
Efld = A*exp(-alpha*(t-betadt).^2);
close all;
figure(1)
plot(t./1.e-9,Efld,'k','LineWidth',2)
axis([0 10 0 1])
set(gca,'FontSize',12,'FontWeight','bold','Xtick',...
[0,2.5,5,7.5,10])
xlabel('t (ns)')
ylabel('E(t) [V/m]')
% grid
%
Fs=[0:(Len-1)]./delt./Len;
EfldFFT = abs(fft(Efld)./Len);
figure(2)
```

```
plot(Fs(1:beta)./1.e9,20.*log10(EfldFFT(1:beta).
/max(EfldFFT(1:beta))),...
'k','LineWidth',2)
axis([0 2 -180 0])
set(gca, 'FontSize',12, 'FontWeight','bold','Xtick',
[0, 0.5, 1, 1.5, 2.0],...
    'Ytick',[-180, -140, -100, -60, -20, 0])
xlabel('f(GHz)')
ylabel('|E(f)| [dB]')
% grid
```

4.2.2 Execution of MATLAB/OCTAVE Code

To execute the MATLAB script, import the script File > Open > FileName.m and run the script just by typing the "FileName" in the MATLAB command prompt "≫." Also, one can execute the script using the graphic user interface (GUI) by clicking the "Run" tab located on the top of the Editor. A MATLAB desktop environment has been depicted in Figure 4.3 for visualization.

Also, it is important to note that the script written in MATLAB is compatible with OCTAVE [38] – a freeware MATLAB clone. A desktop environment of OCTAVE with GUI has been illustrated in Figure 4.4. Simulation can be executed by clicking the run tab located on the top of the "Editor." The ribbons shown

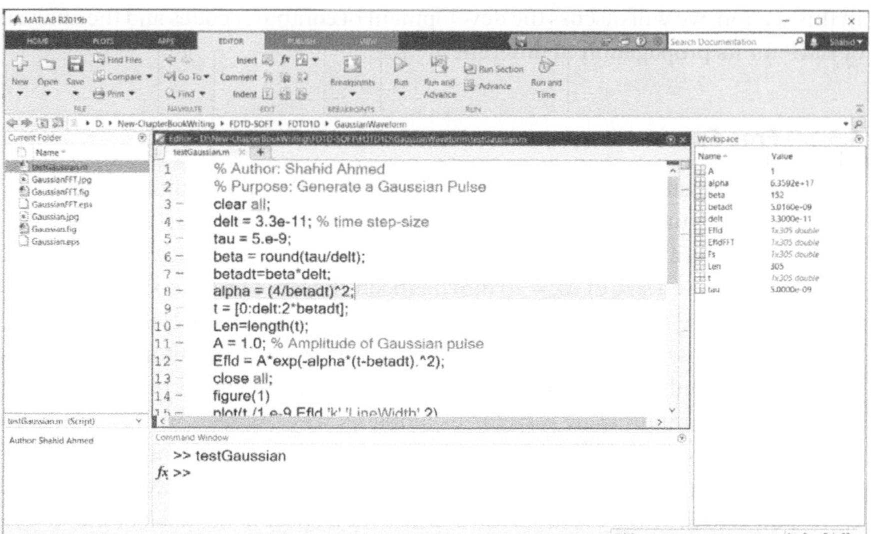

Figure 4.3 A MATLAB [37] desktop GUI environment. (Source: The Mathworks [37] .)

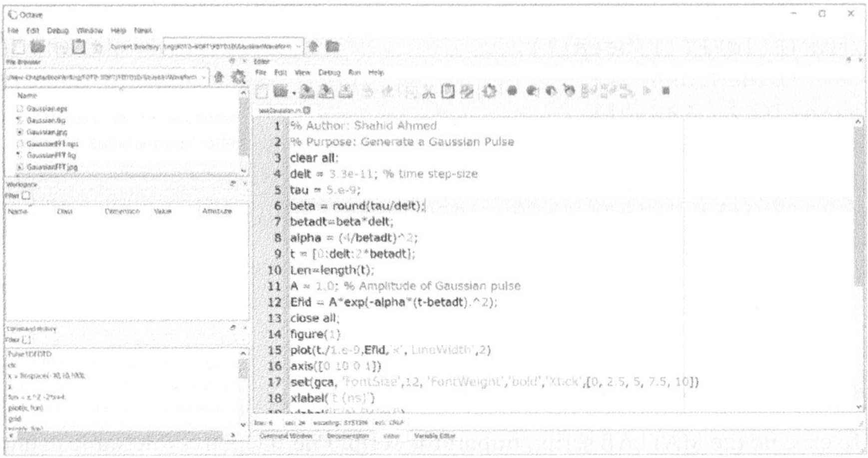

Figure 4.4 An OCTAVE [38] desktop GUI environment. (Ref. [38] .)

in the bottom can be used for different purposes such as "Command Line" for watching the execution/error and "Editor" for writing scripts.

4.3 One-dimension Approach

In this section, we will discuss the development of computer codes and the analysis of EMP for its propagation in different media.

4.3.1 Free Space

Here, the propagation of the EMP in free-space (see the spatial profile of medium properties in Figure 4.5) is simulated by solving the Maxwell's curl equations (3.1) and (3.2) as described in Chapter 3. A one-dimensional space lattice of 500 uniform grid cells with a cell size of $\Delta x = 10$ mm has been chosen. The time-step size of $\Delta t =$ 33.3 ps is dictated by the Courant stability criterion – see Eq. (3.21).

Figure 4.6 shows the propagation of EMPs in free-space monitored at time $t =$ $250\Delta t$, $400\Delta t$, and $550\Delta t$, respectively. Due to the nondispersive properties of air, these pulses maintain their waveforms along its way. As seen in Figure 4.1, the peak of the source pulse has developed at $t = 5$ ns; which corresponds to $\approx 150\Delta t$. This means the pulse propagation in Figure 4.6, for example captured at $t = 400\Delta t$ corresponds to the time evolution of $250\Delta t$ with respect to the peak of the source. This is equivalent to 8.325 ns times the speed of light in free space ($c_0 = 3.0 \times 10^8$ m/s) ≈ 2.5 m, which is indeed the spatial position of the peak as observed in

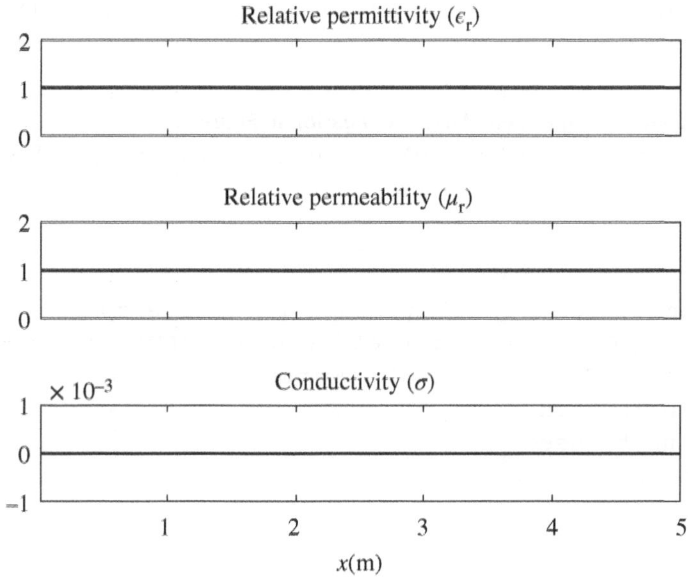

Figure 4.5 Spatial profile of material, showing the free-space medium properties.

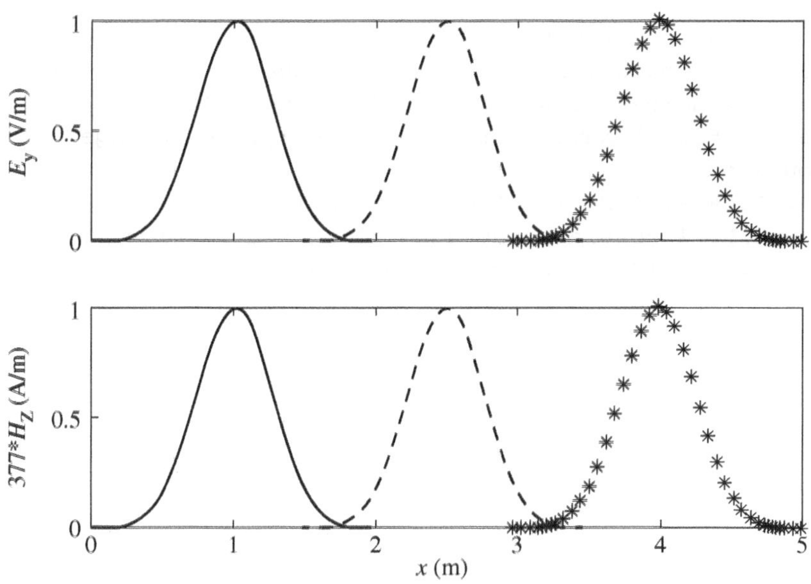

Figure 4.6 Time snapshots at $t = 250\Delta t$, $400\Delta t$, and $550\Delta t$, respectively; showing electromagnetic pulses propagating in free-space along the x-direction. The electromagnetic wave is transverse in nature, which shows y-component of electric-field and z-component of magnetic field.

Figure 4.6. Similarly, we can explain the other two pulses captured at $t = 250\Delta t$ and $400\Delta t$.

4.3.1.1 MATLAB Code Listing #1: EM Wave Propagation in Free-space

The computer code for the simulation of EM wave propagation in free-space has been written in MATLAB is illustrated below.

```
%%%%%%%%%%%%%%%%%%%%%%%%%%%%%%%%%%%%%%%
% Code type: One-dimension
% Purpose: Simulation of EM wave propagation in free-space
%          using Finite-Difference Time-Domain (FDTD) method.
%          Assume Ey and Hz field components along
%          x-direction.
% Units:   SI units used.
%          Distance -- Meter (m)
%          Time -- Second (s)
%          Frequency -- Hertz (Hz)
%
%%%%%%%%%%%%%%%%%%%%%%%%%%%%%%%%%%%%%%%
clear all;
% Define basic parameters
L = 5.0; % Domain length (m), please change as you wish
N = 500; % Number of spatial grids along "L"
NTsteps = 550; % Number of time steps, please change
as you wish
fsrc = 1.0E+9; % Source frequency (Hz)
delx = L/N; % spatial step-size along x-direction (m)
clight0 = 3.0E+8; % Speed of light in free-space (m/s)
delt = delx/clight0; % Time step-size (s) according to
Courant Cond.
eps0 = 8.854E-12; % Permittivity of free space (Farad/m)
mu0 = 4*pi*1.0E-7; % Permeability of free space (H/m)
x = linspace(0,L,N); % X-coordinate of spatial samples
% End of defining basic parameters
%
% Initialization block
ae = ones(N,1)*delt/(delx*eps0); %Scale factor for E-field
am = ones(N,1)*delt/(delx*mu0); %Scale factor for H-field
as = ones(N,1);
epsr = ones(N,1);
mur = ones(N,1);
```

```
sigma = zeros(N,1);
Hz = zeros(N,1);
Ey = zeros(N,1);
% End of initialization block
%
% Define Gaussian pulse of E-field excitation
A = 1.0; % Amplitude (V/m)
tau = 5/fsrc;
alpha = (4.0/tau)^2;

%%%%%%%%%%%%%%%%%%%%%%%%%%%%%%%%%%%%%
% Here we specify the material properties at FDTD grids
% Relative permittivity -- epsr(i)
% Relative permeability -- mur(i)
% Conductivity -- sigma(i)
%
for i=1:N
   epsr(i) = 1; % Relative permittivity
   mur(i) = 1;  % Relative permeability
   sigma(i) = 0.0;
   w1 = 0.5;
   w2 = 1.5;
end
% End of material definition at FDTD grids

ae = ae./epsr;
am = am./mur;
ae = ae./(1+delt*(sigma./epsr)/(2*eps0));
as=(1-delt*(sigma./epsr)/(2*eps0))./(1+delt*(sigma./epsr)
/(2*eps0));

% Plot the medium permittivity, permeability, and
conductivity profiles
figure(1)
subplot(3,1,1);
plot(x,epsr,'-k','LineWidth',2);
% grid on;
set(gca,'FontName','Times','FontSize',14','Xtick',
[1,2,3,4,5],'XtickLabel',[])
title('Relative Permittivity($\epsilon_{r}$)
','interpreter','latex')
subplot(3,1,2);
plot(x,mur,'-k','LineWidth',2);
```

```
set(gca,'FontName','Times','FontSize',14','Xtick',
[1,2,3,4,5],'XtickLabel',[])
title('Relative Permeability ($\mu_{r}$)','interpreter','
latex');
subplot(3,1,3);
plot(x,sigma,'-k','LineWidth',2);
axis([3*delx L min(sigma)*0.9-0.001 max(sigma)*1.1+0.001]);
set(gca,'FontName','Times','FontSize',14')
title('Conductivity ($\sigma$)','interpreter','latex');
set(gca,'FontName','Times','FontSize',14,'Xtick',
[1,2,3,4,5])
xlabel('x(m)')

% E and H-field plots
figure(2);
% Set double buffering on for smoother graphics
set(gcf,'doublebuffer','on');
subplot(211),plot(Ey,'LineWidth',2);
subplot(212),plot(Hz,'LineWidth',2);
grid on;

% Loop for E and H-field's calculations for entire
FDTD grids for total
% time-steps NTsteps
for iter=1:NTsteps

% Define source excitation (Gaussian pulse)
 pulse = A*exp(-alpha*(iter*delt - tau)^2);
% Apply source excitation at 3rd spatial grid
 Ey(3) = Ey(3)+2*(1-exp(-((iter-1)/50)^2))*pulse;
% Absorbing boundary conditions for left-propagating
waves
 Hz(1) = Hz(2);

 for i=2:N-1 % Update H field
      Hz(i) = Hz(i)-am(i)*(Ey(i+1)-Ey(i));
 end
% Absorbing boundary conditions for right
propagating waves
 Ey(N) = Ey(N-1);
 for i=2:N-1 % Update E field
      Ey(i) = as(i)*Ey(i)-ae(i)*(Hz(i)-Hz(i-1));
 end
```

```
% Show the evolution of E and H fields over total
simulation time
figure(2)
subplot(211),plot(x,Ey,'k','LineWidth',2);
axis([3*delx L -1 1]);
set(gca,'FontName','Times','FontSize',14,'Xtick',
[1,2,3,4,5],'XtickLabel',[])
ylabel('E_y (V/m)')
% grid on;
subplot(212),plot(x,377*Hz,'k','LineWidth',2);
axis([3*delx L -1 1]);
set(gca,'FontName','Times','FontSize',14,'Xtick',
[1,2,3,4,5])
ylabel('377*H_z (A/m)')
% grid on;
xlabel('x (m)');
pause(0);
iter

end
%%%%%%%%%%%%%%%%%%%%%%%%%%%%%%%%%%%%%%%%%%%%%%%%%%%%%%%%%%%
```

4.3.2 Data Recording and Visualization

The graph shown in Figure 4.6 has been recorded at time $t = 250\Delta t$, $400\Delta t$, and $550\Delta t$, respectively, by running the simulation for specified values of parameter "Niter" equal to 250, 400, and 550, respectively. The E and H fields are saved in "*.mat" format using the MATLAB command ≫ save EY250 E_y; for E-field at $t = 250\Delta t$. This has been repeated for all other time snapshots for recording E and H fields. Once the complete data have been recorded, a MATLAB script (depicted below) has been run for generating the plot shown in Figure 4.6.

4.3.2.1 MATLAB Script for Visualization: Listing #2

```
%
% Purpose: Example shows how to read E and H field's data
% and generate graphs as shown in Fig. 6.
load 'EY250.mat';
EY250=Ey;
clear Ey;
```

```
load 'EY400.mat';
EY400=Ey;
clear EY;
load 'EY550.mat';
EY550=Ey;
clear EY;
%%%%
load 'HZ250.mat';
HZ250=Hz;
clear Hz;
load 'HZ400.mat';
HZ400=Hz;
clear Hz;
load 'HZ550.mat';
HZ550=Hz;
clear Hz;


close all;
% Create spatial grid along x-direction
L = 5.0; % distance along x-direction in meters
N = 505; % grid points
x = linspace(0,L,N);
subplot(211)
plot(x(1:200),EY250(1:200),'k','LineWidth',2)
hold on
plot(x(150:350),EY400(150:350),'--k','LineWidth',2)
plot(x(300:6:505),EY550(300:6:505),'*k','LineWidth',1)
%
set(gca,'FontSize',14,'FontName','Times','FontWeight',...
'bold','XtickLabel',[])
% xlabel('x (m)')
ylabel('E_{y} (V/m)');
% legend('Inc','Ref')
axis([0 5 0 1])
subplot(212)
plot(x(1:200),377*HZ250(1:200),'k','LineWidth',2)
hold on
plot(x(150:350),377*HZ400(150:350),'--k','LineWidth',2)
plot(x(300:6:505),377*HZ550(300:6:505),'*k','LineWidth',1)
%
```

```
set(gca,'FontSize',14,'FontName','Times','FontWeight',
'bold')
xlabel('x (m)')
ylabel('377*H_{z} (A/m)');
%%%%%%%%%%%%%%%%%%%%%%%%%%%%%%%%%%%%
```

4.3.3 Dielectric Medium

In this section, we will focus on the propagation of an EMP in a lossless and lossy dielectric medium. Before we go for full-wave simulations, let us quantify the magnitudes of the transmitted and reflected pulses using the concept of wave propagation in different media. The reflection (Γ) and transmission (T) coefficients of a wave travelling from a medium of characteristic impedance (η_1) through another medium (η_2) are given by:

$$\Gamma = \frac{\eta_2 - \eta_1}{\eta_2 + \eta_1} \tag{4.2}$$

$$T = \frac{2\eta_2}{\eta_2 + \eta_1} \tag{4.3}$$

The wave impedance for a given medium can be calculated as

$$\eta = \sqrt{\frac{\mu_0 \mu_r}{\epsilon_0 \epsilon_r^*}} \tag{4.4}$$

The complex relative permittivity (ϵ_r^*) can be defined as

$$\epsilon_r^* = \epsilon_r + \frac{\sigma}{j\omega\epsilon_0} \tag{4.5}$$

where σ = conductivity, ϵ_0 = free-space permittivity, ω = angular frequency = $2\pi f$.

Using the medium property as illustrated in Figure 4.7, we can express the reflection and transmission coefficients as follows:

$$\Gamma = \frac{\sqrt{\epsilon_1} - \sqrt{\epsilon_2}}{\sqrt{\epsilon_1} + \sqrt{\epsilon_2}} \tag{4.6}$$

$$T = \frac{2\sqrt{\epsilon_1}}{\sqrt{\epsilon_1} + \sqrt{\epsilon_2}} \tag{4.7}$$

The functional waveform of electric field E propagating along the x-direction can be expressed as

$$E_z(x) = E_0 e^{-\alpha x} e^{-j\beta x} \tag{4.8}$$

where, E_0 = amplitude at $x = 0$; α and β are attenuation and phase constants expressed as

$$\alpha = \frac{\omega}{c_0}\sqrt{\frac{\epsilon_r}{2}}\left[\sqrt{1+\left(\frac{\sigma}{\omega\epsilon_0\epsilon_r}\right)^2} - 1\right]^{1/2} \quad \text{(Np/m)} \tag{4.9}$$

$$\beta = \frac{\omega}{c_0}\sqrt{\frac{\epsilon_r}{2}}\left[\sqrt{1+\left(\frac{\sigma}{\omega\epsilon_0\epsilon_r}\right)^2} + 1\right]^{1/2} \quad \text{(rad/m)} \tag{4.10}$$

4.3.3.1 Lossless Dielectric Medium

In this section, we would like to analyze the propagation of the EMP (see Figure 4.1) hitting a lossless dielectric medium. The one-dimensional spatial profile of the material properties, which describes the medium is illustrated in Figure 4.7. The medium shows a transition at $x = 2.5$ m from the free space to a perfect dielectric with relative permittivity (ϵ_r) of 4; the relative permeability is assumed unity and the conductivity value zero. Figure 4.8 shows the propagation of the EMP in the air-dielectric half-space. The incident pulse ("inc") has been recorded at $t = 300\Delta t$, which corresponds to the travel distance of 1 m in free-space for $\Delta t = 33$ ps. Note that the peak of the incident pulse takes about 6.7 ns to evolve at the point of excitation, which is about 0.2 m of pulse

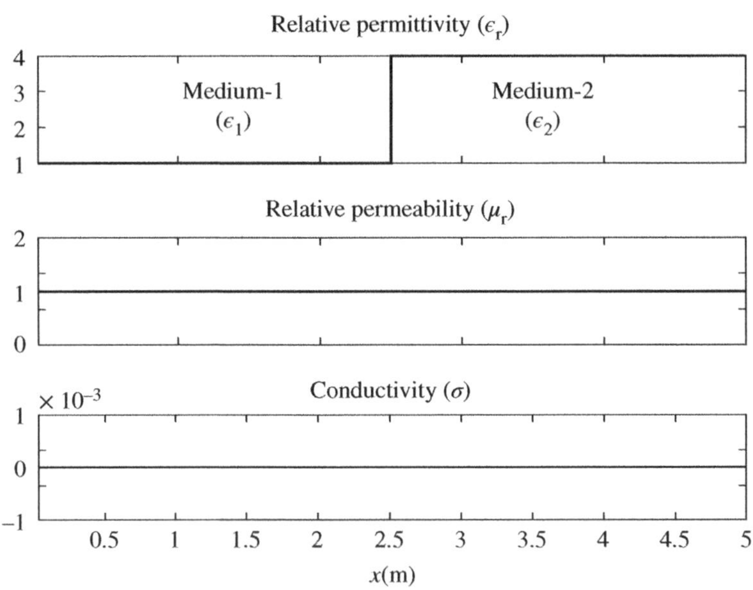

Figure 4.7 Spatial profile of material, showing the lossless dielectric medium properties.

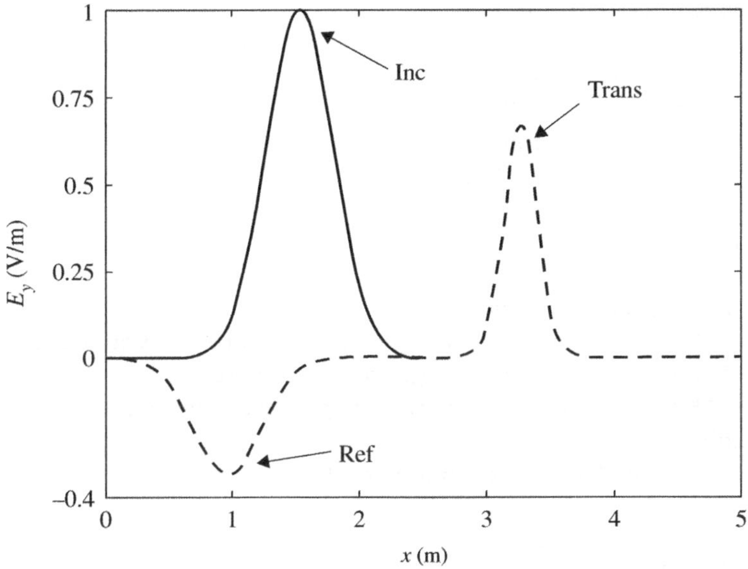

Figure 4.8 The E-field of electromagnetic pulse propagation through a lossless dielectric medium illustrated in Figure 4.7. The incident pulse has been monitored at $t = 300\Delta t$, which after striking the dielectric transmit through and partially reflects.

propagation in free-space. This means the pulse at $t = 300\Delta t$ will travel a distance $d = 1.0$ m + 0.2 m = 1.2 m, which is indeed seen in Figure 4.8. Moreover, it is important to note that the reflected pulse has traveled twice the distance of the pulse transmitted in the dielectric. This is due to the fact that the transmitted pulse goes through the dielectric ($\epsilon_r = 4.0$), where the signal speed ($= c_0/\sqrt{\epsilon_r}$) is one-half of the free space ($c_0 = 3 \times 10^8$ m/s).

Substituting the values of relative permittivity of the two media, $\epsilon_1 = 1$ and $\epsilon_2 = 4$ in Eqs. (4.6) and (4.7), we get $\Gamma = 0.33$ and $T = 0.67$. This means that 33% of the incident wave will be reflected to medium-1 and 67% will transmit through the medium-2 – see Figure 4.8.

4.3.3.2 MATLAB Code Listing #2: EM Wave in Air and Lossless-dielectric Medium

The script for the wave propagation in half-space air-dielectric (lossless) medium is listed below.

```
%%%%%%%%%%%%%%%%%%%%%%%%%%%%%%%%%%
% Code Type: One-dimension
% Purpose: Simulation of EM wave propagation in half-space
% air-dielectric (lossless) medium using Finite-Difference
```

```
% Time-Domain (FDTD) method.
% Assume Ey and Hz field components along x-direction.
% Units:   SI units used.
%          Distance -- Meter (m)
%          Time -- Second (s)
%          Frequency -- Hertz (Hz)
%
%%%%%%%%%%%%%%%%%%%%%%%%%%%%%%%%%%%%%
clear all;
% Define basic parameters
L = 5.0; % Domain length (m), please change as you wish
N = 500; % Number of spatial grids along "L"
NTsteps = 550; % Number of time steps, please change
as you wish
fsrc = 1.0E+9; % Source frequency (Hz)
delx = L/N; % spatial step-size along x-direction (m)
clight0 = 3.0E+8; % Speed of light in free-space (m/s)
delt = delx/clight0; % Time step-size (s) according
to Courant Cond.
eps0 = 8.854E-12; % Permittivity of free space (Farad/m)
mu0 = 4*pi*1.0E-7; % Permeability of free space (H/m)
x = linspace(0,L,N); % X-coordinate of spatial samples
% End of defining basic parameters
%
% Initialization block
ae = ones(N,1)*delt/(delx*eps0); % Scale factor
for E-field
am = ones(N,1)*delt/(delx*mu0); % Scale factor
for H-field
as = ones(N,1);
epsr = ones(N,1);
mur= ones(N,1);
sigma = zeros(N,1);
Hz = zeros(N,1);
Ey = zeros(N,1);
% End of initialization block
%
% Define Gaussian pulse of E-field excitation
A = 1.0; % Amplitude (V/m)
tau = 5/fsrc;
alpha = (4.0/tau)^2;
```

```
%%%%%%%%%%%%%%%%%%%%%%%%%%%%%%%%%%%%%%
% Here we specify the material properties at FDTD grids
% Relative permittivity -- epsr(i)
% Relative permeability -- mur(i)
% Conductivity -- sigma(i)
%
for i=1:N
   if (x(i)>L/2) epsr(i)=4;sigma(i)=0.0;end % Loss-
less Dielectric region
   mur(i) = 1;  % Relative permeability
   w1 = 0.5;
   w2 = 1.5;
end
% End of material definition at FDTD grids

ae = ae./epsr;
am = am./mur;
ae = ae./(1+delt*(sigma./epsr)/(2*eps0));
as=(1-delt*(sigma./epsr)/(2*eps0))./(1+delt*(sigma./epsr)
/(2*eps0));

% Plot the medium permittivity, permeability, and
conductivity profiles
figure(1)
subplot(3,1,1);
plot(x,epsr,'-k','LineWidth',2);
% grid on;
set(gca,'FontName','Times','FontSize',14','Xtick',
[1,2,3,4,5],'XtickLabel',[])
title('Relative Permittivity($\epsilon_{r}$)
','interpreter','latex')
subplot(3,1,2);
plot(x,mur,'-k','LineWidth',2);
set(gca,'FontName','Times','FontSize',14','Xtick',
[1,2,3,4,5],'XtickLabel',[])
title('Relative Permeability ($\mu_{r}$)','interpreter
','latex');
subplot(3,1,3);
plot(x,sigma,'-k','LineWidth',2);
axis([3*delx L min(sigma)*0.9-0.001 max(sigma)*1.1+0.001]);
set(gca,'FontName','Times','FontSize',14')
```

```
title('Conductivity ($\sigma$)','interpreter','latex');
set(gca,'FontName','Times','FontSize',14,'Xtick',
[1,2,3,4,5])
xlabel('x(m)')

% E and H-field plots
figure(2);
% Set double buffering on for smoother graphics
set(gcf,'doublebuffer','on');
subplot(211),plot(Ey,'LineWidth',2);
subplot(212),plot(Hz,'LineWidth',2);
grid on;

% Loop for E and H-field's calculations for entire FDTD
grids for total
% time-steps NTsteps
for iter=1:NTsteps

% Define source excitation (Gaussian pulse)
 pulse = A*exp(-alpha*(iter*delt - tau)^2);
% Apply source excitation at 3rd spatial grid
 Ey(3) = Ey(3)+2*(1-exp(-((iter-1)/50)^2))*pulse;
% Absorbing boundary conditions for left-propagating waves
 Hz(1) = Hz(2);

 for i=2:N-1 % Update H field
      Hz(i) = Hz(i)-am(i)*(Ey(i+1)-Ey(i));
 end
% Absorbing boundary conditions for right
propagating waves
 Ey(N) = Ey(N-1);
 for i=2:N-1 % Update E field
      Ey(i) = as(i)*Ey(i)-ae(i)*(Hz(i)-Hz(i-1));
 end

% Show the evolution of E and H fields over total
simulation time
figure(2)
subplot(211),plot(x,Ey,'k','LineWidth',2);
axis([3*delx L -1 1]);
set(gca,'FontName','Times','FontSize',14,'Xtick',
[1,2,3,4,5],'XtickLabel',[])
```

```
ylabel('E_y (V/m)')
% grid on;
subplot(212),plot(x,377*Hz,'k','LineWidth',2);
axis([3*delx L -1 2]);
set(gca,'FontName','Times','FontSize',14,'Xtick',
[1,2,3,4,5])
ylabel('377*H_z (A/m)')
% grid on;
xlabel('x (m)');
pause(0);
iter

end
%%%%%%%%%%%%%%%%%%%%%%%%%%%%%%%%%%%%%%%%%%%%%%%%%%%%%%%%%%%%%%%%%%%%%%
```

4.3.3.3 Lossy Dielectric Medium

The one-dimensional spatial profile of the material properties, which describes the lossy dielectric medium is illustrated in Figure 4.9. The medium shows a transition at $x = 2.5$ m from the free space to a lossy dielectric with relative permittivity (ϵ_r) of

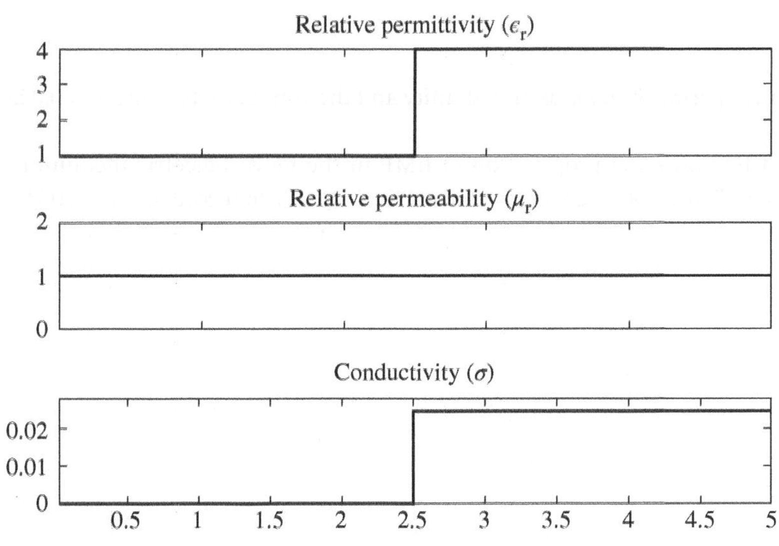

Figure 4.9 Spatial profile of material, showing the lossless dielectric medium properties.

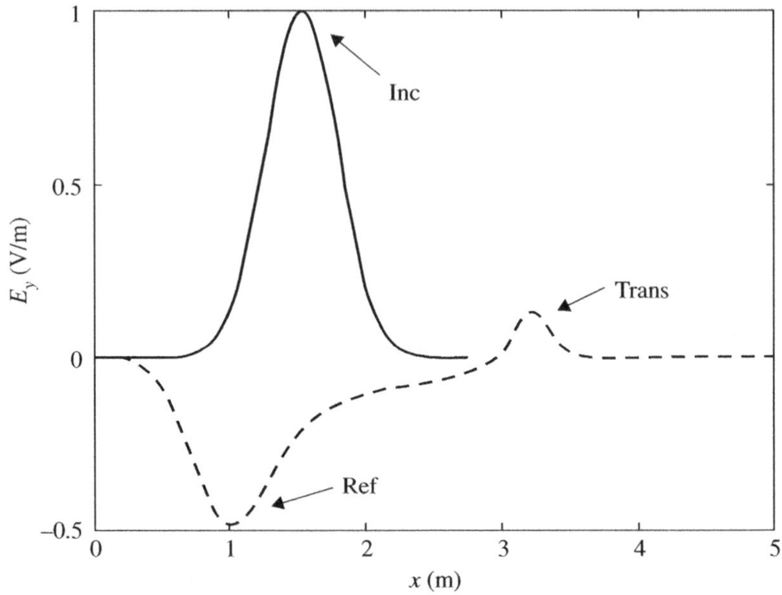

Figure 4.10 The E-field of electromagnetic pulse propagation through an interface with air and lossy dielectric medium (see Figure 4.5 for material profile). The reflected/transmitted pulse has been monitored at $t = 550\Delta t$.

4; the relative permeability is assumed unity and the conductivity value $\sigma = 0.025$ S/m.

Figure 4.10 shows the propagation of EMP in the lossy dielectric medium as illustrated in Figure 4.9. The incident pulse ("inc") has been recorded at $t = 300\Delta t$, which corresponds to the travel distance of 1 m in free-space for $\Delta t = 33$ ps. Note that the peak of the incident pulse takes about 6.7 ns to evolve at the point of excitation, which is about 0.2 m of pulse propagation in free-space. This means the pulse at $t = 300\Delta t$ will travel a distance $d = 1.0\text{ m} + 0.2\text{ m} = 1.2\text{ m}$, which is indeed seen in Figure 4.10. It is important to note that the reflected pulse has traveled twice the distance of the transmitted pulse due to the reduced signal speed ($= c_0/\sqrt{\epsilon_r}$); which is one-half of the free space ($c_0 = 3 \times 10^8$ m/s). Also, the transmitted pulse gets attenuated to approximately 12% in the lossy dielectric ($\epsilon_r = 4.0, \sigma = 0.025$ S/m). Remember that we had 67% transmission in the lossless dielectric medium, which means the pulse amplitude of 0.67. This further attenuated to approximately 0.12 due to the materials' conducting loss as predicted by the analytical calculation shown in Eq. (4.8). Figure 4.11 illustrates the attenuation of the EMP in the lossy dielectric.

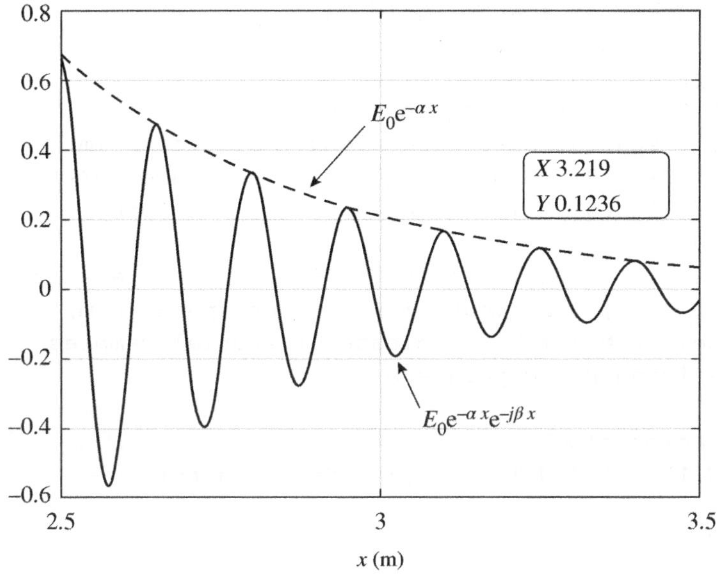

Figure 4.11 Plot shows analytical waveform (see Eq. (4.8)) of a wave in lossy dielectric as a function of distance along the propagation.

4.3.3.4 MATLAB Code Listing #3: EM Wave in Air and Lossy-dielectric Medium

The script for the wave propagation in half-space air-dielectric (lossy) medium is listed below.

```
%%%%%%%%%%%%%%%%%%%%%%%%%%%%%%%%%
% Code Type: One-dimension
% Purpose: Simulation of EM wave propagation in half-space
%          air-dielectric (lossy) medium using
%          Finite-Difference Time-Domain (FDTD) method.
%          Assume Ey and Hz field components along
%          x-direction.
% Units:   SI units used.
%          Distance -- Meter (m)
%          Time -- Second (s)
%          Frequency -- Hertz (Hz)
%
%%%%%%%%%%%%%%%%%%%%%%%%%%%%%%%%%
clear all;
% Define basic parameters
L = 5.0; % Domain length (m), please change as you wish
```

```
N = 500; % Number of spatial grids along "L"
NTsteps = 550; % Number of time steps,
please change as you wish
fsrc = 1.0E+9; % Source frequency (Hz)
delx = L/N; % spatial step-size along x-direction (m)
clight0 = 3.0E+8; % Speed of light in free-space (m/s)
delt = delx/clight0; % Time step-size (s) according to
Courant Cond.
eps0 = 8.854E-12; % Permittivity of free space (Farad/m)
mu0 = 4*pi*1.0E-7; % Permeability of free space (H/m)
x = linspace(0,L,N); % X-coordinate of spatial samples
% End of defining basic parameters
%
% Initialization block
ae = ones(N,1)*delt/(delx*eps0); % Scale factor
for E-field
am = ones(N,1)*delt/(delx*mu0); % Scale factor
for H-field
as = ones(N,1);
epsr = ones(N,1);
mur= ones(N,1);
sigma = zeros(N,1);
Hz = zeros(N,1);
Ey = zeros(N,1);
% End of initialization block
%
% Define Gaussian pulse of E-field excitation
A = 1.0; % Amplitude (V/m)
tau = 5/fsrc;
alpha = (4.0/tau)^2;

%%%%%%%%%%%%%%%%%%%%%%%%%%%%%%%%%%%%%
% Here we specify the material properties at FDTD grids
% Relative permittivity -- epsr(i)
% Relative permeability -- mur(i)
% Conductivity -- sigma(i)
%
for i=1:N
   if (x(i)>L/2) epsr(i)=4;sigma(i)=2.5e-2;end
%Lossy Dielectric region
   mur(i) = 1;  % Relative permeability
```

```
    w1 = 0.5;
    w2 = 1.5;
end
% End of material definition at FDTD grids

ae = ae./epsr;
am = am./mur;
ae = ae./(1+delt*(sigma./epsr)/(2*eps0));
as=(1-delt*(sigma./epsr)/(2*eps0))./(1+delt*
(sigma./epsr)/(2*eps0));

% Plot the medium permittivity, permeability,
and conductivity profiles
figure(1)
subplot(3,1,1);
plot(x,epsr,'-k','LineWidth',2);
% grid on;
set(gca,'FontName','Times','FontSize',14','Xtick',
[1,2,3,4,5],'XtickLabel',[])
title('Relative Permittivity($\epsilon_{r}$)
','interpreter','latex')
subplot(3,1,2);
plot(x,mur,'-k','LineWidth',2);
set(gca,'FontName','Times','FontSize',14','Xtick',
[1,2,3,4,5],'XtickLabel',[])
title('Relative Permeability ($\mu_{r}$)','interpreter
','latex');
subplot(3,1,3);
plot(x,sigma,'-k','LineWidth',2);
axis([3*delx L min(sigma)*0.9-0.001 max(sigma)*1.1+0.001]);
set(gca,'FontName','Times','FontSize',14')
title('Conductivity ($\sigma$)','interpreter','latex');
set(gca,'FontName','Times','FontSize',14,'Xtick',
[1,2,3,4,5])
xlabel('x(m)')

% E and H-field plots
figure(2);
% Set double buffering on for smoother graphics
set(gcf,'doublebuffer','on');
subplot(211),plot(Ey,'LineWidth',2);
subplot(212),plot(Hz,'LineWidth',2);
grid on;
```

```
% Loop for E and H-field's calculations for entire
FDTD grids for total
% time-steps NTsteps
for iter=1:NTsteps

% Define source excitation (Gaussian pulse)
 pulse = A*exp(-alpha*(iter*delt - tau)^2);
% Apply source excitation at 3rd spatial grid
 Ey(3) = Ey(3)+2*(1-exp(-((iter-1)/50)^2))*pulse;
% Absorbing boundary conditions for left-propagating waves
 Hz(1) = Hz(2);

 for i=2:N-1 % Update H field
      Hz(i) = Hz(i)-am(i)*(Ey(i+1)-Ey(i));
 end
% Absorbing boundary conditions for right
propagating waves
 Ey(N) = Ey(N-1);
 for i=2:N-1 % Update E field
      Ey(i) = as(i)*Ey(i)-ae(i)*(Hz(i)-Hz(i-1));
 end

% Show the evolution of E and H fields over total
simulation time
figure(2)
subplot(211),plot(x,Ey,'k','LineWidth',2);
axis([3*delx L -1 1]);
set(gca,'FontName','Times','FontSize',14,'Xtick',
[1,2,3,4,5],'XtickLabel',[])
ylabel('E_y (V/m)')
% grid on;
subplot(212),plot(x,377*Hz,'k','LineWidth',2);
axis([3*delx L -1 2]);
set(gca,'FontName','Times','FontSize',14,'Xtick',
[1,2,3,4,5])
ylabel('377*H_z (A/m)')
% grid on;
xlabel('x (m)');
pause(0);
iter

end
%%%%%%%%%%%%%%%%%%%%%%%%%%%%%%%%%%%%%%%%%%%%%%%%%
```

4.3.3.5 MATLAB Code Listing #4: Analytical Approach for Wave in Lossy Medium

MATLAB script for graphical illustration of the analytical equation (4.8) is listed below.

```
% Analytical calculation of wave propagation in lossy
dielectric
Clear all;
freq = 1.e9; %highest frequency in pulse (Hz)
E0 = 0.67; % Amplitude when wave hits the lossy
dielectric
omega = 2*pi*freq; % angular frequency
epsr = 4.0; % relative permittivity of dielectric block
sigma = 0.025; % conductivity of lossy dielectric (S/m)
clight = 3.0e8;%speed of light in free-space (m/s)
eps0 = 8.854e-12; % free-space permittivity (F/m)
omegabyc0 = omega/clight;
epsrby2 = epsr/2;
eps = epsr*eps0;
omega_eps = omega * eps;
sigmabyomega_eps = sigma/omega_eps;
sigmabyomega_eps_sq = sigmabyomega_eps*sigmabyomega_eps;

% alpha and beta are propagation and phase constants
alpha = omegabyc0*sqrt(epsrby2);
alpha = alpha * sqrt(sqrt(1+sigmabyomega_eps_sq)-1);
beta = omegabyc0*sqrt(epsrby2);
beta = beta * sqrt(sqrt(1+sigmabyomega_eps_sq)+1);

xdis = linspace(2.5,3.5,500);% distance of wave
propagation
% Efld is E(x) as in equation (8)
Efld = E0*exp(-alpha.*(xdis-2.5)).*exp(-j*beta.*(xdis-2.5));
ExpDecay = E0*exp(-alpha.*(xdis-2.5));% only decay
part of Efld

close all
plot(xdis, Efld,'k','LineWidth',2)
hold on;
plot(xdis, ExpDecay,'--k','LineWidth',2)
grid
```

```
set(gca,'FontSize',14,'FontName','Times','FontWeight',
'bold')
xlabel('x (m)')
%%%%%%%%%%%%%%%%%%%%%%%%%%%%%%%%%%%%%%%%%%%%%%%%%%%%%%%%%
```

4.3.4 Perfect Electric Conductor (PEC)

A one-dimensional spatial profile of an air–conductor (PEC) interface where the medium-1 corresponds to free-space ($\sigma_1 = 0, \epsilon_r = 1, \mu_r = 1$) and the medium-2 is a PEC ($\sigma_2 = 5 \times 10^7$ S/m) is illustrated in Figure 4.12. Let us analyze the reflection (Γ) and transmission (T) coefficient for the air–conductor interface, which are governed by the Eqs. (4.2) and (4.3) where the relative permittivity of the medium-1 and medium-2 are defined by the complex permittivity equation (4.5). For medium-1 $\epsilon_1 = 1$, however, for medium-2 $\epsilon_2 = 1 + \frac{\sigma_2}{j\omega\epsilon_0}$. Since $\sigma_2 \gg j\omega\epsilon_0$ we can simplify $\epsilon_2 = \frac{\sigma_2}{j\omega\epsilon_0}$. Now, substituting the values of ϵ_1 and ϵ_2 in Eq. (4.6) we get the reflection co-efficient in the air–conductor interface as

$$\Gamma = \frac{1 - \sqrt{\frac{\sigma_2}{j\omega\epsilon_0}}}{1 + \sqrt{\frac{\sigma_2}{j\omega\epsilon_0}}} \tag{4.11}$$

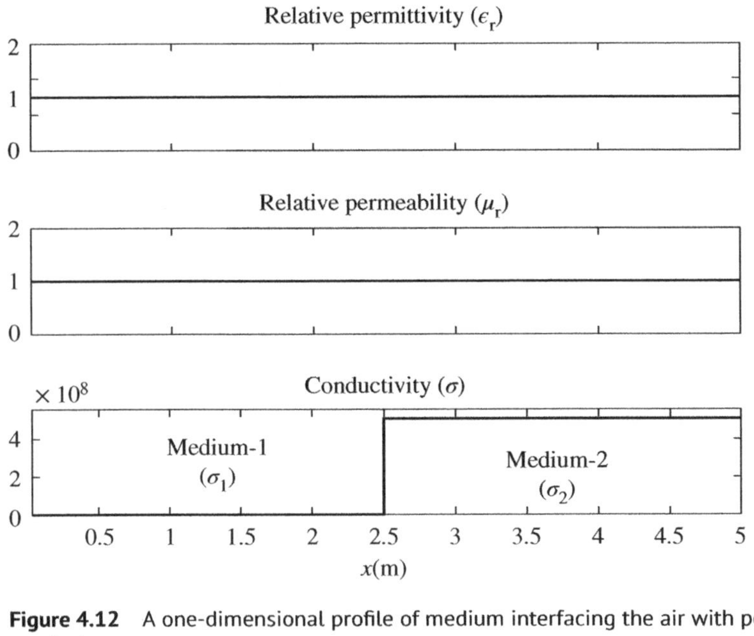

Figure 4.12 A one-dimensional profile of medium interfacing the air with perfect conductor.

As the second term in Eq. (4.11) is much greater than 1, we can further simplify Eq. (4.11) as

$$\Gamma = \frac{-\sqrt{\frac{\sigma_2}{j\omega\epsilon_0}}}{\sqrt{\frac{\sigma_2}{j\omega\epsilon_0}}} = -1 \tag{4.12}$$

Similarly, we can show that the transmission co-efficient (T) for the air–conductor interface turns out to be

$$T \approx 0 \tag{4.13}$$

Therefore, the EMP travelling in the free-space (medium-1) after hitting the medium-2 (PEC) at $x = 2.5$ m totally reflects to the medium-1 (air) – see Figure 4.13. The incident pulse "inc" has been monitored at $t = 300\Delta t$ and the reflected pulse has been recorded at $t = 550\Delta t$.

4.3.4.1 MATLAB Code Listing #5: EM Wave in Air-PEC Half-space

The script for the wave propagation in air–conductor (PEC) half space is listed below.

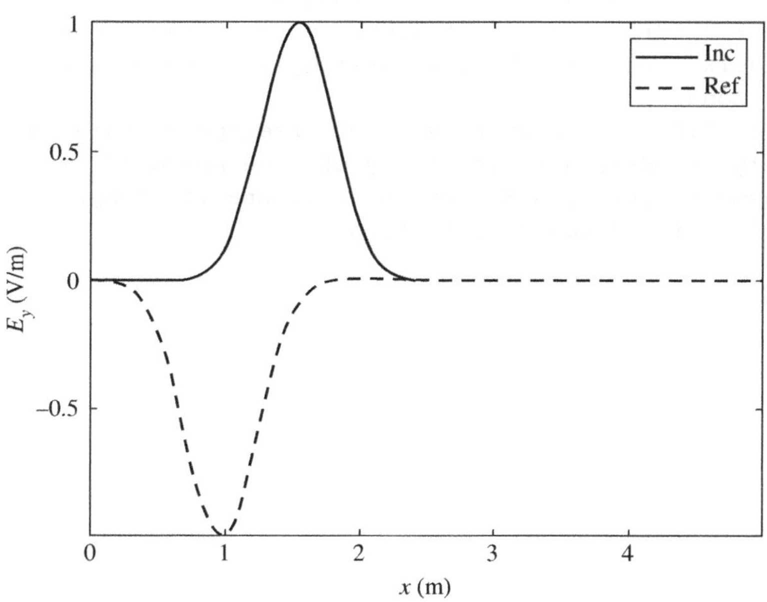

Figure 4.13 Pulse propagation in a medium interfacing air with PEC; the incident wave "inc" after hitting the conductor at $x = 2.5$ m returns to the air medium with opposite polarity.

```
%%%%%%%%%%%%%%%%%%%%%%%%%%%%%%%%%%%%%%
% Code Type: One-dimension
% Purpose: Simulation of EM wave propagation in half
%          air-PEC region using Finite-Difference
%          Time-Domain (FDTD) method. Assume Ey and
%          Hz field components along x-direction.
% Units:   SI units used.
%          Distance -- Meter (m)
%          Time -- Second (s)
%          Frequency -- Hertz (Hz)
%
%%%%%%%%%%%%%%%%%%%%%%%%%%%%%%%%%%%%%%
clear all;
% Define basic parameters
L = 5.0; % Domain length (m), please change as you wish
N = 500; % Number of spatial grids along "L"
NTsteps = 550; % Number of time steps, please change as
you wish
fsrc = 1.0E+9; % Source frequency (Hz)
delx = L/N; % spatial step-size along x-direction (m)
clight0 = 3.0E+8; % Speed of light in free-space (m/s)
delt = delx/clight0; % Time step-size (s) according to
Courant Cond.
eps0 = 8.854E-12; % Permittivity of free space (Farad/m)
mu0 = 4*pi*1.0E-7; % Permeability of free space (H/m)
x = linspace(0,L,N); % X-coordinate of spatial samples
% End of defining basic parameters
%
% Initialization block
ae = ones(N,1)*delt/(delx*eps0); % Scale factor for E-field
am = ones(N,1)*delt/(delx*mu0); % Scale factor for H-field
as = ones(N,1);
epsr = ones(N,1);
mur= ones(N,1);
sigma = zeros(N,1);
Hz = zeros(N,1);
Ey = zeros(N,1);
% End of initialization block
%
% Define Gaussian pulse of E-field excitation
A = 1.0; % Amplitude (V/m)
```

```
tau = 5/fsrc;
alpha = (4.0/tau)^2;

%%%%%%%%%%%%%%%%%%%%%%%%%%%%%%%%%%%%%%
% Here we specify the material properties at FDTD grids
% Relative permittivity -- epsr(i)
% Relative permeability -- mur(i)
% Conductivity -- sigma(i)
%
for i=1:N
   if (x(i)>L/2) sigma(i) = 6.0e7; end % PEC region
   mur(i) = 1;  % Relative permeability
   w1 = 0.5;
   w2 = 1.5;
end
% End of material definition at FDTD grids

ae = ae./epsr;
am = am./mur;
ae = ae./(1+delt*(sigma./epsr)/(2*eps0));
as=(1-delt*(sigma./epsr)/(2*eps0))./(1+delt*(sigma./epsr)
/(2*eps0));

% Plot the medium permittivity, permeability,
and conductivity profiles
figure(1)
subplot(3,1,1);
plot(x,epsr,'-k','LineWidth',2);
% grid on;
set(gca,'FontName','Times','FontSize',14','Xtick',
[1,2,3,4,5],'XtickLabel',[])
title('Relative Permittivity($\epsilon_{r}$)
','interpreter','latex')
subplot(3,1,2);
plot(x,mur,'-k','LineWidth',2);
set(gca,'FontName','Times','FontSize',14','Xtick',
[1,2,3,4,5],'XtickLabel',[])
title('Relative Permeability ($\mu_{r}$)','interpreter
','latex');
subplot(3,1,3);
plot(x,sigma,'-k','LineWidth',2);
axis([3*delx L min(sigma)*0.9-0.001 max(sigma)*1.1+0.001]);
```

```
set(gca,'FontName','Times','FontSize',14')
title('Conductivity ($\sigma$)','interpreter','latex');
set(gca,'FontName','Times','FontSize',14,'Xtick',
[1,2,3,4,5])
xlabel('x(m)')

% E and H-field plots
figure(2);
% Set double buffering on for smoother graphics
set(gcf,'doublebuffer','on');
subplot(211),plot(Ey,'LineWidth',2);
subplot(212),plot(Hz,'LineWidth',2);
grid on;

% Loop for E and H-field's calculations for entire FDTD
grids for total
% time-steps NTsteps
for iter=1:NTsteps

% Define source excitation (Gaussian pulse)
 pulse = A*exp(-alpha*(iter*delt - tau)^2);
% Apply source excitation at 3rd spatial grid
 Ey(3) = Ey(3)+2*(1-exp(-((iter-1)/50)^2))*pulse;
% Absorbing boundary conditions for left-propagating waves
 Hz(1) = Hz(2);

 for i=2:N-1 % Update H field
      Hz(i) = Hz(i)-am(i)*(Ey(i+1)-Ey(i));
 end
% Absorbing boundary conditions for right
propagating waves
 Ey(N) = Ey(N-1);
 for i=2:N-1 % Update E field
      Ey(i) = as(i)*Ey(i)-ae(i)*(Hz(i)-Hz(i-1));
 end

% Show the evolution of E and H fields over total
simulation time
figure(2)
subplot(211),plot(x,Ey,'k','LineWidth',2);
axis([3*delx L -1 1]);
set(gca,'FontName','Times','FontSize',14,'Xtick',
```

```
[1,2,3,4,5],'XtickLabel',[])
ylabel('E_y (V/m)')
% grid on;
subplot(212),plot(x,377*Hz,'k','LineWidth',2);
axis([3*delx L -1 2]);
set(gca,'FontName','Times','FontSize',14,'Xtick',
[1,2,3,4,5])
ylabel('377*H_z (A/m)')
% grid on;
xlabel('x (m)');
pause(0);
iter

end
%%%%%%%%%%%%%%%%%%%%%%%%%%%%%%%%%%%%%%%%%%%%%%%%%%%%
```

4.4 Summary

In this chapter, we have gained insight of EMP propagation through various media – free-space and materials (dielectric and perfect conductor). Computer codes have been developed in MATLAB, which is compatible with its freeware clone package OCTAVE. Also, this chapter provides great details for writing scripts for visualizing the simulation output data. Analytical approach has been introduced to correlate the simulation results.

Exercises

4.1 In this chapter, the computer codes for the pulsed electromagnetic wave propagation in air-material half-space have been listed. Run the programs, execute the postprocessing scripts, and reproduce the results illustrated in Figures 4.1 through 4.13.

4.2 The impedance that a medium offers to an incident EMP can be manipulated through the variation of material properties (ϵ_r, σ). Study the electromagnetic pulse wave phenomena by computing the reflection (Γ), transmission (T), and attenuation (α) coefficients by varying the medium properties (ϵ_r, σ). Use analytical equations to gain physical insight.

4.3 The Gaussian waveforms have been chosen for representing the EMP waveform. Repeat Exercise 4.2 for EMPs having Gaussian waveforms of different rise-times, derivatives of Gaussian waveforms, modulated Gaussian such as Gaussian times $\sin(\omega t)$, Gaussian times $\cos(\omega t)$, where $\omega = 2\pi f$. Choose f as maximum frequency in the desired EMP or any frequency of your interest.

5

Simulation of Capacitor Bank

5.1 Introduction

It is necessary to set up the capacitor and switch within the finite-difference time-domain (FDTD), along with the transverse electromagnetic (TEM) cell. Since the capacitor and switch can have a fairly complex geometry, it becomes necessary to make idealizations in their geometries and dimensions. However, assuming that the geometry is reasonably well represented, this method offers the advantage of taking into account distributed parameter effects within the capacitor and switch. These effects can often be important due to the short time-scales of this event, involving risetimes as low as 1–2 ns [9].

Details of the overall idea behind a self-consistent analysis of an electromagnetic pulse (EMP) simulator will be presented in Chapter 6 – the discussion would focus on the electromagnetic field variation within the TEM structure. In this chapter, we present details of the methodology used for setting up a parallel-plate capacitor, and its charging and discharging, using the FDTD method.

The FDTD method involves the numerical solution of Maxwell's curl equations [32]. Hence, it does not explicitly include charge dynamics. The presence of charges in the FDTD grid is known by the divergence of electric field from a Gaussian surface bounding the object of interest [30].

This chapter is organized as follows: Section 5.2 gives details of the computational model for the capacitor and closing switch, including the charging and discharging process. Results for charging and discharging are presented in Section 5.3. In Section 5.4, we describe the use of the method-of-moments (MOM) to cross-check the FDTD results. Subsequent sections deal with specific details of performance. The effect of boundary conditions on the electric field distribution is examined in Section 5.5. The conclusive remarks of the chapter is summarized in Section 5.6.

Electromagnetic Pulse Simulations Using Finite-Difference Time-Domain Method, First Edition.
Shahid Ahmed.

5.2 Details of Model

5.2.1 Description of Geometry

We have set up an air-gap parallel-plate capacitor with square plates, inside the FDTD domain. The plates are made of perfect electrical conductor (PEC). Figure 5.1 shows a schematic of the arrangement. Apart from the plates, switch and resistive connection, the rest of the FDTD domain is assumed to consist of free space, i.e. it is an electrical insulator and has a dielectric constant ϵ_0. A second-order outer radiation boundary condition (ORBC) has been used [32]. Figure 5.1 shows square plates of side w_p, separated by a gap g_p. The aspect ratio of the plate, A_s, is defined as w_p/ g_p. We have examined capacitors with a wide range of aspect ratios. The limit of infinite A_s would correspond to an ideal parallel-plate capacitor, with a capacitance given by $\epsilon w_p{}^2/g_p$, where ϵ is the dielectric constant.

The number of computational cells in the three directions, and the mesh sizes, vary from problem to problem. However, an example will illustrate typical numbers. For the case of a capacitor with $A_s = 20$, corresponding to w_p of 2 m and g_p of 0.1 m, a typical FDTD domain consists of 262 × 122 × 122 cells in the x, y, and z directions, with cell sizes Δx, Δy, and Δz of 0.5, 3.34 and 3.34 cm, respectively.

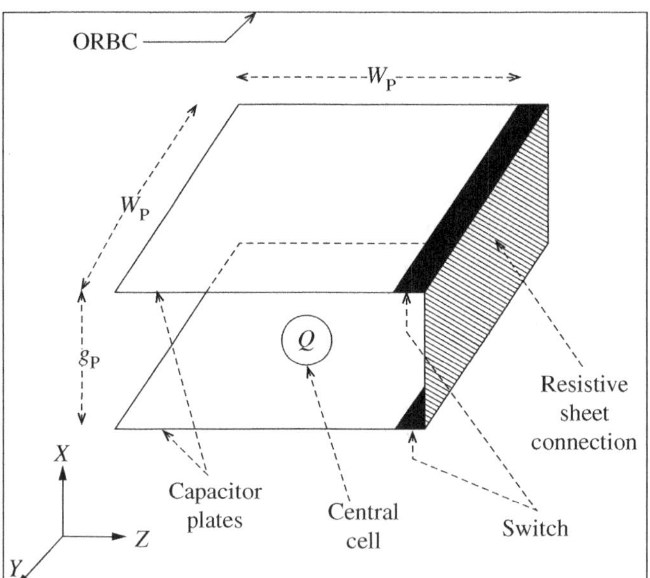

Figure 5.1 Schematic of parallel plate capacitor, showing the switch and a resistive sheet connection between the two plates.

This corresponds to a time-step Δt of 16.3 ps according to the Courant stability criterion [32]. The capacitor is symmetrically placed at the center of the FDTD domain. This means that there is a spacing of 31 cells between the capacitor edge and the domain boundary in the y and z directions and 121 cells in the x direction.

5.2.2 Method of Charging

We could, in principle, have started the FDTD simulation with an initial electric field distribution $\vec{E}(\vec{r})$ corresponding to the equilibrium distribution in the capacitor. However, this requires prior knowledge of $\vec{E}(\vec{r})$, not just within the capacitor but throughout the FDTD domain. A search of the available literature did not yield any analytical form valid for arbitrary aspect ratio. Hence, it was decided to start with an uncharged capacitor.

Electrostatic energy can be stored in a capacitor by charging it to its rated voltage level. Different techniques have been reported in the literature for depositing charge within the FDTD domain. For example, Wagner and Schneider [30] have used a single-element filamentary current source at a particular grid location for studying grid capacitance effects.

In our simulation, capacitor charging has been accomplished by applying a time-dependent electric field $E_x(t)$ in the space between the two plates. Any temporal waveform of E_x, starting at zero and reaching the desired peak value, could be used here. The only constraint is that the highest frequency with significant power content should not exceed the maximum frequency that can be handled accurately by the FDTD mesh. We have chosen a half-Gaussian waveform given by

$$E_x(t) = \begin{cases} E_0 e^{-\alpha(t-\tau)^2} & \text{if } 0 \le t \le \tau \\ 0 & \text{if } t < 0 \end{cases} \tag{5.1}$$

where τ is the characteristic time of charging and $\alpha = (4/\tau)^2$. This choice of α is explained in [32] for a full Gaussian waveform. This waveform means that the excitation is only applied until the Gaussian reaches its peak value.

The FDTD algorithm evolves the electric and magnetic fields in the entire domain as a function of time. For the computational cells lying in-between the top and bottom capacitor plates, the electric field evolved by the FDTD equations is overwritten by the above form, for $0 \le t \le \tau$. FDTD implementation of the impressed electric field in the grid space can be coded as follows:

```
if (t .ge. 0 .and. t .le. tau) then
do 40 i = imin, imax - 1
  do 40 j = jmin + 1, jmax - 1
    do 40 k = kmin + 1, kmax - 1
40        Ex(i,j,k) = Ex(t)
```

where “imin” and “imax” correspond to the location of the top and bottom plates (i.e. the separation between the plates), however, the j and k index values represent the dimension of the plates.

This means that we violate Maxwell’s equations over some interval of space and time. Gauss’ law (Eq. (5.2)), then implies that a finite charge is deposited on the plates, even though there is no physical connection between them. The peak applied electric field, E_0, depends upon the desired voltage between the two plates. The method for choosing τ is discussed below.

Accurate calculations using the FDTD technique require that there be at least 10 computational cells within a wavelength at the highest frequency of interest [32]. Hence, the smallest wavelength λ_{min} that can be accurately handled by the mesh is governed by the biggest computational cell of the FDTD domain. We must, therefore, choose τ such that, at the highest frequency that can be handled by the mesh, $|E_x(\omega)|$ falls by several orders of magnitude as compared to its peak value. Now, for a full Gaussian pulse, $|E_x(\omega)|$ falls by seven orders at a frequency $f_{\text{h}} \simeq 6/\tau$ [32]. An example will illustrate this calculation of τ.

Consider a parallel-plate capacitor having square plates of side 2 m in the y- and z-directions, a plate separation of 0.1 m in the x-direction, and cell sizes Δx, Δy, and Δz of 0.5, 3.34 and 3.34 cm, respectively. We then have $\lambda_{\text{min}} = 10 \times \Delta y = 0.334$ m, since $\Delta y = \Delta z$. The highest frequency that can be handled by FDTD mesh is given by $f_{\text{h}} = c/\lambda_{\text{min}} = 898$ MHz. This, in turn, allows a minimum permissible $\tau = 6/f_{\text{h}} = 6.68$ ns.

It must be noted that this is only a lower bound on τ – we could always opt for higher values. However, recall that the FDTD timestep is given by the Courant criterion, which depends only on cell sizes and the speed of light through the medium. Higher values of τ would, therefore, require longer computational times for charging the capacitor, with no compensating benefit. Hence, in this work, we have used the minimum τ value obtained by the above method.

This method of charging yields a nonuniform charge density distribution on the plates, which fluctuates in time before stabilization. This is discussed in a later section.

5.2.3 Method for Calculating FDTD Charge and Capacitance

An FDTD simulation does not explicitly account for charge dynamics, since it only evolves Maxwell’s curl equations. The presence of charges is only observed, therefore, through the divergence of the electric field [30]:

$$\vec{\nabla} \cdot \vec{E} = \frac{\rho}{\epsilon_0} \tag{5.2}$$

where ρ is the charge density.

To measure the total charge associated with either plate of the capacitor, we construct an imaginary bounding box, called the "Gaussian surface," which encloses the plate of interest. This surface could be of any shape – for convenience, it is taken to have the shape of a cuboid, with each of its six faces lying normal to either the x-, y- or z-axes. We then apply the integral form of Gauss's law (Eq. (5.2)) on each face of the bounding box, to calculate the flux coming out of that face. For example, the net flux Γ_x flowing out of the two faces normal to the x-direction is obtained by adding up the fluxes through the positive and negative x-faces, Γ_{x+} and Γ_{x-}, respectively, with the proper sign:

$$\Gamma_x = \Gamma_{x+} - \Gamma_{x-} \tag{5.3}$$

The sign convention is chosen such that the flux flowing out normal to a surface is positive. Integration of Γ_x over the x-directed surface of the cuboid yields the charge, Q_x, associated with the x-directed flux. Q_y and Q_z can be calculated in the same way.

The magnitude of total charge Q_{FDTD} on the plate is given by

$$Q_{\mathrm{FDTD}} = Q_x + Q_y + Q_z \tag{5.4}$$

The capacitance C_{FDTD} is now determined from the ratio of the total charge Q_{FDTD} on either plate to the potential difference V_{FDTD} between the plates, i.e.

$$C_{\mathrm{FDTD}} = Q_{\mathrm{FDTD}}/V_{\mathrm{FDTD}} \tag{5.5}$$

$$V_{\mathrm{FDTD}} = \int_{+\mathrm{plat}}^{-\mathrm{plat}} \vec{E} \cdot \vec{dl} \tag{5.6}$$

where the line integral is taken from the positive to the negative plate. The example of FDTD codes for calculating the potential difference between the capacitor plates separated along the x-direction is illustrated below.

```
Voltage = 0.0d0
Do 2 i = imin,imax
    Voltage = Voltage + EX(i,j,k) * Delx
2   Continue
```

In the above FDTD code for voltage calculation, it has been considered that the other components, e.g. E_y and E_z are negligibly small. However, the effects of all the E-field components can be taken into account by the line integral along the x-direction illustrated below.

```
Voltage = 0.0d0
Do 2 i = imin,imax
    Voltage = Voltage + (EX(i,j,k) + EY(i,j,k) + EZ(i,j,k))* Delx
2   Continue
```

The FDTD codes for calculating the charge deposited on the plates can be obtained from the following example:

```
j = (jmin + jmax)/ 2
k = (kmin + kmax)/2
  Do i = 1,nx
    Net Flux =(EX(i+1,j+1,k+1)-EX(i,j+1,k+1))*Dely*Delz
                    +(EY(i+1,j+1,k+1)-EY(i+1,j,k+1))*Delx*Delz
                      +(EZ(i+1,j+1,k+1)-EZ(i+1,j+1,k))*Delx*Dely
  Enddo
  charge = Net Flux * eps0
```

where jmin, jmax and kmin, kmax are the minimum and the maximum extension of the plate along y- and z-directions, respectively.

The capacitance obtained in this way has been cross-checked by letting the capacitor discharge through a known resistance R, and measuring the characteristic discharge time. If inductive effects are kept small, the e-folding time for the decay of total charge on a plate is the RC time, which yields the capacitance. Details of this procedure are given in Section 5.2.5.

5.2.4 FDTD Model of Closing Switch

A detailed description of the simplified model used for the fast closing switch has already been presented in Chapter 6. A brief description is presented here for the sake of completeness.

The idealized closing switch consists of two thin resistive strips, each having a length of one computational cell in the z-direction, placed at one end of the top and bottom plates. The arrangement is shown in Figure 5.1.

The strip material has a time-dependent resistivity, which is kept infinite during the charging of the capacitor. This means that no charge can move through the sheet termination while the capacitor is being charged. The temporal waveform for switch conductivity, from the fully open to the fully closed state, is assumed to be given by a "half-Gaussian":

$$\sigma = \sigma_c \exp\left[-\alpha_s(t - \tau_s - \tau_0)^2\right], \quad \tau_0 \le t \le \tau_s + \tau_0 \tag{5.7}$$

where τ_s is a characteristic time for closure, τ_0 is the time when switching starts, $\alpha_s = (4/\tau_s)^2$ and σ_c is the switch material conductivity in the fully closed state. As in the case of the capacitor-charging waveform, other functional forms could also have been used. However, a half-Gaussian waveform is better suited to model switching, despite its limitations as regards continuity of the second derivative at the peak [39].

The FDTD code for "closing switch" is illustrated below.

```
taus = 7.0e-9
aa = (4.0d0 / taus)**2
if (t .ge. (tau0 + taus)) then
        sigma(t)=sigma0
else
        sigma(t)=sigma0*exp(-aa*(t-(tau0+taus))^2)
endif
```

where σ_0 is the switch conductivity that can be determined by the material properties (resistivity ρ) and the dimension (area $[A]$ and length $[l]$) of the switch. This can be calculated by

$$\sigma_0 = \frac{1}{\rho} = \frac{l}{R \times A} \tag{5.8}$$

Note that the switch closure time τ_s is usually take much longer time than the capacitor charging time τ_0 in order for the transient to settle down. As the conductivity $\sigma(t)$ is time-dependent, it should be updated at each time-step.

In a capacitor charging-discharging problem, τ_0 should be large enough so that the charge density distribution on the capacitor plates equilibrates before discharge is started. We find that $\tau_0 \simeq 5\tau$ generally yields satisfactory results. Here, τ is the time constant used for capacitor charging (Eq. (5.1)). τ_s must be chosen so as to avoid the generation of frequencies higher than those that can be handled by the FDTD mesh [32].

5.2.5 Discharging a Charged Capacitor

The parallel-plate capacitor is charged as described in Section 5.2.2. Switching starts at $\tau_0 = 5\tau$ and is completed by $\tau_{sc} = \tau_0 + \tau_s$. Since the switch resistance is finite even during the switching process, capacitor discharge starts for $t > \tau_0$. However, the assumed half-Gaussian waveform of conductivity implies that the conductivity becomes high only toward the end of the switching period. This, in turn, implies that the rate of discharge becomes significant only after t comes fairly close to τ_{sc}.

The capacitor discharges through a series combination of the switch and sheet resistances, yielding a total resistance R_{eff}. The switch conductivity in the fully closed state, σ_c, is chosen such that the sheet resistance, R_{sheet}, dominates by far. The resistivity of the sheet resistance, ρ_{sheet}, is chosen such that

$$R_{\text{sheet}} = \rho_{\text{sheet}} \frac{L_{\text{sheet}}}{A_{\text{sheet}}} \tag{5.9}$$

where L_{sheet} and A_{sheet} are the effective length and cross-sectional area of the load resistance, respectively.

If inductive effects are neglected, the characteristic time for discharging, τ_{dis}, is given by

$$\tau_{dis} = R_{eff}C_{FDTD} \tag{5.10}$$

A sheet resistance has been chosen in order to minimize the inductance. In the FDTD simulation, the total charge on either of the plates, $Q_{plate}(t)$, is monitored as a function of time. τ_{dis} is obtained by fitting an exponential curve to $Q_{plate}(t)$, starting at $t = \tau_{sc}$. Since R_{sheet} is known, the best fit value of capacitance, C_{fit}, can be calculated. This capacitance serves as a cross-check on the value calculated by an alternative method in Section 5.2.3.

5.3 Results and Discussion

5.3.1 Charge Deposition on Plates

Figure 5.2 shows the charge deposition in corresponding cells on the top and bottom plates. As expected, the charges are equal in magnitude and opposite in sign. The capacitor plates have a width w_p of 2 m and separation g_p of 0.1 m.

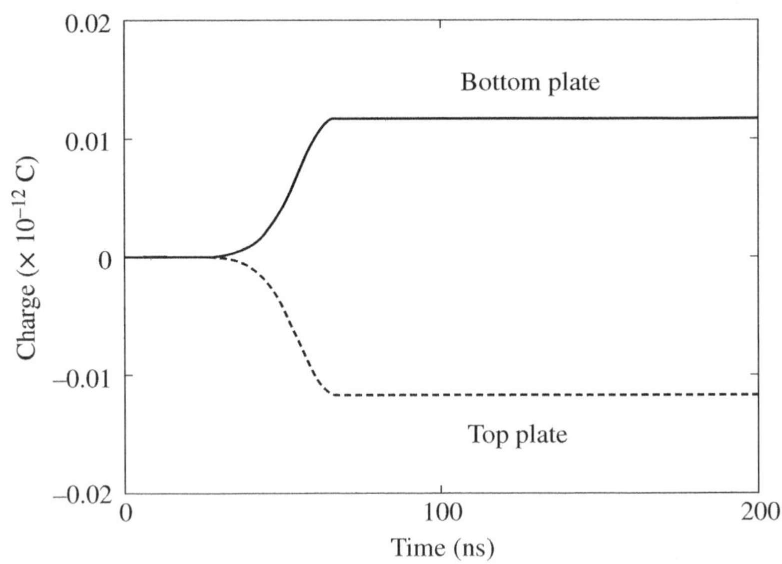

Figure 5.2 Charge deposition on corresponding computational cells of the upper and lower plates, for a capacitor with $w_p = 2$ m and $g_p = 0.1$ m.

The computational domain consists of 222 × 122 × 122 cells in the x, y, and z directions, with cell sizes of 0.5, 3.34, and 3.34 cm in the three directions, respectively.

5.3.2 Stabilization of Charge Density Distribution

As discussed earlier, charge is deposited on the plates by violating Maxwell's equations in the computational region between the two plates, over the charging period τ. During this period, the applied electric field E_x is assumed to be uniform in the space between the plates. Now, in a charged parallel-plate capacitor of finite size, there are always fringing fields which increase in magnitude as we approach the edges of the plates. This means that E_x varies to some extent between the central region of the capacitor and its edges. The mismatch between the actual E_x distribution and the distribution imposed during charging means that for $t > \tau$, the electric field throughout the FDTD domain will change in time, and gradually relax to its equilibrium distribution. Since we calculate the charge density from Gauss' law, its distribution over the plates is also expected to settle down in the same way. This is indeed observed, as shown in the following example:

Consider a capacitor having $A_s = 20$, set up within an FDTD mesh described in Section 5.2.1. Figure 5.3 shows the charge level in the central cell of the bottom plate – this location is indicated in Figure 5.1. The charge is seen to undergo damped oscillations, with an oscillatory period of ~7.3 ns. This is fairly close to one

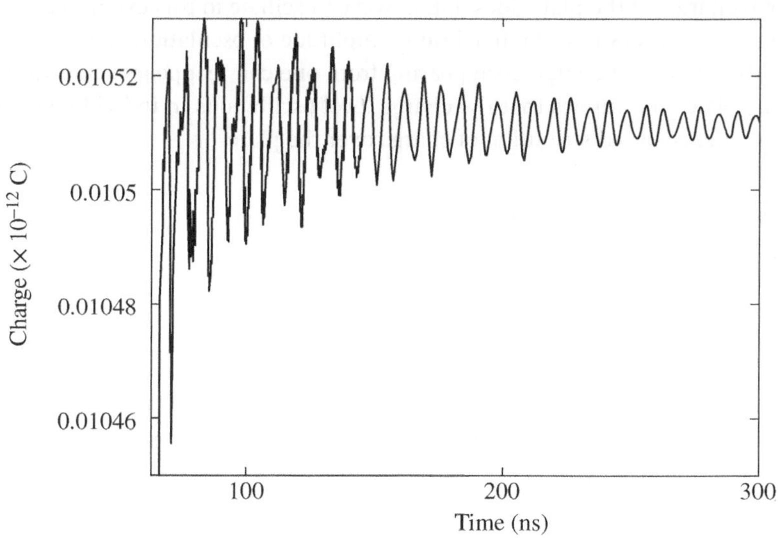

Figure 5.3 Charge stabilization in the central cell of the bottom plate. The location of this cell is indicated in Figure 5.1.

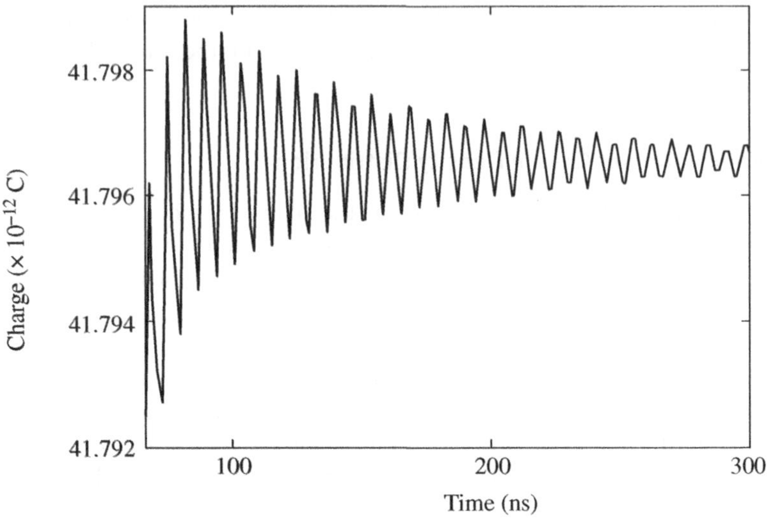

Figure 5.4 Total charge on bottom plate.

transit time through the plate width, i.e. 6.7 ns. Charge stabilization occurs after about 20–25 transits. The period of oscillation indicates that there is an oscillatory, but damped, flow of charge within the plate. The amplitude of oscillation is ~0.2% for $t > 80$ ns.

The total charge on the plate does not, however, oscillate to this extent. For $t >$ 80 ns, Figure 5.4 shows that the maximum amplitude of oscillation is only 4.8×10^{-3}%. This means that charge disappearing from one cell reappears somewhere else on the plate and implies that the charge level oscillations in individual cells are a consequence of redistribution of charge over the plate.

5.3.3 Determination of Characteristic Discharge Time

Figure 5.5 shows the charging, stabilization, and discharge waveforms for the central cell on each plate. Following the procedure described in Section 5.2.3, we get a capacitance of 434.8 pF, charged to 98 mV, and discharging through a series combination of a 0.02Ω switch and a 100Ω load resistance.

To determine the RC-time for discharge, we fit an exponentially decaying curve to the discharge data, starting a little after $t = \tau_{sc}$, i.e. after the switch resistance becomes constant. This portion of the data and the best-fit curve are shown in Figure 5.6. The e-folding time from curve fitting is $\tau_{dis} = 44.01$ ns, which matches within 1.2% with the RC value calculated from the abovementioned values of R and C. This validates the procedure described in Section 5.2.3 for calculation of capacitance.

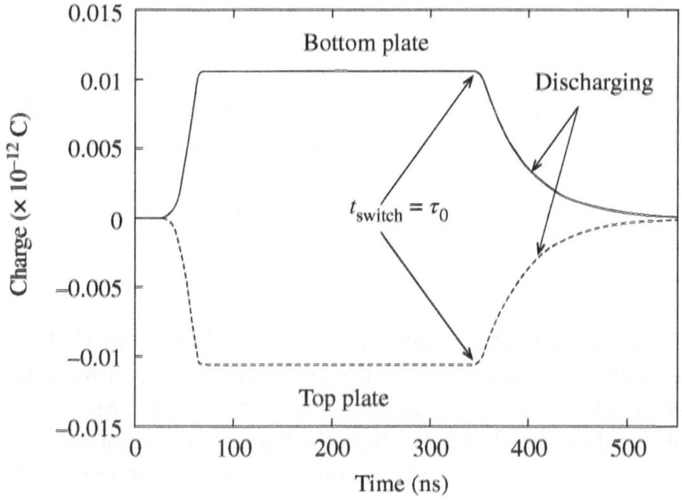

Figure 5.5 Charge variation with time in the central cells of both plates.

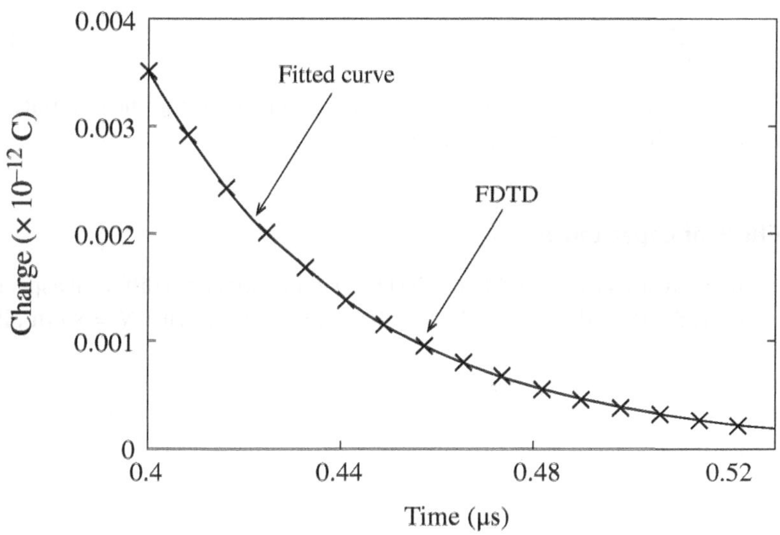

Figure 5.6 Discharge of capacitor along with the best-fit exponential Curve.

5.4 Cross-check of FDTD Results Using Method-of-Moments

The charge distribution on the plates and the resulting capacitance, have been cross-checked using the MOM [40]. Details of the method can be obtained from [40]. However, a few important features are given below for the sake of completeness.

The capacitor plates, assumed to be infinitesimally thin in the x-direction, are divided into a rectangular computational mesh in the y-z plane. It is assumed that the charges in all rectangles are concentrated at their respective centers, hereafter called the "mesh points." Each of these point charges contributes to the electrostatic potential at all mesh points. The objective is to determine the value of charge at each mesh point such that all points on a given plate have the same potential. The calculation of potential on either plate must take into account contributions from both plates.

Since we have square plates, these are divided into an $N \times N$ square mesh. From symmetry considerations, we need only calculate the charges at a quarter of the mesh points, i.e. $N^2/4$ points in all. For accuracy, N must be chosen large enough so that the mesh size is significantly smaller than the inter-plate separation.

We thus have $N^2/4$ linear equations with an equal number of unknowns. The solution yields the charge at each mesh point. The total charge on each plate, Q_{MOM}, is obtained by summing these charges. The capacitance is given by

$$C_{\text{MOM}} = \frac{Q_{\text{MOM}}}{V_{\text{MOM}}} \tag{5.11}$$

The charge density in each cell can now be calculated by dividing the computed charge at each mesh point by the surface area.

5.4.1 Check of Capacitance

We now compare results from MOM and FDTD for capacitors with different aspect ratios A_s, viz. 1, 5, 10, and 20. The MOM calculation is done with $N = 80$ in all cases.

The capacitor having $A_s = 20$ is studied using the computational mesh described in Section 5.2.1. Due to various numerical constraints, the FDTD calculations for capacitors of aspect ratios 1, 5, and 10 are done with a computational domain consisting of $110 \times 90 \times 90$ cells in the x, y, and z directions. This mesh has cell sizes $\Delta y = \Delta z = 6.67$ cm for all aspect ratios, while Δx has values of 20, 4, and 2 cm for $A_s = 1$, 5 and 10, respectively. In all cases, we have $10 \times 30 \times 30$ cells in the capacitor region and gaps of $50 \times 30 \times 30$ cells between the capacitor edges and the boundaries in the x-, y-, and z-directions.

Table 5.1 compares results from the two techniques. Also shown in Table 5.1 is the theoretical capacitance for an infinite parallel-plate capacitor, given by

$$C_{\text{thy}} = \frac{\epsilon_0 w_{\text{p}}^2}{g_{\text{p}}} \tag{5.12}$$

Table 5.1 Comparison between FDTD and MOM capacitance for different aspect ratios.

A_s	C_{FDTD} (pF)	C_{MOM} (pF)	$100(1 - C_{FDTD}/C_{MOM})$ (%)	C_{thy} (pF)
1	48.4	61.7	21.6	17.7
5	136.1	145.1	6.2	88.5
10	241	245.8	1.95	177
20	430.8	431.3	0.1	354

Three points may be noted. First, as A_s increases, the results from both methods gradually approach C_{thy}. This is what we expect, since C_{thy} is valid for infinite aspect ratio. Second, there is reasonable agreement between MOM and FDTD, except at $A_s = 1$, despite the fact that the two methods rest on different assumptions. Third, the agreement between FDTD and MOM clearly becomes better as we go to higher A_s. This systematic convergence is explained in Section 5.4.3.

5.4.2 Edge Effects on Charge Density Distribution

We expect the electric field and the charge density to be higher near the edges of the plates as compared to the central regions. This effect should be particularly marked at the four corners of a square plate.

This effect is indeed observed in the FDTD simulations. Figure 5.7 shows the distribution of charge density ρ over the surface of the bottom plate of a capacitor with $A_s = 20(w_p = 2$ m, $g_p = 0.1$ m). The same FDTD mesh has been used

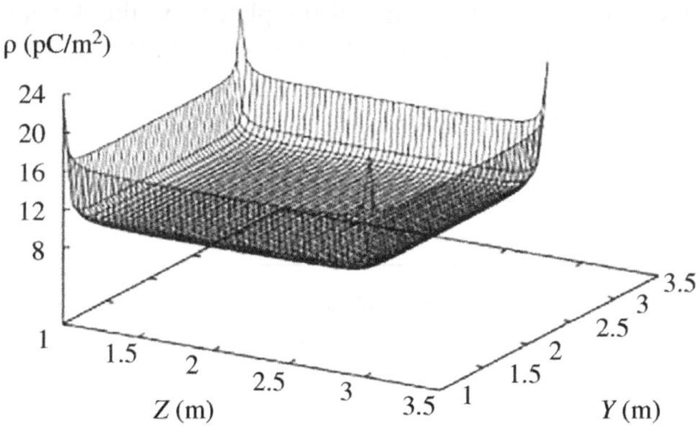

Figure 5.7 Charge-density distribution over the surface of the bottom plate, computed using the FDTD method. pC refers to picocoulombs.

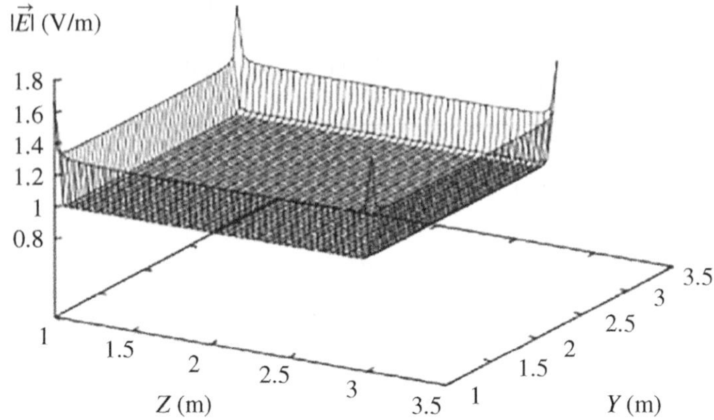

Figure 5.8 Distribution of electric field amplitude $|\vec{E}|$ over the surface of the bottom plate. The distribution has been computed using the FDTD method.

as in Section 5.2.1. The corresponding electric field amplitude, $|\vec{E}|$, is shown in Figure 5.8. As expected, $|\vec{E}|$ is almost uniform over the central region of the plate but rises sharply near the corners.

5.4.3 Check of Charge Density Distribution

For the same capacitor as in Section 5.4.2, the charge density distribution computed using FDTD and MOM has been compared in Table 5.2.

Three different mesh sizes have been examined, corresponding to $N = 20$, 40, and 60 in the MOM mesh. Now, the MOM mesh is two-dimensional and only covers the plates. The FDTD mesh, on the other hand, is three-dimensional and must also allow some distance between the edges of the plates and the domain boundaries. Hence, for the three cases, we require FDTD meshes of $110 \times 60 \times 60$, $110 \times 80 \times 80$, and $262 \times 120 \times 120$ cells, with Δx, Δy, and Δz values of (1, 10, 10) cm, (1, 5, 5) cm, and (0.5, 3.34, 3.34) cm, respectively.

Table 5.2 Comparison of surface charge density from FDTD and MOM, for $A_s = 20$.

N_{MOM}	ρ_{FDTD} (pC/m^2)		ρ_{MOM} (pC/m^2)	
	Low	High	Low	High
20×20	9.55	13.62	9.51	15.25
40×40	9.64	17.84	9.17	21.37
60×60	9.52	23.26	9.24	27.82

In Table 5.2, the symbols "high" and "low" refer to charge densities at the corners and the center of the plate, respectively. The charge densities are in picocoulomb per square meter (pC/m^2).

Three conclusions can be drawn. First, there is reasonable agreement between the calculated charge densities at the center of the plate. Also, these values are relatively insensitive to the mesh resolution. Second, the charge density at the corners increases when we increase the spatial resolution. This is consistent with the fact that charge density is inversely related to the radius of curvature; since the plates have right-angled corners, i.e. zero radius of curvature, the charge density would theoretically become infinite at the corners. The finer the mesh we use, the closer we get to the corner, hence, the increase in the peak charge density. Third, there is significant deviation between the "high" values computed by the two methods.

The deviations in the "high" values could be due to differences in the assumptions underlying the two methods. For example, in the FDTD method, we have calculated the charge density from $\vec{E}$ by assuming that the components of $\vec{E}$ vary linearly in all directions, thereby implying a constant charge density in each computational cell. In MOM, on the other hand, the charge is assumed to be localized at each mesh point. This difference may not be important in regions where $\vec{E}$ changes slowly, such as the plate center, but becomes important near the corners.

We can now explain the observation in Section 5.4.1 that the capacitances computed using MOM and FDTD show progressively better agreement as the aspect ratio is increased. Table 5.3 compares the charge densities computed using the two methods for different aspect ratios. The FDTD computational domain used in these calculations is described in Section 5.4.1. The MOM calculation is done with $N = 30$ for $A_s = 1, 5, 10$ and $N = 60$ for $A_s = 20$.

We observe that for all values of A_s, the central charge densities yielded by FDTD and MOM differ by less than ~3%. However, the charge densities in the corner cells are significantly higher with MOM. The mismatch decreases as A_s increases. This means that MOM will yield a higher charge on the plate, and hence a higher capacitance, for low values of A_s.

Table 5.3 Comparison of FDTD and MOM charge density for different aspect ratios.

Aspect ratio	ρ_{fdtd} (pC/m^2)		ρ_{MOM} (pC/m^2)	
(A_s)	Low	High	Low	High
1	15.54	38.12	15.65	145.22
5	10.36	27.9	10.08	43.45
10	9.77	21.25	9.45	27.26
20	9.52	23.26	9.24	27.82

5.5 Effect of Boundary Condition

A second-order ORBC has been used in the simulations [32]. This boundary condition assumes that the fields radiated by the object are plane waves traveling normal to the domain boundaries. This assumption is valid only when the object-to-boundary distance is at least of the order of a wavelength.

It is difficult to rigorously satisfy this condition when the problem involves a wide range of frequencies. On the one hand, even if the object being simulated is small in size, the total domain size would be governed by the lowest frequency. On the other hand, the mesh size would have to be small enough to handle the highest frequency of interest. This combination would require unacceptably large numbers of computational cells.

Furthermore, it is clearly impossible to satisfy this condition for very low frequencies, especially zero frequency. We therefore expect some error in our calculations, especially in the long-time response. Specifically, for the capacitor problem, the final charge density distribution is likely to be affected by the proximity of capacitor plates to the domain boundary.

Luckily, for the capacitor-charging problem, a simple solution suggests itself. It is known that the fringing fields fall off with distance from the plates, with a characteristic decay length of the order of the plate separation. If the field strength at the boundary were small, any errors introduced by the application of ORBC would have a smaller impact upon the charge distribution. Hence, the solution is to maintain a distance between the plates and domain boundaries equal to several times the plate separation.

We have studied the boundary effect for a capacitor with $A_s = 20$, i.e. $w_p = 2$ m and $g_p = 0.1$ m. The distance between the plates and the domain boundaries in the y and z directions, i.e. D_y and D_z, are held fixed at 1 m, i.e. $D_y/g_p = D_z/g_p = 10$. This is because any further increase in this ratio does not yield significant changes in the results. The third parameter, D_x/g_p, is varied from 3 to 6. The FDTD domain has mesh sizes of 0.5, 3.34, and 3.34 cm in the x, y, and z directions, respectively. In each case, the FDTD capacitance (C_{FDTD}) is compared with the MOM capacitance (C_{MOM}). Figure 5.9 shows that the deviation decreases steadily as D_x/g_p is increased, with a mismatch of less than 1% for $D_x/g_p = 5$.

The convergence between MOM and FDTD results can be explained by modification of the electric field distribution when the boundaries are closer to the object. Figures 5.10 and 5.11 show the contours of $|\vec{E}|$ in the $x - z$ plane for the cases with $D_x/g_p = 3$ and $D_x/g_p = 6$, respectively. The contours are shown at the central value of y in the domain, i.e. 2 m.

For the case with $D_x/g_p = 6$, we see that the contours are fairly symmetric about the central values of x and z in the domain, i.e. 0.65 and 2 m, respectively. However, there is marked asymmetry in the case with $D_x/g_p = 3$. This shows that the

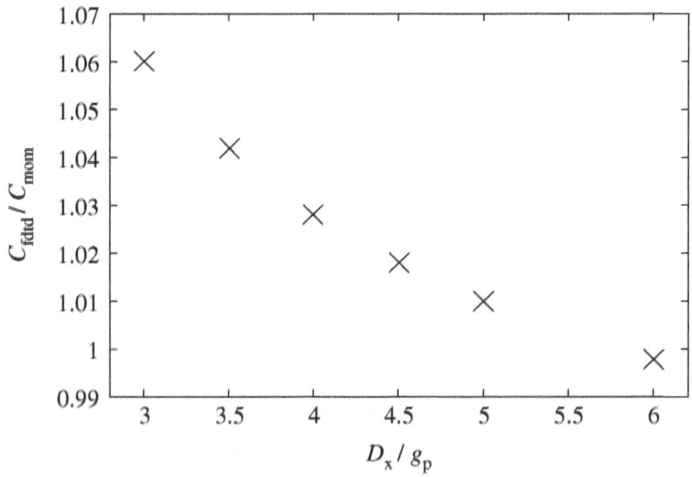

Figure 5.9 Effect of distance to boundary on the capacitance for $D_y/g_p = D_z/g_p = 10$.

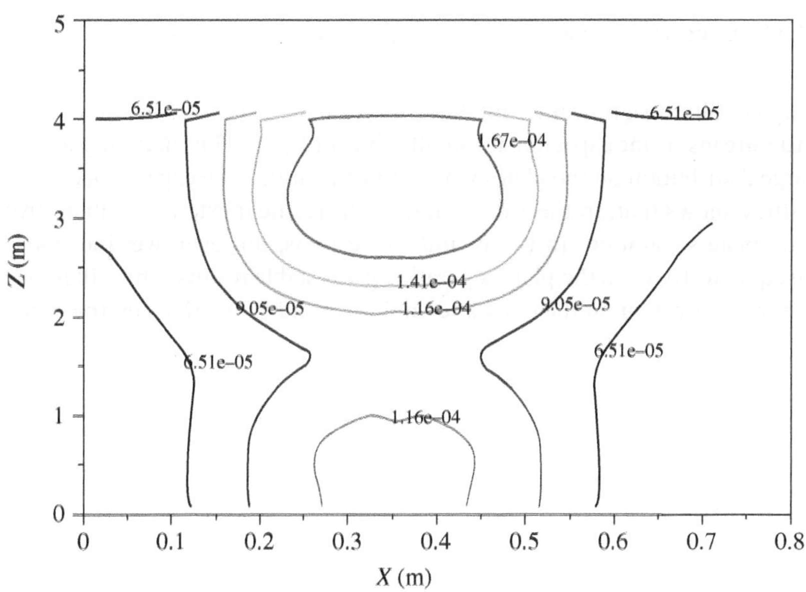

Figure 5.10 Contours of $|\vec{E}|$ for $D_x/g_p = 3$, $D_y/g_p = D_z/g_p = 10$.

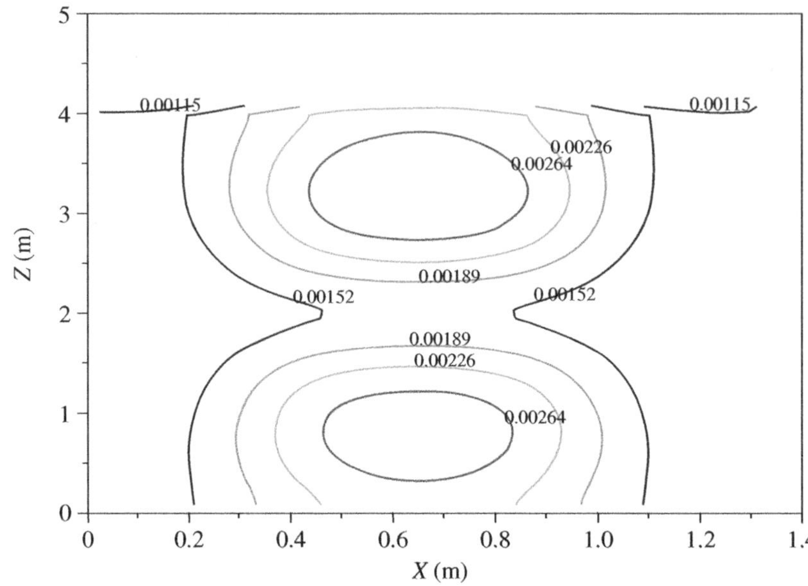

Figure 5.11 Contours of $|\vec{E}|$ for $D_x/g_p = 6$, $D_y/g_p = D_z/g_p = 10$.

boundary condition significantly modifies not just the fringe fields but also the field structure inside the capacitor, for small values of D_x/g_p. This, naturally, affects the charge distribution on the plates which, in turn, affects the capacitance.

Our study shows that, in the x direction, the fringe-field extends to about five times the plate separation. In the y- and z-directions, however, we find that a spacing equal to 10 times the plate separation gives stable results. This difference between the x direction on the one hand, and the y and z directions on the other, can be understood as follows. The perfectly conducting plates of the capacitor are likely to greatly reduce fringe field effects in the x-direction, which is normal to the plane of the plates. Hence, a smaller gap can be tolerated in the x direction.

5.6 Summary

We have performed a three-dimensional FDTD analysis of the charging and discharging of an air-gap, parallel-plate capacitor with square plates. Charging of the plates has been achieved by violating Maxwell's equations in the space between the plates over a certain period.

The simulation gives reasonably accurate results for the 3-D structure of electric field within a charged capacitor and the fringing fields in its vicinity. The capacitance values so obtained generally show a good match with the MOM, except when the width of the plates and the inter-plate separation are comparable.

The study has helped identify the critical numerical issues that must be taken into account. First, we find that the FDTD domain must be large enough to allow a gap between the plates and the domain boundaries that is five to ten times the plate separation. This has also been explained in terms of the fringing fields. Second, the temporal waveform of the charging electric field should be chosen such that it does not excite frequencies higher than the highest frequency that the FDTD mesh can handle. A similar restriction applies to the temporal waveform of the closing switch resistance, when the capacitor is being discharged. Third, after the charging phase is over, some time must be allowed for the electric fields to equilibrate.

FDTD modeling of pulsed experiments is normally done with only the load, e.g. an antenna, set up inside the FDTD domain, the evolution of the pulsed energy source being modeled by some other method. The present study shows one way to set up the pulse-power driver within the same domain.

The study has been done for a single, simple geometry, viz., a parallel-plate capacitor with an air gap. However, the methodology can readily be extended to more complex systems. It can be particularly useful for examining distributed-parameter effects inside capacitors and other fast energy storage and transmission devices.

Exercises

5.1 The capacitors used in short-pulse experiments tend to be compact, which implies the use of dielectrics with high dielectric constant ϵ. The geometry may also be different, e.g. tubular instead of parallel-plate. The difference in dielectric material and geometry means that transmission-line effects, charge distribution, and fringing-fields will be different.

5.2 As shown in Section 5.5, for a capacitor with aspect ratio $A_s = 20$, the distance of computational boundary from plates with respect to the gap between plates (D_x/g_p) has been progressively increased from three to six, holding $D_y/g_p = D_z/g_p = 10$. A good match of FDTD results with MOM was obtained only for D_x/g_p = 5 to 6. Perform simulations for capacitors with lower A_s and analyze the fringe field, charge distribution. The symbols have their usual meaning as defined in the chapter.

5.3 The use of second-order outer radiation boundary condition (ORBC) has an inherent property of absorption of only those waves which are plane and normal to the boundary. The wavefronts tend to become planar when the separation of the observation point (the boundary in this case) from the source becomes of the order of a few wavelengths. This condition may not be satisfied by longer wavelength (low-frequency) signals. Consider other boundary conditions such as perfectly matched layer (PML) to show the effect of the longer wavelength signals.

6

Bounded Wave Simulator for Electromagnetic Pulses

6.1 Introduction

Electromagnetic pulse (EMP) simulators are widely used for testing the susceptibility of electronic equipment to EMP. A bounded-wave EMP simulator, in its simplest form, consists of two electrically conducting triangular plates, making up a transverse electro-magnetic (TEM) structure, separated by a parallel plate region [10]. A capacitor bank, charged to high voltage, is discharged through a fast closing switch into the front triangular plate. The current flows through the structure to the rear triangular plate, then through a matching termination, and back through a conducting ground plane. The front plate, which displays a near-constant impedance over a wide frequency range, plays a significant role in determining the EMP waveform, while the middle and rear plates serve to guide the signal [10]. The object to be tested is mounted in the bounded volume of the parallel-plate region. The discharging capacitor produces an intense, rapidly varying electromagnetic field, covering a wide frequency range, in the vicinity of the test object [9].

6.1.1 Organization of This Chapter

We have divided the study of a bounded-wave EMP simulator into two parts. In Part I, consisting of Sections 6.2–6.8, we consider the performance of the capacitor-switch-TEM structure, with a resistive termination placed at the end of the tapered section. This means that the test volume and test object are not included, i.e. we only examine the formation and propagation of an EMP. In Part II, consisting only of Section 6.9, we model the structure shown in Figure 6.1, where the termination is placed at the end of a parallel-plate extension, and a

Electromagnetic Pulse Simulations Using Finite-Difference Time-Domain Method, First Edition.
Shahid Ahmed.

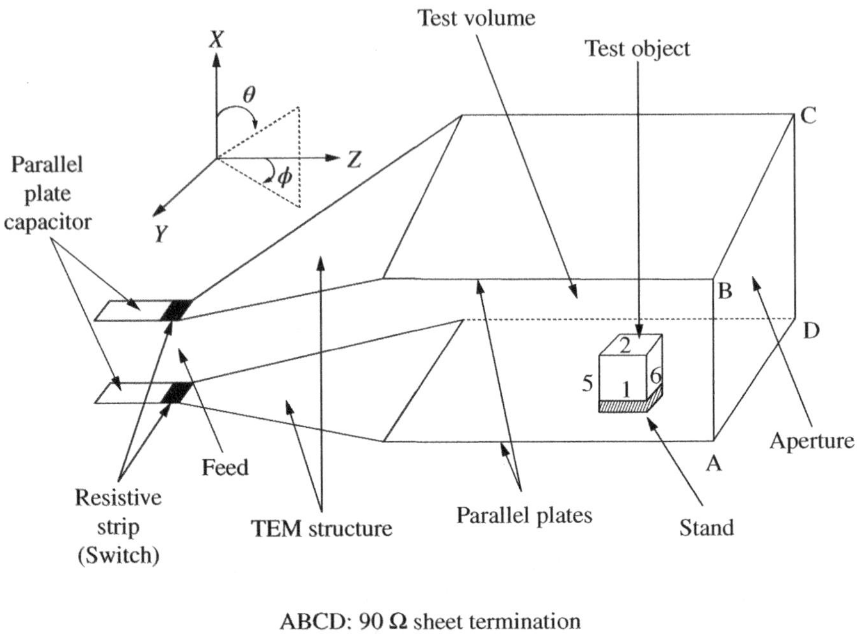

Figure 6.1 Schematic of EMP simulator with parallel-plate test volume and resistive sheet termination.

simple test object is placed within that extension. Here we study the interaction of the EMP with a test object.

This chapter is organized as follows: Section 6.2 describes the simulator examined in this study and provides some information about the computational model. Section 6.3 describes the procedure for confirming that we have set up the proper geometry of the TEM structure. An finite-difference time-domain (FDTD) model of the fast closing switch is described in Section 6.4. In Section 6.5, we explain the procedure for choosing a suitable distance between the simulator and the FDTD domain boundary. In Section 6.6, we present results of electromagnetic field structure within the bounded space, while current flow through the simulator plates is discussed in Section 6.7. The prepulse seen before switching is discussed in Section 6.8. Simulator performance in the presence of the test object is examined in Section 6.9. The results obtained using the FDTD code need to be validated against known results for simpler problems. Some of the validation checks are discussed in Section 6.10. Finally, the conclusions appear in Section 6.11.

6.2 Geometry and Computational Model

Figure 6.1 shows a schematic of the system being modeled. For purposes of illustration, the parameters of the pulser and the TEM structure are chosen to be similar to those of the EMP simulator described in [9]. The TEM structure of [9] has a characteristic impedance of 90 Ω and a matching resistive termination at the aperture. It is driven by an 82 picofarad (pF) peaking capacitor charged to 1 megavolt (MV). Only limited data is available about that simulator. To facilitate the analysis, and in some cases, due to lack of data, we have made certain idealizations, which are given in Section 6.2.1. Note that the objective of this study is to illustrate the utility of 3-D FDTD calculations for such simulators, rather than an exact match with experiment. The method can readily be used to handle more realistic designs.

6.2.1 Idealizations

First, the test volume of the simulator is not shown in [9]. However, such simulators typically have large test volumes. Inclusion of the complete test volume would require a very large increase in the computational domain, while the mesh size would be restricted to small values due to the need to handle high frequencies. Computer hardware limitations do not presently allow us to handle the full problem. Hence, for different studies, we have assumed test section lengths typically lying between 3 and 9 m. Even with this restriction, we are unable to use a mesh size finer than 1.85 cm, which permits a maximum frequency of 1.6 GHz. The fastest 10–90% risetime obtained in our simulations is 2.2 nanosecond (ns), which can be handled by this mesh. However, the fastest experimental risetime (20–90%) reported by Schilling et al. [9] is 1 ns. A finer mesh would be necessary to handle such risetimes.

Second, the peaking section in [9] consists of several water capacitors. We have taken a single air-gap, parallel-plate capacitor, but with similar capacitance. The difference in dielectric material and geometry means that transmission-line effects within the capacitor will not be handled realistically. However, this should affect only very short time-scale phenomena.

Third, no details of the fast closing switch are available from [9]. Hence, we have assumed a simple model which is described in Section 6.4. This means that the computed prepulse and the early-time part of the field in the TEM structure may not match experimental values.

Fourth, details of the geometry of the termination are not available. It is common, however, to use a number of chains of resistances in parallel, connecting the upper and lower plates. The parameters of these chains are chosen so as to provide the appropriate matching resistance. This arrangement effectively acts like a "resistive sheet," which also reduces radiation leakage from the aperture of the bounded volume. For this study, therefore, we have assumed a single resistive sheet. In certain studies, however, we have also studied the effect of using alternate geometries, such as two resistive rods in parallel.

6.2.2 Geometry

Figure 6.2 shows a schematic of the TEM horn geometry, illustrating some of the symbols used later. In all cases, w refers to the width of the plate in the y-direction and H to the separation between plates in the x-direction. "S" refers to the slant length, i.e. the actual length of an edge of the top plate between two specified points. Subscripts "f" and "a" refer to the feed and aperture, respectively.

The parameters known from Ref. [9] are the plate width at the aperture $w_a = 2.32$ m, the angle $\alpha = 22°46' \approx 22.77°$ and angle $\beta \approx 30°$. Here, $\beta/2$ is the angle of elevation of the top plate measured with respect to the ground plane. Now, these two angles are related through

$$\alpha = 2\tan^{-1}\left[\frac{w}{2H}\sin(\beta/2)\right] \tag{6.1}$$

Hence, we obtain $w/H = 1.56$. Since we already know w_a, this yields the height at the aperture $H_a = 1.49$ m.

We next consider the feed. The height at the feed H_f is chosen to be $\simeq$0.055 m, since it must match with the heights of the capacitor and switch as set up within the FDTD mesh. Since we desire to have a nearly constant w/H ratio, the width at the feed works out to be $\simeq$0.086 m. However, the FDTD mesh size imposes a constraint, so that the actual value used for w_f is 0.074 m.

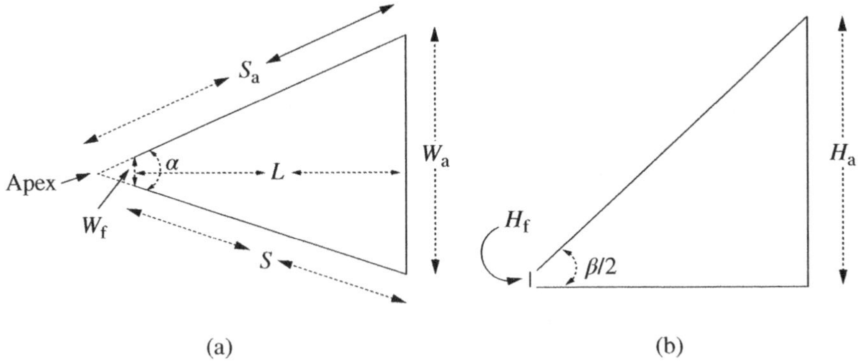

Figure 6.2 Schematic showing the top (a) and side (b) view of simulator.

We next consider the slant length S_a between the apex and the aperture, as shown in Figure 6.2. We get $S_a = w_a / [2\sin(\alpha/2)] \approx 5.88$ m. Proceeding in the same way, we get the projected length of the top plate on the y–z plane, viz. $L = 5.37$ m, and $S = 5.67$ m.

6.2.3 FDTD Model

Figure 6.3 shows a cross-sectional elevation and Figure 6.4 shows the top view of the FDTD mesh. The peaking capacitor, which consists of a set of water capacitors in [9], is idealized as a single, parallel-plate, air-gap capacitor in our model. Both plates of the capacitor and the simulator are assumed to be made of perfect electrical conductor (PEC). The peaking capacitor is charged to ~1 MV. Our simulation starts with an uncharged capacitor and charging is accomplished by the application of an x-directed electric field in the region lying between the capacitor plates. This electric field has a temporal waveform following a half-Gaussian

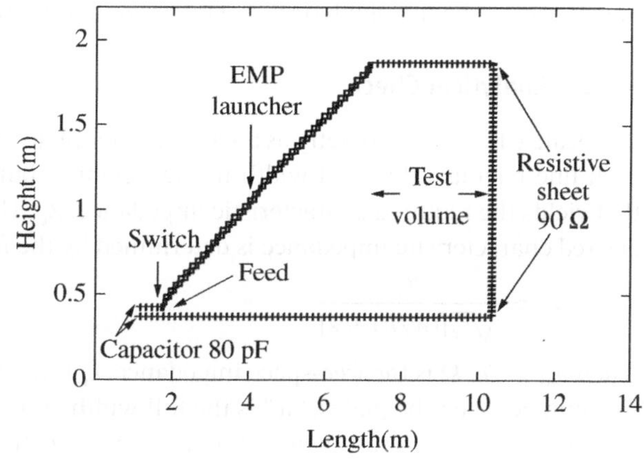

Figure 6.3 Cross-sectional elevation of FDTD mesh.

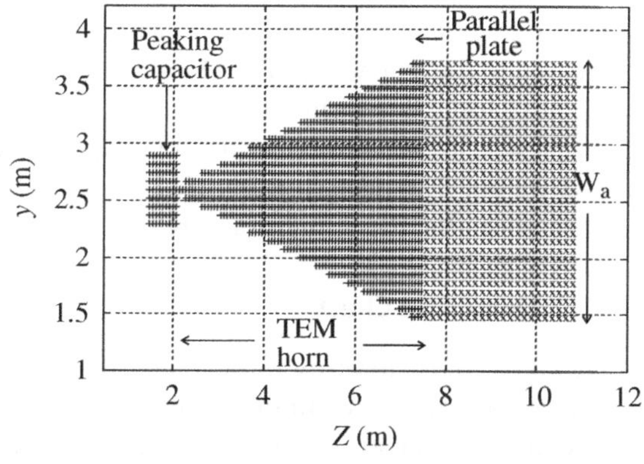

Figure 6.4 Top view of FDTD mesh.

in time. This waveform will be briefly described in Section 6.8. A detailed study of the charging procedure is reported in Chapter 5.

The computational mesh changes according to the needs of each problem. A typical mesh consists of $122 \times 156 \times 365$ cells in the x-, y-, and z-directions, with uniform cell sizes Δx, Δy and Δz of 1.85 cm. This corresponds to a time-step of 35.6 ps according to the Courant criterion [32]. The active region, consisting of the capacitor, switch, and TEM structure, is placed within this domain with a uniform spacing of at least 20 cells to the domain boundaries.

6.3 Validation of TEM Structure Geometry

The TEM structure, in its simplest form, consists of two triangular plates, each having the shape of an isosceles triangle, separated by some angle as shown in Figure 6.1. Before we proceed with detailed time-dependent simulations, it is necessary to confirm that we have indeed set up a 90 Ω structure. This has been done in two ways – the first analytical and the second numerical.

6.3.1 Analytical Check

We idealize the horn geometry as a set of parallel-plate transmission lines in series, each line having a different width w and height h, but maintaining a w/h ratio that yields the required characteristic impedance, Z_c. The ratio w/h that gives the desired characteristic impedance is determined by the formula [41]

$$Z_c \simeq \frac{\eta_0}{\sqrt{\epsilon_r}[(w/h) + 2]} \tag{6.2}$$

where $\eta_0 = 377\ \Omega$ is the free-space impedance, ϵ_r is the relative permittivity of the medium between the plates, "w" is the full-width and "h" the half-height, i.e. the height above the ground plane, at a given cross section of the horn. It must be noted that this formula is only approximate. An estimate of the error is obtained as follows: For $w/h \approx 4.8$, and $\epsilon_r = 1$, we get $Z_c = 55.4\ \Omega$, an increase of ~10.8% as compared to the true value of 50 Ω reported in [18].

We next apply this formula to the TEM structure examined in this study. This has $w/h = 1.56$, yielding $Z_c = 105.9\ \Omega$. Applying the correction factor obtained above, we get $Z_c = 95.6\ \Omega$, fairly close to the reported 90 Ω [9]. This is the first confirmation that the dimensions are reasonably correct.

6.3.2 Numerical Check

We would next like to verify, numerically, that the structure set up in the 3-D FDTD mesh, consisting of the simulator plates and termination, actually has an input impedance around 90 Ω over the frequency range of interest.

In principle, it is possible to determine the input impedance by applying a sinusoidal voltage waveform, $V_0(t)$, having the desired frequency ω, at the input to the TEM structure and then "measuring" the resulting input current $I_0(t)$. We find that $I_0(t)$ normally exhibits a transient and then settles down to a "pure" sinusoid having a phase difference with reference to $V_0(t)$. The ratio of the voltage and current amplitudes gives the magnitude of the input impedance Z, while the phase angle gives its breakup into real and imaginary parts.

Since we need to determine the impedance over a wide range of frequencies, we would need to repeat this process for a large number of frequencies. A more convenient method, and one widely used in the FDTD literature, is to apply $V_0(t)$ having the form of a Gaussian pulse [32]:

$$E_x(t) = \begin{cases} E_0 e^{-\alpha(t-\tau)^2} & \text{if } 0 \leq t \leq 2\tau \\ 0 & \text{if } t < 0 \text{ or } t > 2\tau \end{cases} \tag{6.3}$$

where $\alpha = (4/\tau)^2$, E_0 is the peak field and the characteristic time $\tau = 6/f_{\text{maxcell}}$. Here, f_{maxcell} is the maximum frequency that can be handled by the FDTD mesh, given by [32]

$$f_{\text{maxcell}} = \frac{c}{10\Delta}$$

where Δ is the largest of Δx, Δy and Δz.

The input current $I_0(t)$ is measured at the feed location using Ampere's circuital law, $\int \vec{H} \cdot \vec{dl}$. The input impedance $Z_{\text{in}}(\omega)$ is then given by

$$Z_{\text{in}}(\omega) = \frac{V_0(\omega)}{I_0(\omega)} = -\frac{g_{\text{f}} \times E_x(\omega)}{I_0(\omega)} = R_{\text{in}}(\omega) + jX_{\text{in}}(\omega) \tag{6.4}$$

where g_{f} is the gap between top and bottom plates at the feed. $V_0(\omega) = g_{\text{f}} E_x(\omega)$, a complex number in general, is the Fourier transform of the applied input voltage. The impedance consists of an input resistance $R_{\text{in}}(\omega)$ and an input reactance $X_{\text{in}}(\omega)$.

Figure 6.5 shows the frequency response of the designed TEM structure from 30 to 550 MHz for the case of a matched sheet termination. We observe strong spikes in the low-frequency range. At higher frequencies, these spikes tend to die out, so that the mean input resistance is almost constant around a value ~90 Ω. The same trends have been reported in experimental measurements of input impedance for TEM horns [24], [42].

The same trend can be observed from Figure 6.6, which shows the input impedance plot for the TEM structure with two parallel resistive rods acting as the termination, each rod having a resistance of 180 Ω. This is the second confirmation that the geometry is correct.

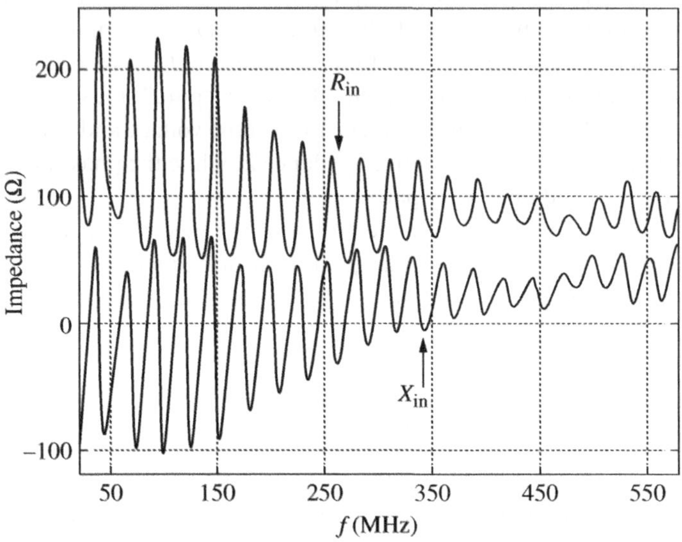

Figure 6.5 Input impedance of TEM structure with a matched 90 Ω sheet termination.

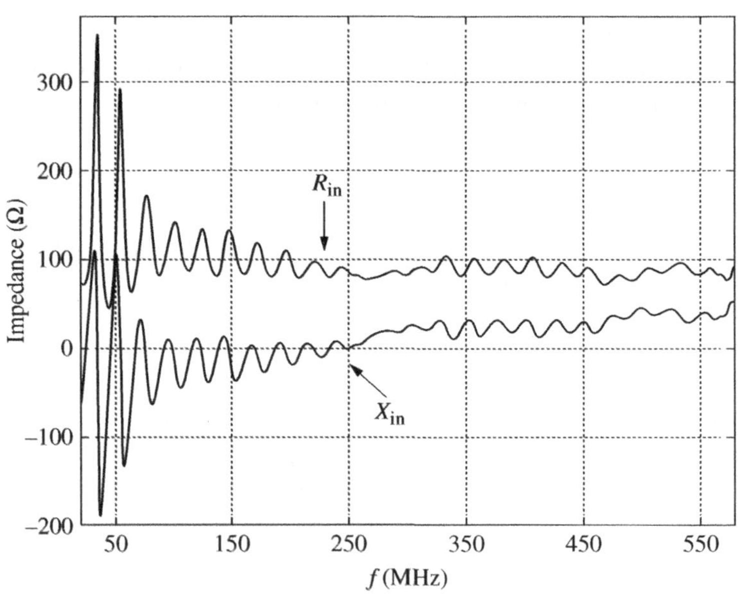

Figure 6.6 Input impedance of TEM structure with a matched rod termination, consisting of two 180 Ω resistive rods in parallel.

6.4 FDTD Model of Closing Switch

A simplified model has been used for the closing switch, as shown in Figures 6.1 and 6.3. This consists of two resistive strips, each with a length of one computational cell, separating the upper and lower capacitor plates from the corresponding plates of the TEM structure. The strip material has a time-dependent resistivity, which is initially kept infinite during the charging of the capacitor. This means that no charge can move to the simulator structure while the capacitor is being charged. However, a prepulse is still possible due to capacitive coupling – this is examined in Section 6.8.

The actual temporal variation of switch resistivity would depend upon the physics of the breakdown process. Since detailed modeling of the switch is beyond the scope of this work, we must choose an approximate waveform that is physically justifiable. It is well known that many fast closing switches, involving breakdown of a gas or liquid, exhibit a formative time-lag [43]. Starting from the time at which a strong electric field is applied, there is a slow build-up of the avalanche breakdown process, during which the effective resistance of the switch remains high. Following this time lag, the avalanche builds up rapidly, leading to rapid voltage collapse across the electrodes.

A waveform that yields this kind of behavior, from the fully open to the fully closed state, is given by the "half-Gaussian"

$$\sigma = \sigma_c \exp\left[-\alpha_s(t - \tau_s - \tau_0)^2\right], \quad \tau_0 \leq t \leq \tau_s + \tau_0 \tag{6.5}$$

where τ_s is a characteristic time for closure, τ_0 is the time when switching starts, $\alpha_s = (4/\tau_s)^2$ and σ_c is the switch material conductivity in the fully closed state. τ_s is chosen so as to avoid the generation of frequencies higher than those that can be handled by our FDTD mesh [32]. σ_c is chosen such that, in the fully closed state, the switch resistance becomes very small in comparison with the characteristic impedance of the TEM structure.

We see that the conductance remains small until "t" comes fairly close to $\tau_s + \tau_0$, followed by a small interval exhibiting rapid growth in σ. Hence, the "time lag" is reproduced to some extent.

Another advantage with the half-Gaussian waveform is that its first derivative falls to zero in the fully closed state, which should reduce the spurious generation of high frequencies.

Let us now consider another property of this waveform. We see that, at the start of switching, i.e. $t = \tau_0$ in Eq. (6.5), there is a discontinuity in the conductance, which is taken to be zero during capacitor charging. However, the magnitude of the discontinuity is small, being ~7 orders of magnitude smaller than the peak conductance in the fully closed state. This step-change would, naturally, excite all frequencies. Now, the FDTD mesh can accurately handle frequencies up to some upper bound determined by mesh sizes. However, since the step-change has a small magnitude, it is numerically acceptable [32].

A question that naturally arises is whether some other waveform might not be superior to the half-Gaussian in some or all of these areas. From the point of view of the temporal derivatives of σ at $t = \tau_s + \tau_0$, an error function waveform would be better, since both the first- and second-derivatives would become zero in the fully closed state. The error function, however, suffers from two major disadvantages. The first is that it does not exhibit the "time delay" discussed above. Indeed, it would yield a $\mathrm{d}\sigma/\mathrm{d}t$ waveform that has its maximum at $t = \tau_0$ and steadily decreases as we move toward the fully closed state. Second, it has a problem around the time switching starts, as explained below.

The characteristic frequency f_c for variation of σ can be estimated from

$$f_c = \frac{1}{\sigma}\frac{\mathrm{d}\sigma}{\mathrm{d}t} \tag{6.6}$$

where $\sigma = \sigma_0 \times \mathrm{erf}(x)$ and $x = (t - \tau_0)/\tau_s$. Substituting for σ, we get

$$f_c = \frac{2}{\tau_s\sqrt{\pi}} \cdot \frac{\mathrm{e}^{-x^2}}{\mathrm{erf}(x)} \tag{6.7}$$

This frequency is a function of time. When switching starts at $t = \tau_0$, we have $\sigma = 0$, hence, this would yield an infinitely high frequency.

To get around this problem, we could start at a finite value of σ, as in the half-Gaussian model, by redefining $x = (t - \tau_0 + \delta\tau_s)/\tau_s$. This is where a major difference between the half-Gaussian and error function waveforms become apparent. In the Gaussian form, at small σ/σ_0 values, both the function and its derivative are small, so that the characteristic frequency remains reasonable. With erf(x), however, the derivative is initially large and progressively decreases in time. Hence, f_c can become very large when switching starts. For example, starting with $\sigma = 10^{-7}\sigma_0$, i.e. $\delta = 8.863 \times 10^{-8}$, and $\tau_s = 19.7$ ns, we get $f_c = 5.7 \times 10^{14}$ Hz, which is far too high.

We could further increase the initial value of σ/σ_0 to overcome this problem. Let us say we want to restrict f_c to below 400 MHz, the limit of the computational mesh in one problem. This requires $\sigma/\sigma_0 = 0.15$ at $t = \tau_0$. This, however, raises its own problems – the initial jump in σ is now considerable, and the resulting generation of high frequencies introduce substantial errors into the FDTD calculation.

Figure 6.7 shows the effect of choosing $\sigma/\sigma_0 = 10^{-7}$ and 0.15, respectively, on the observed E-field waveform inside the TEM structure. There is significant high-frequency noise, which is not observed with the half-Gaussian waveform (shown in Section 6.6). We believe that this noise is due to the high frequencies generated by the error function.

We conclude, therefore, that a half-Gaussian waveform is better suited to model switching, despite its limitations as regards continuity of the second derivative at the peak.

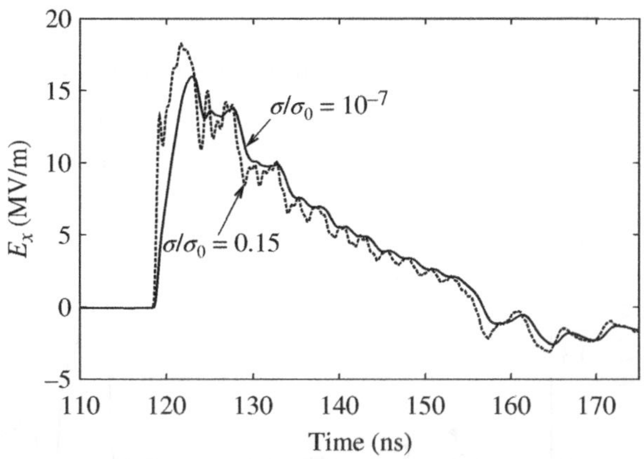

Figure 6.7 Vertical component of electric field at TEM structure feed for different values of $\sigma(t = \tau_0)/\sigma_0$, using the error function waveform for switching.

These simulations yields the temporal behavior of the input current to the TEM structure and the 3-D distribution of electromagnetic fields within the structure.

6.5 Choice of Distance to Domain Boundary

A second-order outer radiation boundary condition (ORBC) has been used in the simulations [32]. ORBC is applicable only if the distance between the simulator and the domain boundaries is at least of the order of a wavelength [32]. This condition becomes difficult to satisfy when the simulation involves an ultra-wideband spectrum. On the one hand, the total domain size would be governed by the lowest frequency. At the same time, the mesh size must be dictated by the highest frequency of interest. This combination would require unacceptably large numbers of computational cells. It is clearly impossible to satisfy this condition for very low frequencies, e.g. for the long-time response. It has been reported in [18] that good agreement is obtained between simulation and experiments when the above condition is satisfied for the highest frequency of interest.

We have performed Fourier analysis of $E_x(t)$ at the feed point to identify the highest frequency of interest, f_{max}, for different switching times. This is defined as the frequency where $|E_x(f)|$ falls to 0.001 of its peak value. The distance to the domain boundary has then been chosen appropriately.

6.6 Electric Field within TEM Structure

We now present results for the case of a charged capacitor discharging through the TEM structure terminated by a perfectly matched resistive sheet having a resistance of 90 Ω. In these simulations, the main component of the electric field is

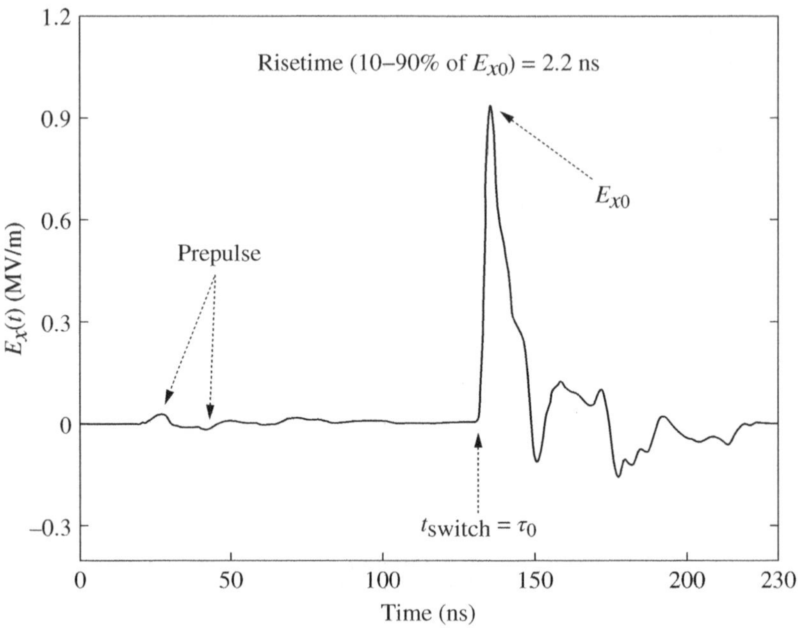

Figure 6.8 Vertical component of electric field inside TEM structure at a point in-between the feed and aperture.

taken as E_x, where "x" is the height, as shown in Figure 6.3. Figure 6.8 illustrates the general nature of $E_x(t)$ at a point in-between the feed and the aperture. The figure shows both the prepulse (before switching) and the main pulse (after switching). An arrow marks the time τ_0 when switching starts. A detailed discussion of the prepulse is presented in Section 6.8. In the rest of this section, we focus on the main pulse.

6.6.1 Effect of Switch Closure Time

Given the size and cost of EMP simulators, it is often desirable to use the same TEM structure for susceptibility testing over rather different frequency ranges. One such example comes from Ref. [9], where a pulser yielding a ~7 ns risetime was replaced by one with a ~1 ns risetime. A faster risetime would certainly permit testing at higher frequencies, but may also have unacceptable side effects.

For the case of a TEM structure with a 90 Ω resistive termination, we have studied the effect of varying the switching time. Figure 6.9 shows the main pulse at the feed point for three different values of τ_s, viz., 19.7, 14.8 and 7.4 ns. Runs with faster switching times require a finer FDTD mesh, so the capacitance C_0 and charging voltage V_0 of the parallel-plate capacitor are slightly different in the three cases, viz.

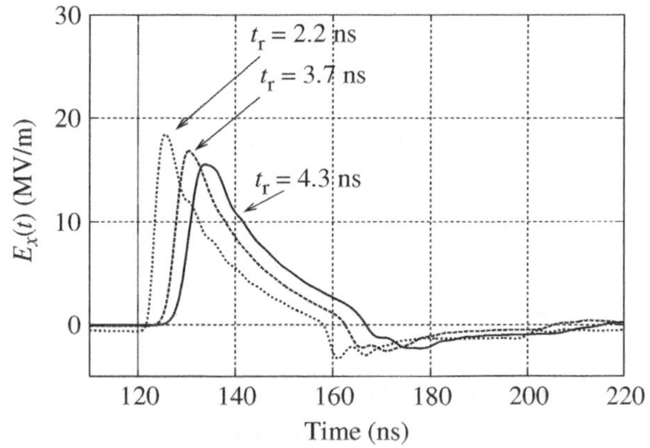

Figure 6.9 Electric field waveform at feed point for different switching times. Case with matching 90 Ω resistive termination. The prepulse is omitted for clarity.

$V_0 = 917$, 911 and 906 kV and the corresponding $C_0 = 85$, 82, and 80 pF, respectively. We observe a 10–90% risetime for $E_x(t)$, i.e. $\tau_E = 4.3$, 3.7, and 2.2 ns in the three cases. These correspond to $0.22\tau_s$, $0.25\tau_s$, and $0.30\tau_s$, respectively.

This relation between τ_s and τ_E can be understood as follows: The switch conductivity waveform of Eq. (6.5) implies that σ rises from $0.2\sigma_c$ to σ_c in $\sim 0.32\tau_s$. It is reasonable to assume that the flow of charge from the capacitor to the TEM structure only becomes significant when σ is high enough, say $\sigma > 0.2\sigma_c$. It is this flow of charge that builds up the electric field inside the TEM structure. Hence, a risetime in the range 0.2–$0.3\tau_s$ for $E_x(t)$ in the TEM structure is understandable. It is not presently clear why the ratio τ_E/τ_s should increase as we reduce the risetime. It is likely that the inductance of the connection region between the capacitor and TEM structure, including the switch, starts playing a role for very fast risetimes.

The deleterious effect of faster risetimes on radiation leakage from the simulator are examined in detail in Chapter 8. The effect on mode structure within the simulator are examined in Chapter 7.

6.6.2 Pulse Fidelity

EMP simulators are used for producing a nuclear electromagnetic pulse (NEMP) like environment in a finite-sized laboratory system. The objective is to subject a test object to a wideband pulse having a desired frequency spectrum. Now, the pulsed-power source, consisting of the capacitor, switch, and connections, may provide an excitation with the appropriate frequency content. However, it is equally important to ensure that the TEM structure maintains pulse fidelity, i.e. does not significantly change the relative importance of different frequencies.

At a simple level, maintenance of pulse fidelity can be checked by observing the risetime of the primary pulse at different locations along the TEM structure.

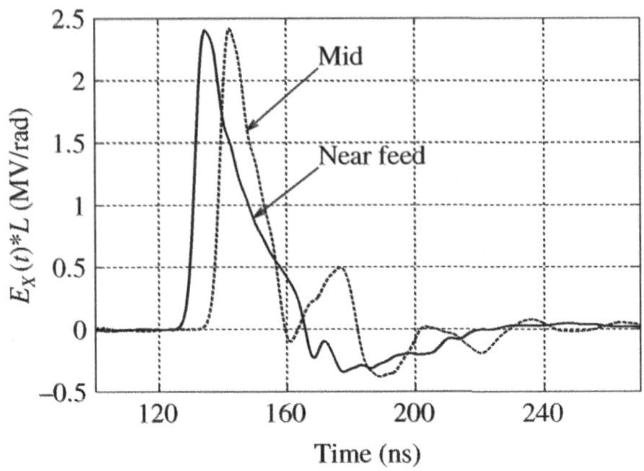

Figure 6.10 Vertical electric field $E_x * L$ at two locations along boresight. Case with matched sheet-termination and $\tau_s =$ 19.7 ns.

Figure 6.10 shows the product $E_x(t) * L$ in the main pulse at two locations along the boresight of the TEM structure, viz., near the feed point and midway between the feed and aperture, for the case with $\tau_s = 19.7$ ns. Here "L" is the distance of the observation point from the apex of the TEM structure. If we had a perfectly spherical wavefront, the electric field magnitude would scale inversely with distance from the apex. Hence, multiplication by L provides a reasonable scale factor.

We note that the rise-time and pulse-width are approximately the same at both points, confirming that fidelity is maintained to a reasonable extent. Maintenance of "reasonable" fidelity does not, by itself, imply that the wavefront is perfectly spherical. The attempt here is simply to scale the two E_x waveforms for convenience in comparison.

6.7 Flow of Current through Simulator Plates

For the case with $\tau_s = 19.7$ ns, Figure 6.11 shows the temporal waveform of the z-directed current flowing through the lower plate near the aperture. This current has been obtained through the path-integral $\int \vec{H} \cdot \vec{dl}$. The observed time-delay of 16.5 ns between the peak current in Figure 6.11 and the peak $E_x * L$ near the feed in Figure 6.10 is consistent with the transit time between these points.

6.8 Prepulse

A prepulse, caused by the displacement current through the switch capacitance, can be significant when the charging time of the source capacitor is comparable to the switch closure time. Such a situation can arise in practical simulators, as in

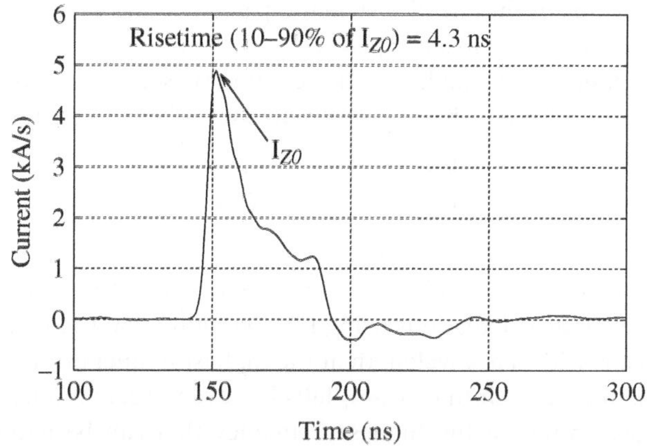

Figure 6.11 *z*-Directed current flowing through lower plate near the aperture. Case with matched termination and $\tau_s = 19.7$ ns.

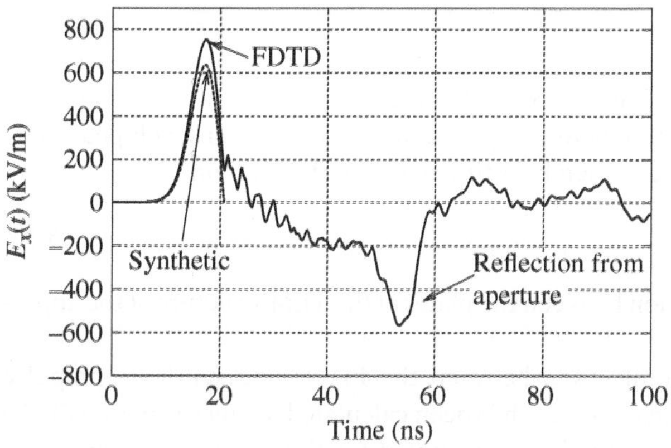

Figure 6.12 Electric field prepulse.

Ref. [9], where the charging time of the peaking section is ~10 ns, only five to ten times higher than the switch closure time.

Figure 6.8, from an FDTD simulation, shows the existence of a prepulse in this system for the case of $\tau_s = 7.4$ ns. An enlarged view is shown in Figure 6.12, for the case with $\tau_s = 19.7$ ns. The first peak is the main prepulse, which is followed by a reflection from the aperture. The time delay between the main (positive) and reflected (negative) peaks corresponds to the double transit time between the feed and aperture, a distance of 10.8 m.

It is of interest to estimate the prepulse from simple considerations and compare it with the FDTD waveform. This is reported below.

The displacement current I_D is given by the relation

$$I_D(t) = C\frac{dV(t)}{dt} \tag{6.8}$$

where C is the effective switch capacitance and $V(t)$ is the charging voltage applied between the capacitor plates.

Since we do not know the actual $V(t)$ used in the experiments [9], we have arbitrarily chosen a half-Gaussian-like waveform given by

$$V(t) = \begin{cases} V_0 e^{-\alpha(t-\tau)^2} & \text{if } 0 < t \leq \tau \\ 0 & \text{if } t < 0 \end{cases} \tag{6.9}$$

where $\alpha = (4/\tau)^2$, and τ is the characteristic time of charging. This waveform means that the excitation is only applied until the Gaussian reaches its peak value. Accurate calculations using FDTD require that there be at least 10 computational cells within a wavelength at the highest frequency of interest [32]. We, therefore, choose τ such that $|V(\omega)|$ falls by seven orders of magnitude as compared to its peak value at the highest frequency that can be handled by our FDTD mesh. We note that for a full Gaussian pulse, this decay by seven orders occurs at a frequency $f_h \simeq 6/\tau$.

Given $I_D(t)$, the electric field prepulse $E_p(t)$ inside the structure can be estimated as follows: The input impedance Z_{in} (90 Ω) of the structure remains constant over some frequency band because of its ultra-wideband nature. $E_p(t)$ developed at any point inside the structure can then be estimated from the relation

$$E_p(t) = \frac{I_D(t) Z_{in}}{\text{gap}} \tag{6.10}$$

where "gap" is the separation between the plates of the TEM structure at the appropriate point.

Figure 6.12 shows a comparison between the FDTD and synthetic prepulse waveforms. The switch capacitance C has been calculated, using a separate FDTD calculation, as 2.25 pF. Here we have accounted for the fact that the switch in our model actually consists of two resistive strips in series, which halves the capacitance. Three points must be noted. First, the synthetic waveform should only be compared with the forward-traveling FDTD prepulse and not with the reflected portion. Second, the synthetic prepulse falls to zero after the charging waveform reaches its peak. This is because, in our simulation, the charging waveform is not applied beyond this point. Third, we have taken a delay ~1 ns between dV/dt of the charging waveform and the synthetic curve. This is because in the FDTD simulation, the electric field is being measured ~0.3 m away from the closest point on the capacitor.

There is a fairly good agreement between the FDTD-computed main prepulse and the synthetic prepulse. This shows that the computation can be generalized to predict the prepulse from more complex switch geometries.

6.9 Effect of Test Object

The test object is placed within the test volume formed by a parallel-plate extension of the tapered structure. The presence of the test volume and object is likely to affect two things. First, the total energy leakage from the system would depend upon the length of the parallel-plate section and also the size, shape, and material of the test object. It is essential to know the magnitude of this change from the point of view of electromagnetic interference (EMI) with equipment placed in the vicinity. Second, the test object is likely to affect the electromagnetic field structure in its vicinity, especially in terms of the frequency spectrum. Since the simulator is supposed to subject the object to a particular kind of spectrum, any change in this distribution is of interest.

Test objects of practical interest often have complex features like internal cavities, small apertures, complicated geometries, and multiple materials. Hence, the two issues mentioned above cannot be studied by simpler methods – a 3-D FDTD calculation is necessary.

A detailed study of changes in radiation leakage due to the presence of a test object will be presented in Chapter 8. In this section, we focus on determining the effect of the test object on the field structure inside the simulator. We also compare the electromagnetic environment "seen" by the object with that it would have experienced during free-space illumination.

In an actual simulator, the parallel-plate extension is likely to be fairly large. Due to FDTD mesh constraints, we have limited ourselves to an extension 3 m long, having a width equal to that of the aperture in the tapered section, as shown in Figure 6.1. This defines a test volume with the dimensions $3 \times 2.32 \times 1.49$ m in the x, y and z directions, respectively.

Test objects come in a variety of geometries and materials. Our purpose here is only to estimate the kind of modification a test object can produce. Hence, we have examined a simple shape, viz., a perfectly conducting solid cube (see Figure 6.1). The side of the cube, h_{to}, is allowed to take on three values, viz., 0.25 m, 0.5 m, and 1.0 m, respectively. We define an "interaction parameter" I_{p} $(=h_{\mathrm{to}}/h_{\mathrm{pp}})$, where h_{pp} $(= 1.49$ m$)$ is the separation between the parallel plates.

Figure 6.13a,b shows the vertical component of the electric field just outside the centers of the x-directed faces 1 and 2. The electric field strength increases with increase in I_{p} from 0.17 to 0.67. The peak values scale approximately as the inverse of the gap between the object and the parallel-plates. Figure 6.13c,d shows the vertical component of $\vec{E}$ just outside faces 3 and 4, which are oriented normal to the y-direction. The field strength falls off with increase in I_{p}, indicating that E_x

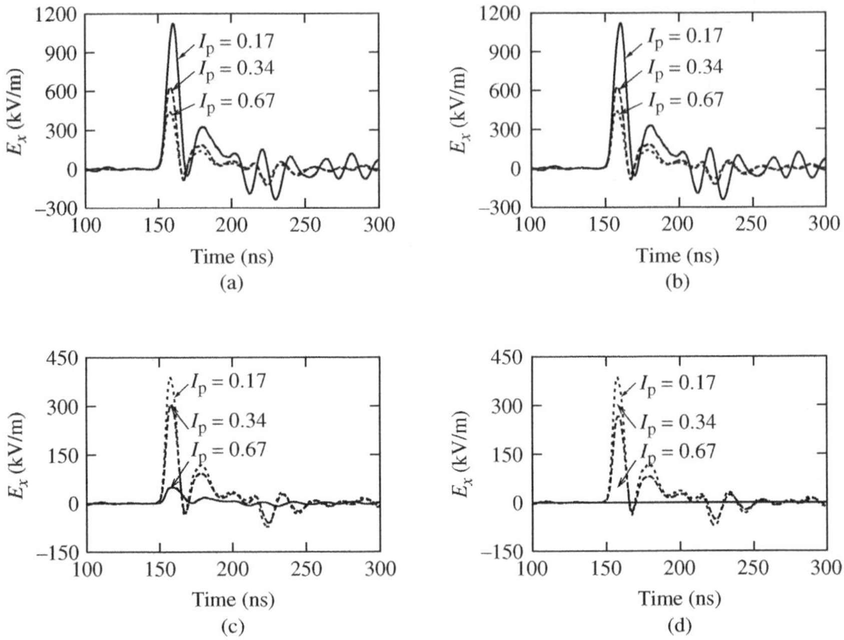

Figure 6.13 Comparison of vertical component of electric field near different faces of the test object. Case with $\tau_s = 19.7$ ns.

gets effectively shorted. A similar trend in the electric field amplitudes has been found outside faces 5 and 6.

For the above values of I_p, there is no distortion in the vertical component of electric field inside the tapered section of the simulator. However, the $E_x(t)$ waveform near the test object is considerably distorted, with risetimes of 4.6, 4.76, and 6.1 ns for $I_p = 0.17, 0.34$, and 0.67, respectively. This means that a larger test object will be subjected to significantly lower frequencies, for identical simulator and pulser parameters. The simulator considered here can be used for testing objects with a maximum dimension of ~0.5 m without significant change in the power spectrum.

An EMP simulator is meant to subject objects to fields similar to those that would be encountered when the pulse comes through free-space [8]. This necessarily requires the application of an E-field waveform at the TEM structure feed, similar in frequency content to the desired EMP waveform. However, even if the feed waveform is appropriate, scattering from various TEM cell components results in the object experiencing fields that are different from the desired form. It is important to quantify this difference.

We have compared the $E_x(t)$ waveforms, measured just above surface#2 (defined in Figure 6.1), for illumination in free space (FS) and inside the

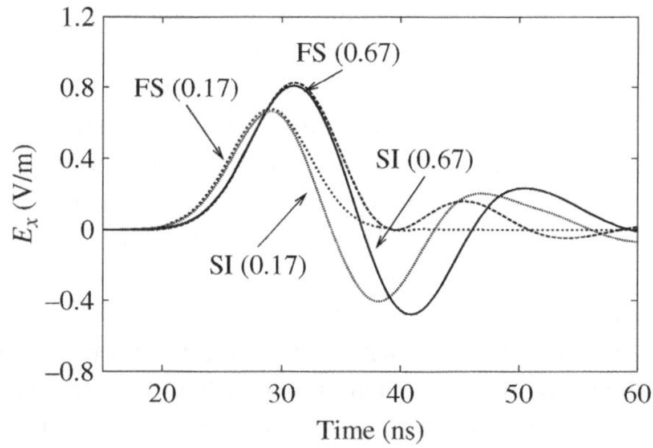

Figure 6.14 Scattered $E_x(t)$ measured just above surface#2 for illumination in free space (FS) and inside the simulator (SI). Curves are shown for $l_p = 0.17$ and 0.67.

simulator (SI). The results are shown in Figure 6.14 for $l_p = 0.17$ and 0.67. In all cases, the applied $E_x(t)$ is taken to be a Gaussian with unit amplitude and $\tau = 19.7$ ns. The SI results have been suitably scaled. As expected, we observe reasonable agreement between the FS and SI results up to the first peak. For both objects, subsequent scattering of fields off the TEM structure, especially the termination, produces negative peaks in the simulator waveforms which are not present in the free-space curves. We also observe that the SI response for the smaller object has a smaller 10–90% risetime, indicating greater high-frequency content. This is consistent with our observations earlier in this section.

One point is noteworthy. The applied Gaussian, with $\tau = 19.7$ ns, has a 10–90% risetime $\simeq$5.7 ns. The 10–90% risetimes observed above, for $l_p = 0.17$ and 0.67, are ~5.9 and 6.2 ns, respectively. This means that the smaller object essentially follows the risetime of the applied pulse, while the larger object exhibits a slower response. This is because the measured $E_x(t)$ has contributions from all points on the object, and these points are further away in the case of the large object. If the applied Gaussian had a faster risetime, e.g. 2 ns, the finite transit-time through the smaller object would also have played some role, further enhancing the difference between risetimes for the two objects.

These results show that 3-D FDTD simulations have an important role to play in predicting the role of test object geometry and material composition on the electromagnetic fields to which they are subjected.

6.10 Validation Checks for FDTD Analysis

The results obtained using the FDTD code need to be validated against known results for simpler problems. A variety of checks have been done, some of which are summarized below.

The general performance of the code has been confirmed by calculating the frequency-dependent input impedance $Z_{in}(\omega)$ of a dipole antenna, for the parameters published in [44]. In this simulation, we have assumed that $2\pi r_0/\lambda \ll 1$, where r_0 is the radius of the dipole and λ the wavelength. This condition implies that there is no variation of current across the cross-section of the dipole. The condition has been imposed by incorporating a wire-mesh model in the FDTD simulation [32]. We find that the FDTD results match published results [44] fairly well.

We have been calculating the radiated far-field using the near-to-far field transform described in a number of references, such as [32, 33, 45, 46]. This implementation has been confirmed in three different problems. First, we have reproduced the far-field FDTD calculation of a 50 Ω TEM horn analyzed by Shlager et al. [18]. Second, we have obtained a reasonable match with the radiation pattern of a thin dipole reported in [32]. Third, the computed radar-cross-section (RCS) of a plate made out of PEC has been matched with published results [35, 47].

Finally, we have checked the proper setting up a of parallel-plate capacitor in the FDTD domain, its charging procedure and its discharge through a known resistance. The charge density distribution and capacitance have been cross-checked through the method-of-moments (MOM). Details of the check are reported in Chapter 5.

6.11 Summary

We have performed self-consistent, three-dimensional, time-domain calculations for a bounded-wave EMP simulator, including a realistic geometry and test object. The simulations yields the detailed 3-D evolution of electromagnetic fields within the structure, within the test object and in its immediate vicinity. To our knowledge, this is the first time such a detailed study has been done.

As expected, the pulse travels along the length of the simulator with a nearly constant pulse-width and risetime. The peak amplitude falls in inverse relation to the height of the TEM structure and is constant in the parallel-plate test region of the simulator.

Shorter closing times for the switch yield shorter risetimes for the electric field within the TEM structure. This is just what one would expect. However, as closure times become smaller, the risetime does not decrease proportionally, possibly due to inductive effects from the connections and switch.

For the parameters of [9], we expect a significant prepulse. This is indeed seen in the simulations. The prepulse waveform has been explained fairly well quantitatively, using simple estimates of capacitive coupling across the switch and the known charging waveform across the capacitor. To our knowledge, this is the first detailed explanation of a prepulse through 3-D simulation. This effort opens the way to prediction of prepulse in situations with complex geometries of the capacitor bank, switch and load.

Placement of a test object within the test section significantly modifies the electric field "seen" by the object. The vertical field E_x experienced by the x-directed faces of the object increases with object size, while E_x outside the y- and z-directed faces shows the reverse trend. $E_x(t)$ on the surface of the object shows progressively larger risetimes as the object size is increased, i.e. larger objects will be subjected to somewhat lower frequencies. However, this increase in risetime becomes significant only beyond some critical object size. This finding has practical significance because it imposes an upper limit on the object size that can be tested in a given simulator.

Ideally, an EMP simulator should subject a test object to an electric field waveform similar to that it would have experienced in free-space illumination. However, we find that, regardless of the size of the test object, the waveforms match only up to the first peak in the electric field. Multiple scattering off the simulator structure and test object produces features in the simulated waveform that is markedly different from the free-space case.

These findings show that 3-D FDTD simulations have an important role to play in predicting the role of test object geometry and material composition on the electromagnetic fields to which they are subjected.

Exercises

6.1 A prepulse forms during the charging state of a capacitor when the switch is fully open. Study the effects of different switch configurations, medium properties, etc. Can we suppress the prepulse in order to avoid false switching due to unwanted breakdown in the medium?

6.2 Examine the performance of an EMP simulator through the possible termination types such as resistive loading on the simulator plates and admittance sheet at the final end.

6.3 Design an EMP simulator for an object-under-test (DUT) of cylindrical geometry and high dielectric permittivity. Study the simulator performance as compared to free-space illumination of the DUT.

7

Electromagnetic Modes Inside Bounded Wave Simulators

7.1 Introduction

In an "Ideal" nuclear electromagnetic pulse (NEMP) simulator, the pulse incident on the test object should be a "pure" transverse electromagnetic (TEM) wave with a near-planar wavefront, as in free-space propagation. In a bounded-wave simulator, however, scattering from different parts of the simulator structure, reflection from the termination, and the finite size of the structure, can lead to the excitation of higher order transverse electric (TE) and transverse magnetic (TM) modes. These modes arise from small, but nonzero, electric field components E_y and E_z inside the simulator shown in Figure 6.1. Furthermore, the presence of the test object itself modifies the fields to which it is subjected. Interpretation of the test results from an electromagnetic pulse (EMP) simulator should thus take into account the strength of higher order modes.

The degree to which the TEM mode is "polluted" by high-order modes depends strongly upon the design of the simulator, the frequency spectrum of interest, and the geometry and material composition of the test object. Given the number of variables, no simple estimate is possible. Hence, a detailed analysis becomes necessary. In this chapter, we concentrate to self-consistently determine the mode structure inside the simulator, in the presence of a test object.

7.1.1 Choice of Method for Modal Analysis

Modal analysis of bounded-wave devices has been performed by other workers using waveguide theory and the method of conformal mapping [28, 29]. However, such analyses have been limited to single-frequency excitation (instead of broadband), for parallel-plate waveguides (instead of a TEM-horn), and do not account for the test object. We have generalized the study to a real-life EMP simulator, having broad-band excitation in the presence of an arbitrary test object. This is

Electromagnetic Pulse Simulations Using Finite-Difference Time-Domain Method, First Edition.
Shahid Ahmed.

accomplished by performing singular value decomposition (SVD) analysis [48] of 3-D, time-domain finite-difference time-domain (FDTD) results.

Modal analysis could also be done by other spectral techniques, such as Fourier spectral analysis and Hilbert spectral analysis [49]. Fourier analysis, with respect to space, of FDTD-computed $\vec{B}(\vec{r},t)$ and $\vec{E}(\vec{r},t)$ inside the simulator, would yield the spatial modes at a particular time. The analysis would then have to be repeated at each time point in order to determine the temporal evolution of these modes. The method would naturally be computationally expensive.

Use of SVD offers the following advantages [48]. Firstly, this method is based on the diagonalization of a rectangular matrix. Therefore, one can easily analyze the data collected by as many channels as desired. Secondly, in a single analysis, SVD gives complete spatial as well as temporal mode information. Thirdly, this method is based upon the analysis of the correlation in space rather than time, which easily filters out the noise, i.e. any feature of one channel uncorrelated with the others. We have, therefore, opted for the SVD technique to analyze FDTD results.

7.1.2 Organization of This Chapter

Section 7.2 describes the computational model used in this study, focusing on the SVD method and its application to modal analysis. Section 7.3 examines the mode structure inside the simulator in the absence of a test object, and Section 7.4 reports on the results with a test object present. A physical explanation for electric field enhancement in the neighborhood of a test object is given in Section 7.5. The chapter is finally concluded with a summary in Section 7.6.

7.2 Details of Model

A detailed description of the FDTD analysis of a bounded-wave simulator has already been presented in Chapter 6. Since this chapter focuses on modal analysis of FDTD results, this section mainly presents the methodology of the SVD technique. However, we do present a few points related to the FDTD model that are relevant for data analysis.

7.2.1 FDTD Model

The length of the parallel-plate section is taken to be ~9 m and the 90 Ω simulator is terminated by a matched resistive sheet.

Figure 7.1 shows the FDTD mesh and the locations used for field measurement. Hereafter, we will interchangeably use the words "location" and "channel", for consistency with the terminology used in experimental data analysis. These channels lie in the *xz*-plane and are located at the mid-point of the simulator

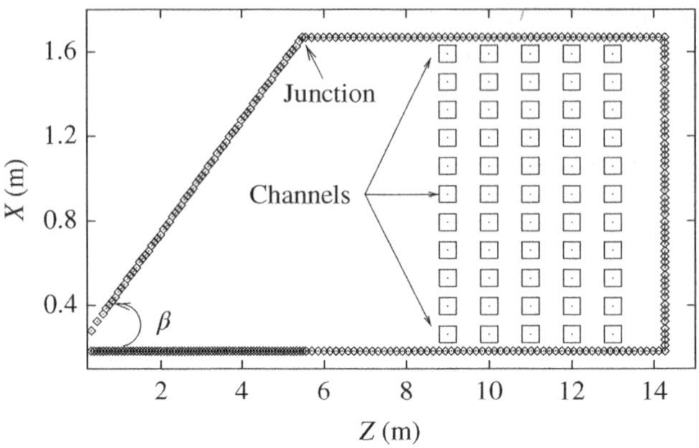

Figure 7.1 Side view of FDTD mesh of EMP simulator, showing locations (channels) for field measurement. Channels lie in the *xz*-plane and are centered with respect to the plate width in the *y*-direction. "Junction" refers to the junction of the tapered and parallel-plate sections.

in the y-direction (plate width). The computational mesh changes according to the needs of each problem. A typical mesh consists of $270 \times 70 \times 244$ cells in the x-, y-, and z-directions, with uniform cell sizes Δy and Δz of 7.4 cm and Δx of 0.9 cm. This corresponds to a constant time-step of 28.7 ps according to the Courant criterion [32]. The active region, consisting of the capacitor, switch, TEM structure, and the test volume, is placed within this domain with a uniform spacing of at least 20 cells to the domain boundaries.

Figure 7.2 shows the temporal variation of the three electric field components at an arbitrary location inside the tapered section (TEM horn) of the simulator. The figure illustrates their relative magnitudes. Since E_x is, by far, the dominant component within the capacitor plates, we would expect it to dominate within the bounded volume as well. This can be seen from Figure 7.2. There is a small, but still significant, longitudinal component E_z, while the transverse component E_y is negligibly small.

Figure 7.3 shows the temporal variation of the three magnetic field components at the same location. The dominant component is H_y, while H_x and H_z are negligibly small. This is what we would expect from the geometrical structure of the simulator, where the z-directed "sheet" currents in the two plates would mainly set up H_y.

7.2.2 Qualitative Discussion of Mode Structure

The simulator shown in Figure 7.1 has conducting plates at two x-locations and is open in the y-direction. Hence, we expect periodicity only in the x-direction, i.e. mode analysis need only be performed with respect to the x-direction.

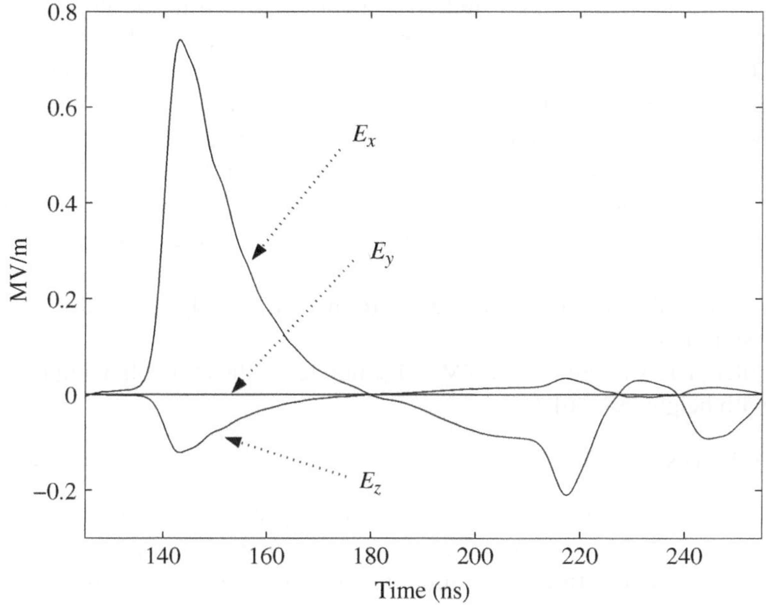

Figure 7.2 Typical level of electric field components at an arbitrary location inside the tapered section of the simulator.

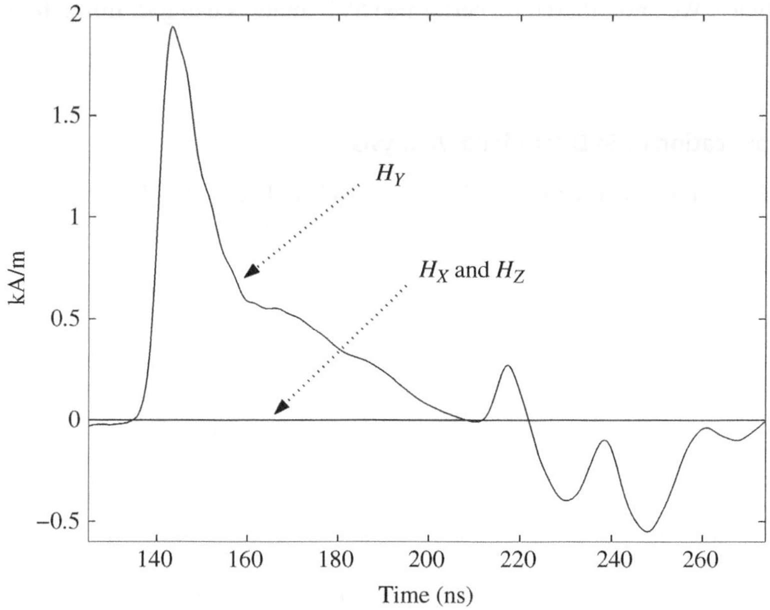

Figure 7.3 Typical level of magnetic field components at an arbitrary location inside the tapered section of the simulator.

Before we start analyzing FDTD results, it is important to identify the modes that are likely to exist within this system. For convenience, we have used the terminology of modes in parallel-plate waveguides, viz. TEM, TE, and TM, corresponding to transverse electromagnetic, transverse electric, and transverse magnetic, respectively. In a parallel-plate waveguide, with the wave propagating in the longitudinal (z-) direction, the TE modes can have nonzero E_y, but not E_z [50]. TM modes, on the other hand, would have nonzero E_z [50]. Now, we have already seen in Figure 7.2 that there is a significant E_z inside the simulator, while E_y is negligible. Also, Figure 7.3 shows that only H_y is significant, while H_x and H_z are negligible. Hence, it is reasonable to suppose that, apart from TEM, mainly TM modes exist within the simulator.

In a parallel-plate waveguide, the TM_m eigenmode exhibits the following field variation with height "x" [50]:

$$E_x = E_{x0} \cos \left[m\pi x/h \right] \tag{7.1}$$

where "m" is the mode number and "h" is the inter-plate separation. The TEM and TM modes support several electromagnetic field components, apart from the dominant E_x. The mode structure would, of course, be expected to show up in more than one component. At the outset, it is necessary to decide which component should be considered for SVD. Now, E_x exists in the TEM as well as the TM modes. Since we are interested in the relative importance of these modes, E_x is the natural choice. We have, therefore, performed SVD analysis using E_x throughout this study.

7.2.3 Application of SVD for Modal Analysis

A detailed description of the SVD method is available in Ref. [51] and several other texts. A brief description of the method and its application to the modal analysis of FDTD results is given below.

Let us consider a physical quantity "x" simultaneously measured at "m" different positions (channels) sampled at "n" different times with a sampling interval t_s. The matrix representation of the above observation can be generally expressed by a rectangular array, $x_{ij} = x_j[(i-1)t_s]$, where the row index "i" refers to time and the column index "j" to the channel. The SVD of the matrix x_{ij} is expressed as $\mathbf{X} = \mathbf{USV^T}$, where superscript "T" represents the transpose of a vector. Now, the matrix x_{ij} has been decomposed into three parts – time, amplitude, and space. The temporal component $\mathbf{U}$ is an $n \times m$ unitary matrix containing the eigenvalues or principal values, the amplitude $\mathbf{S}$ is an $m \times m$ diagonal matrix called the singular values and the spatial component $\mathbf{V}$ is an $m \times m$ unitary matrix containing the eigenvectors of $\mathbf{X}$.

This technique is widely used in the field of digital data processing in order to effectively filter the noise, i.e. any feature of one channel uncorrelated with the others [48]. The details are discussed in [48].

Note that **U** and **V** are simply the basis functions used in the analysis, and have nothing to do with the amplitude of a mode. Only the singular value **S** represents the amplitude. This has been confirmed by performing the following exercise.

Consider an artificially generated signal having the form given in Ref. [48]

$$X_{ij} = (1/\sqrt{N})\sum_{l} a_l \cos[2\pi m_l(j-1)/M$$
$$-2\pi \nu_l t_s(i-1)] \tag{7.2}$$

This is a superposition of cosinusoids of mode number m_l, having the frequencies ν_l with amplitudes a_l, respectively, sampled at M equispaced channels with a timestep t_s. The indices "i" and "j" have their usual meaning. We have performed SVD for a superposition of three cosinusoids having the same m_l but different a_l and ν_l. We find that doubling the amplitudes a_l doubles the singular values, leaving the basis vectors unchanged. Hence, in the rest of this chapter and in Chapter 9, we treat the variation of S-values of different modes as representative of their respective strengths.

7.2.4 Validation of SVD Results

Before SVD is applied to data for the simulator, it is necessary to verify the calculations against known solutions in simple cases. We have verified the consistency of the SVD method in two ways. Firstly, we have reconstructed the original signal from the decomposed components, using the relation $\mathbf{X} = \mathbf{USV}^{\mathrm{T}}$. Secondly, we have performed SVD analysis of FDTD output for single-frequency excitation of a simple parallel-plate waveguide with a matching termination. This analysis yields the theoretically expected shapes of the eigenmodes as well as the temporal waveforms **U**. Thirdly, with multifrequency excitation, i.e. a sum of sinusoids at different frequencies, the appropriate **U** is recovered. Fourthly, we observe the expected excitation of modes with progressively higher mode numbers as the applied frequency is increased.

7.2.5 Sample Calculation

Our first calculation of the mode structure has been done for the case where the switching time τ_s is 19.7 ns. SVD is performed on $E_x(t)$ measured at 25 channels lying between the top and bottom plates inside the test volume. These locations lie at a distance of ~1.5 m in the z-direction, measured from the junction – see Figure 7.1. Each channel $E_x(t)$ has 11 000 time points with a sampling time of 28.7 ps.

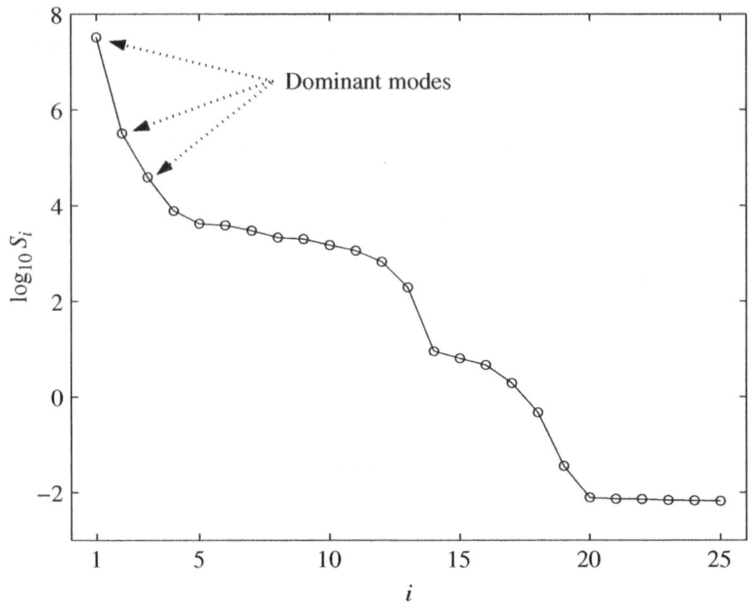

Figure 7.4 Log of singular value S_i shown as a function of serial number i. "i" is simply a serial number, and should not be confused with the mode number.

We have applied SVD to the matrix $\mathbf{E_x}$ formed by these measurements. The singular values of the decomposed components are shown in Figure 7.4. There are 25 singular values. Note that the SVD procedure yields singular values arranged in descending order. Hence "i" is merely a serial number, and should not be confused with the mode number, which must be determined by inspection of the eigenvectors **V** corresponding to each "i."

Figure 7.4 shows that there are three dominant modes (eigenmodes). Figure 7.5a–c shows the eigenvectors corresponding to these modes, in descending order of S-value. The eigenvector in Figure 7.5a is almost constant, showing a maximum of ~1.2% variation across the height of the test volume (x-direction). This clearly corresponds to the TEM mode. Since the S-value of this mode is more than two orders of magnitude higher than its closest mode, we conclude that the TEM mode is dominant, by far, at this particular location "z." This is just what we would expect from a TEM cell.

Figure 7.5b corresponds to the second S-value of Figure 7.4. This eigenvector is symmetric about zero and has only one zero crossing. This is consistent with the TM_1 mode of a parallel-plate waveguide. An important point may be noted here. Since SVD has been performed on E_x data, we should not expect it to fall to zero at the top and bottom plates. Only E_z would fall to zero at the plates. This is why

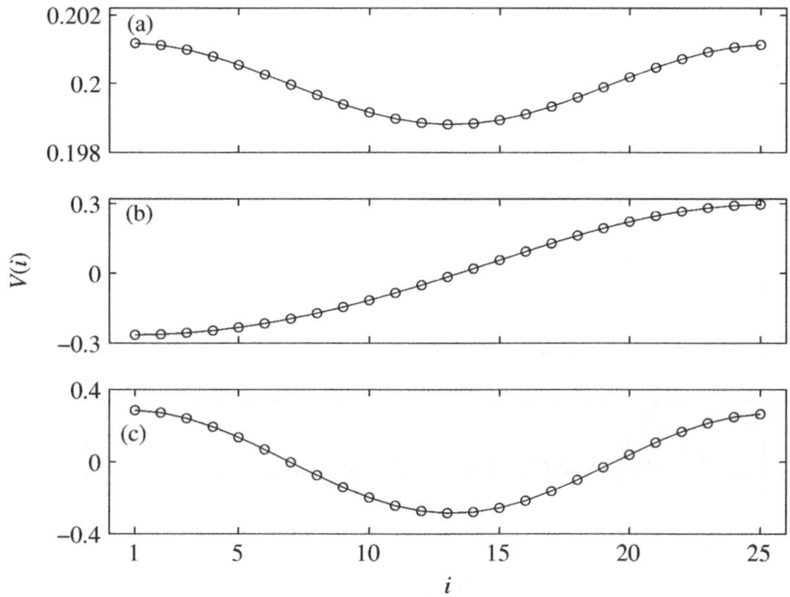

Figure 7.5 Eigenvectors of the three dominant modes indicated in Figure 7.4. Cases (a–c) correspond to TEM, TM_1, and TM_2 modes, respectively. Each curve represents the variation of mode amplitude in the x-direction. Abscissa "i" represents the x-location of the ith channel.

we are using the number of zero-crossings in the eigenvector for identification of the mode number.

Figure 7.5c shows two zero-crossings and matches the structure of the TM_2 mode.

The temporal evolution of the TEM, TM_1, and TM_2 modes is shown in Figure 7.6a–c. As expected, the dominant TEM mode is similar to the original electric field waveform $E_x(t)$. However, the higher order modes have entirely different waveforms.

7.3 Modal Analysis of Simulator Without Test Object

In this section, we examine the mode structure inside the simulator in the absence of a test object.

The discharge of the capacitor through the switch launches a broad range of frequencies into the simulator, as shown in Figure 7.7 for a position very close to the feed. The power spectrum is rapidly decaying, having significant energy content

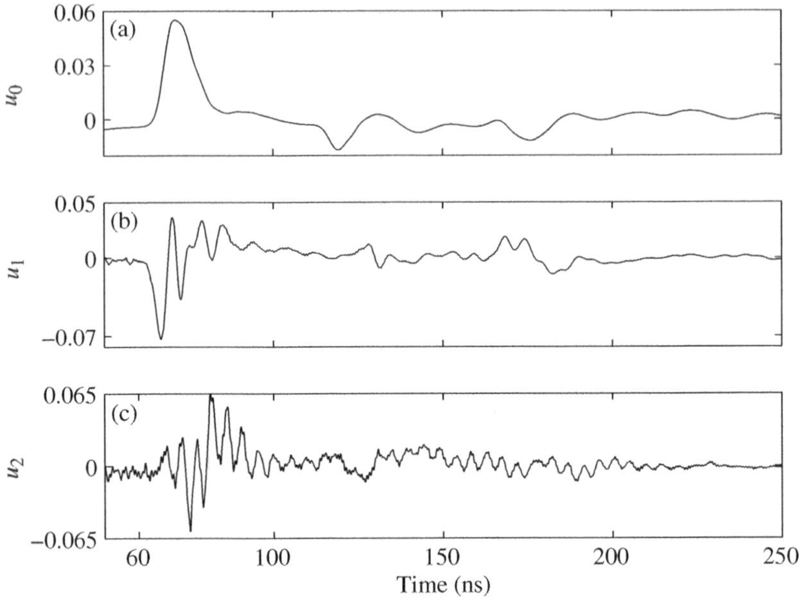

Figure 7.6 Temporal evolution of the three dominant modes. Cases (a–c) correspond to TEM, TM_1, and TM_2 modes, respectively.

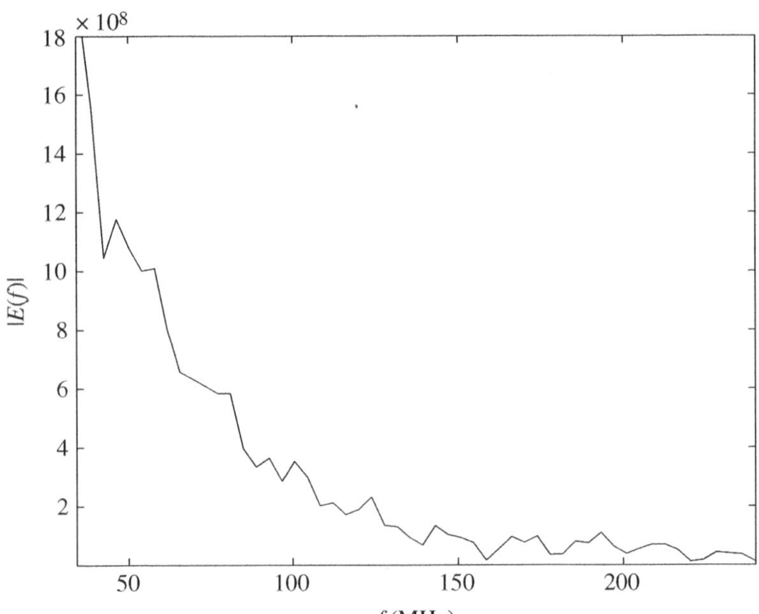

Figure 7.7 Fourier spectrum of vertical component of *E*-field near the feed of the TEM structure.

only up to ~240 MHz. Now, the cutoff condition for the TM_m mode is given by

$$f_c(z) = \frac{mc}{2h(z)} \tag{7.3}$$

where "c" is the speed of light in free space and $h(z)$ is the height between the plates, which is a function of z. The mode is evanescent for frequencies below f_c and propagates at higher frequencies. In the region of the simulator near the feed, $h(z)$ is rather small, hence f_c is high near the feed and decreases as we move toward the test section. This means that the strength of a given TM_m mode should generally increase as we move from the feed toward the test section. Also, since the power spectrum is more-or-less fixed, it is more difficult to excite higher order modes at a given z-location due to their higher cutoff frequencies.

SVD analysis has been performed on three-dimensional, time-dependent E_x data obtained from the FDTD analysis for the simulator without a test object.

A word of caution is necessary before we present the results of this analysis. We have already seen an expression for $E_x(x)$ of the TM_m eigenmode in Eq. (7.1). That expression is, strictly speaking, valid only for an infinitely long parallel-plate waveguide. In the simulator examined here, this condition is violated in two ways. Firstly, in the tapered section, the top plate is inclined with respect to the bottom one, so it is not a parallel-plate arrangement. The bottom plate would "short-out" E_z, as in a parallel-plate geometry, but the top plate would exactly short-out neither E_x nor E_z. This means that we should expect the actual eigenmode shape to deviate from the simple cosinusoidal form given above, especially in the tapered section. Secondly, even the parallel-plate section is not infinite, being terminated by a resistive sheet. Hence, even in the parallel-plate section, we would expect some deviation near the termination.

Given these two points, we expect the following general trend. There should be significant deviations in eigenmode shape all through the tapered section. Moving in the longitudinal (z-) direction within the parallel-plate section, we expect the eigenmode to progressively approach the "ideal" form. Near the termination, the deviation should increase once again. However, given that the termination is a matched one, we expect the point of "best" match to lie closer to the termination than to the junction.

In the rest of this chapter, whenever we refer to a particular mode "m," it actually refers to the mode that most closely resembles that particular mode number. For example, $m = 0$ would refer to a mode which is nearly a constant in the x-direction.

These trends are indeed seen in Figures 7.8–7.11, which show the evolution of the eigenmodes V_0, V_1, and V_2, respectively, as functions of longitudinal position "z." In the case of the $m = 0$ mode, there is increasing distortion as we approach the junction, probably due to the discontinuity. There is significant improvement over a small length soon after entering the parallel-plate section, followed by a region of near-constant eigenmode shape.

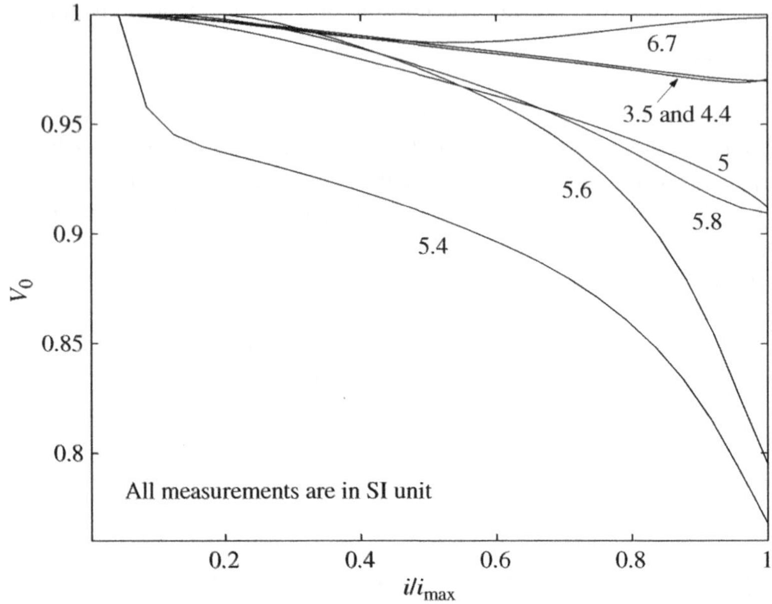

Figure 7.8 Longitudinal (z-) variation of $m = 0$ eigenmode inside the simulator, over the range $z = 3.5–6.7$ m. Note the large distortion in shape inside the tapered section.

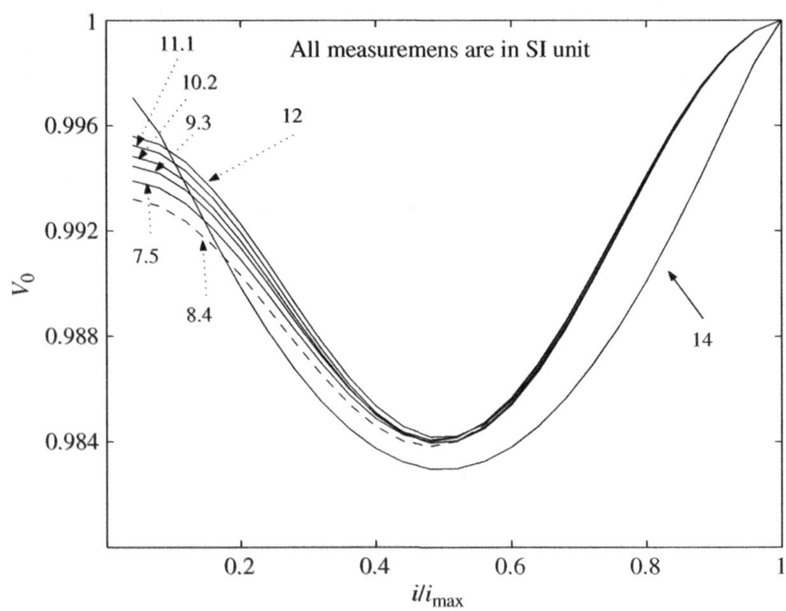

Figure 7.9 Longitudinal (z-) variation of $m = 0$ eigenmode inside the simulator, over the range z =7.5–14 m.

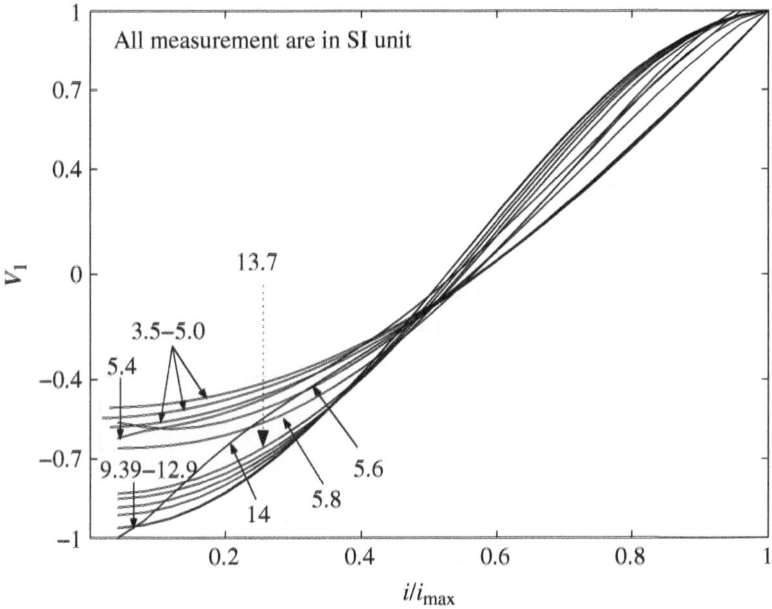

Figure 7.10 Longitudinal (z-) variation of $m = 1$ eigenmode inside the simulator. The numbers next to the curves denote the z-location in meters.

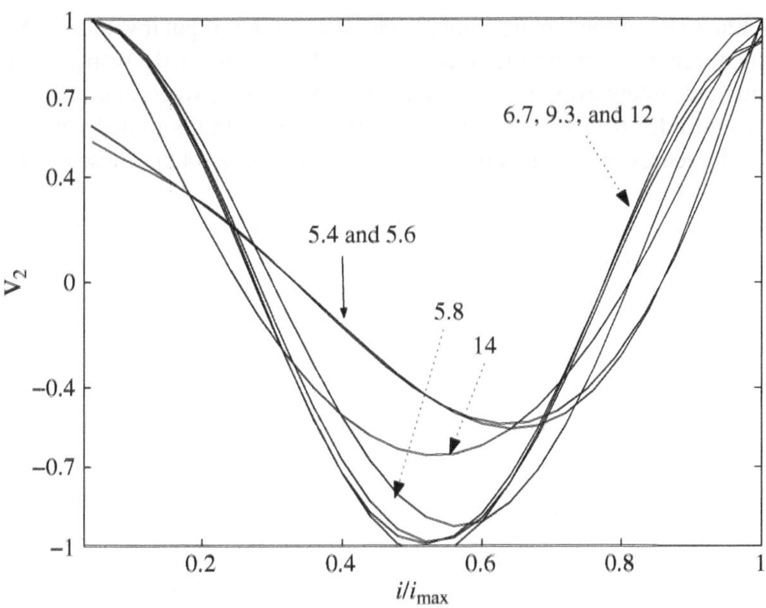

Figure 7.11 Longitudinal (z-) variation of $m = 2$ eigenmode inside the simulator. The numbers next to the curves denote the z-location in meters.

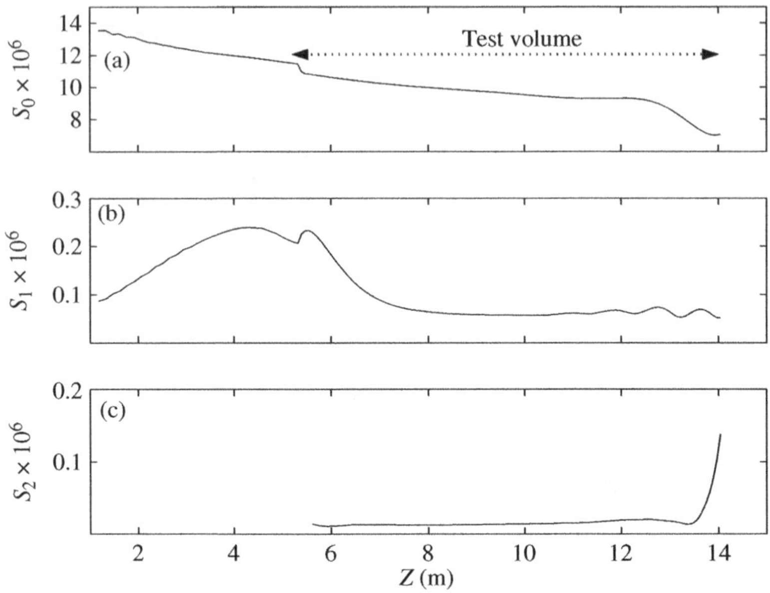

Figure 7.12 Longitudinal (z-) variation of singular values of important modes inside the simulator. Cases (a–c) correspond to TEM, TM_1, and TM_2 modes, respectively.

Figure 7.12a–c shows the longitudinal (z-) variation of the singular values S_0, S_1, and S_2 of the three most important modes, viz. TEM, TM_1, and TM_2, respectively.

Two points are noteworthy in Figure 7.12. Firstly, S_0 is between one and two orders of magnitude stronger than the others throughout the bounded volume. Secondly, S_1 increases monotonically between $z \sim 2$ m and $z \sim 4$ m. These can be understood as follows.

The height $h(z)$ between the plates increases linearly with z between the feed and the junction, as shown in Figure 7.1. Between the junction and the aperture, $h(z)$ remains constant. Equation (7.3) then implies that the cutoff frequency for any given mode steadily falls as we move from the feed to the junction.

Given the power spectrum of Figure 7.7, we expect that the TEM mode, which propagates at all frequencies, must be the dominant mode. At small values of "z," f_c is high for $m = 1$, so that only a small part of the launched energy can exist in the TM_1 mode. As "z" (and $h(z)$) increases, a progressively larger fraction of the energy can exist in the TM_1 mode, leading to an increase in S_1. An example will illustrate the point. For $z \sim 2$ and 3 m, $h(z)$ is ~ 0.65 and 0.9 m, respectively, yielding f_c values of 230 and 170 MHz for TM_1. It is clear from Figure 7.7 that a decrease in f_c from 230 to 170 MHz allows access to a significantly larger fraction of the power, strengthening TM_1.

The TM_2 mode is not well formed in the tapered section because its cutoff frequency is far too high.

The singular values S_0 (S_1) show a rapid decrease (increase) near the junction. This is probably due to a sudden change in the geometry, which produces a discontinuity in E_x. A small reflection at the junction has also been observed in a computational analysis [52].

All three singular values are fairly constant throughout the test volume, with the exception of a small region near the termination. There is a significant decrease in S_0 near the termination, which is explained below.

Figure 7.13 shows the spatiotemporal variation of E_x inside the test volume near the termination. There is a progressive reduction in the peak amplitude of the pulse as we approach the termination, a consequence of reflections from the termination. Note that the reflection has a negative amplitude. At z-locations far from the termination, the first pulse of the forward-traveling wave completes its rise-and-fall well before the reflected wave from the termination reaches that location. Hence, the rise-and-fall of the main peak is unaffected. As we move closer to the termination, the reflections arising out the "rising" part of the main pulse reach the location earlier. This means that the main peak remains unaltered, but there is a faster decay following this peak. Very close to the termination, the reflected pulse arrives even before the main peak is reached, leading to a reduction in the

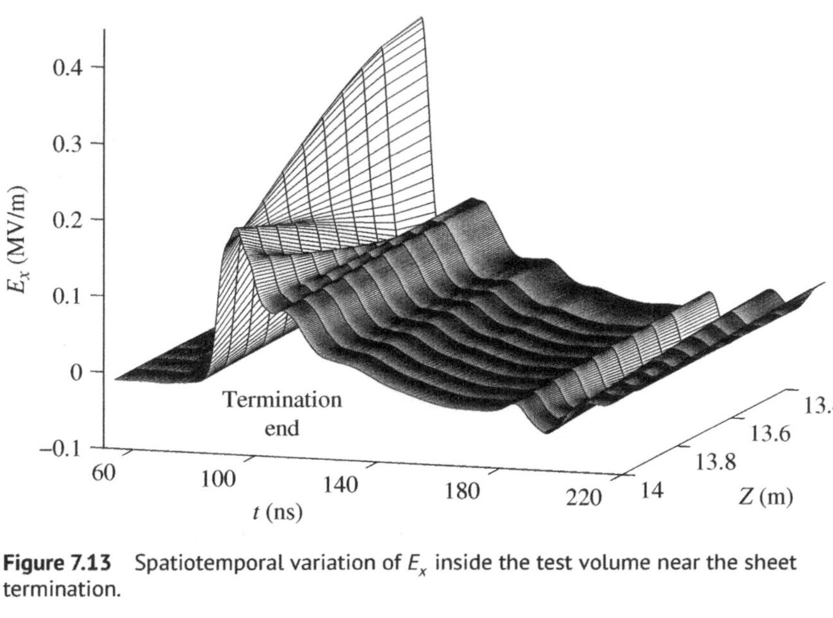

Figure 7.13 Spatiotemporal variation of E_x inside the test volume near the sheet termination.

primary peak itself. Since E_x is the dominant electric field component, and TEM the dominant mode, the same trend is observed in the singular value S_0.

Figure 7.12 also shows that S_2 increases rapidly near the termination. This increase in the magnitude of $m = 2$ can be explained in terms of the variation of current along the height of the termination and the magnetic field H_y produced by this current. Figure 7.14 shows the variation of instantaneous current flowing through the sheet termination, as a function of height "x" from the bottom to the top end, for the case with switching time $\tau_s = 19.7$ ns. At any given x-location, the current is calculated through the line integral $\int \vec{H} \cdot \vec{dl}$, the integral being taken along a loop enclosing the sheet termination. This loop closely encloses the sheet, so that conduction current makes the major contribution. The integral is calculated at a time when the current through the sheet is close to its maximum. We see that the current is maximum at both ends, with a minimum lying near the middle of the sheet. The asymmetry in the current is probably due to the difference between the top and bottom plates. This x-directed current produces a corresponding variation in H_y, which is shown in Figure 7.15. The H_y waveform resembles a constant function with a superimposed $m = 2$ variation. Furthermore, the curvature and amplitude of H_y becomes stronger as we move toward

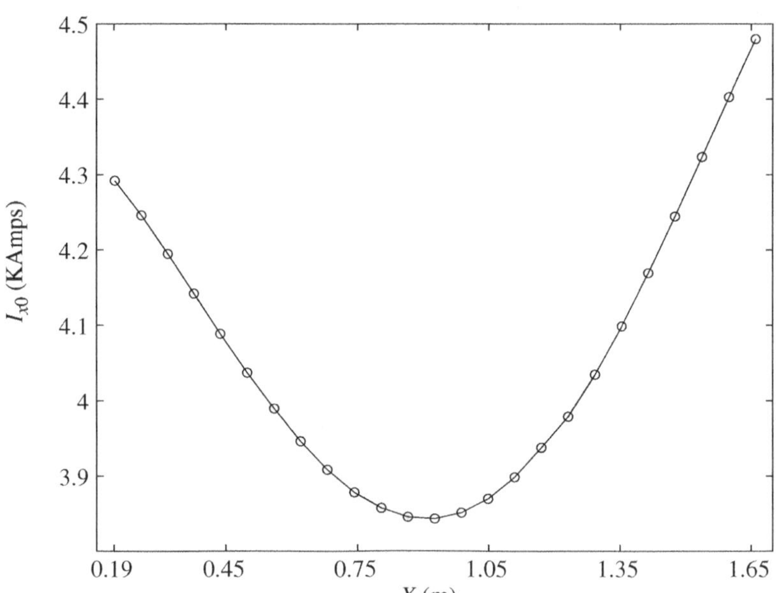

Figure 7.14 Instantaneous x-directed current through the sheet termination as a function of its height from the bottom to the top end. Case with switching time $\tau_s = 19.7$ ns.

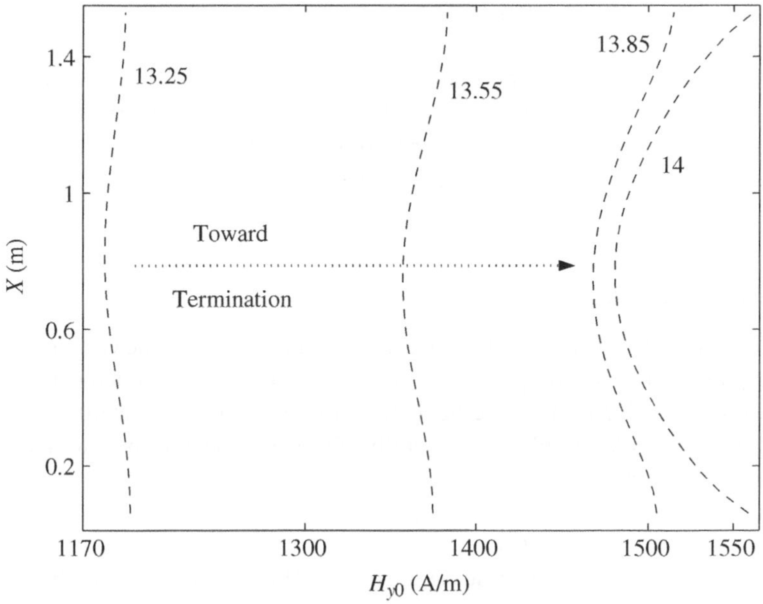

Figure 7.15 Instantaneous magnetic field $H_y(x)$ between the parallel plates. Curves are shown at different z-locations near the sheet termination. Case with switching time $\tau_s = 19.7$ ns. At each x-location, H_y is recorded at the time corresponding to peak magnetic field at the center of xy-plane in the test volume.

the termination. This corresponds to a progressively increasing $m = 2$ mode as we approach the termination.

A simple quantitative check will be in order at this point. From Figure 7.15, we see that H_{y0} varies between ~1560 and ~1480 A/m at $z = 14$ m. This corresponds to an $m = 2$ mode of amplitude ~ 40 A/m, superimposed on an $m = 0$ mode of strength ~ 1520 A/m. This implies a ratio of amplitudes of 40/1520 ≃2.6%. This matches fairly well with the ratio $S_2/S_0 \simeq 2\%$ at $z = 14$ m.

7.4 Modal Analysis of Simulator With Test Object

In an EMP simulator, the presence of the test object itself modifies the fields to which it is subjected. The extent of modification depends upon the dimensions and material composition of the object. We have already reported on the distortion produced by test objects of different sizes in Chapter 6. In particular, we found that the simulator examined here could be used for testing perfectly-conducting cube-shaped objects with a maximum dimension of ~0.5 m, without introducing significant distortion. We also found that the frequency spectrum "seen" by the object also changes with object size.

These changes in the spatial and temporal structure of the field imply a corresponding change in the mode structure. In this Section, we examine the effect of the test object on the mode structure. For purposes of illustration, we have performed modal analysis in the presence of a perfectly conducting solid cube of side 0.5 m.

7.4.1 Qualitative Analysis

Before starting the modal analysis, it is instructive to take a look at the variation of the two major electric field components, E_x and E_z, through the bounded volume. Now, we expect the electromagnetic field structure to deviate significantly from that for an empty simulator only in the vicinity of the test object. Hence, we focus our attention on that region.

The electric field E_x is zero inside the test object, but finite in the gap between the parallel plates and the object, as shown in Figure 7.16. This gives rise to strong gradients of E_x in the region surrounding the object. The variation of E_x in the x–z

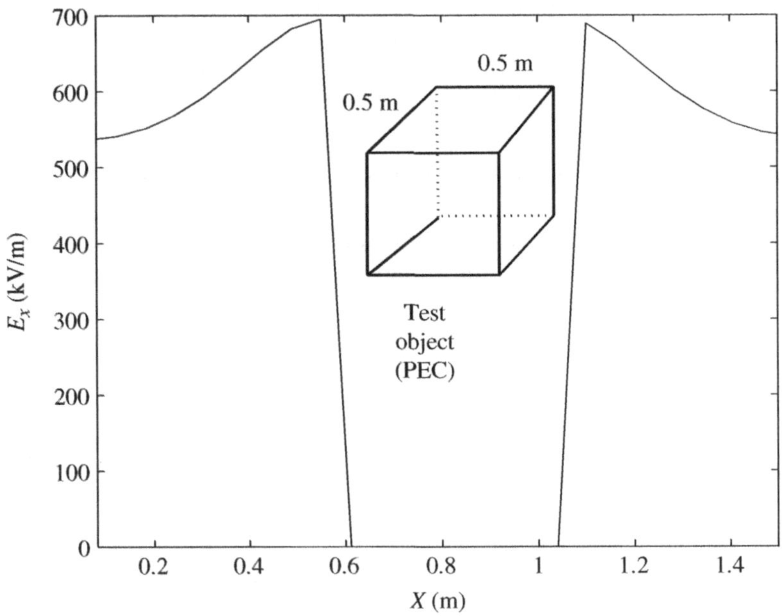

Figure 7.16 Variation of instantaneous electric field E_x with x, with (y, z) coordinates located at the center of the test object, at a time corresponding to the peak in electric field near the object.

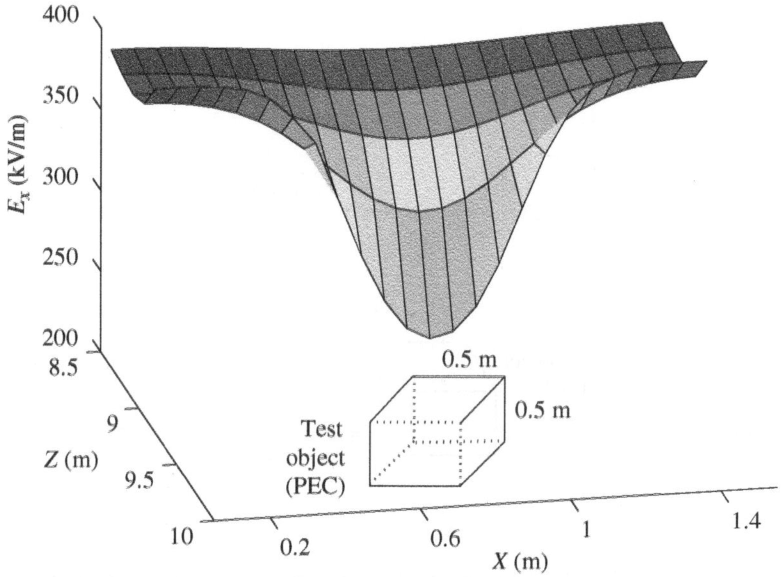

Figure 7.17 Variation of E_x in the xz-plane in the vicinity of the test object. For each z-location, reference time is chosen corresponding to peak electric field at the center of xy-plane. The y-value is taken at the center of the simulator.

plane, in the vicinity of the object, is shown in Figure 7.17. At each z location, the E_x value has been taken at a time corresponding to the peak electric field at the center of the x–y plane. Visual inspection shows that the curve $E_x(x)$ at any given z-location can be considered as a superposition of $m = 0$ (constant) with several higher order modes. The higher order component clearly increases rapidly as we approach the object. In fact, over the range of z spanned by the object, the discontinuity in E_x (Figure 7.16) can only be represented by an infinite number of modes.

It is possible to interpret the rapid change in TM modes in terms of a divergence-free electric field. Since E_y is negligible, we have

$$\partial E_x/\partial x + \partial E_z/\partial z = 0 \tag{7.4}$$

This means that any gradient in E_x in the x-direction must be balanced by a gradient of E_z in the z-direction, and the two must have opposite signs. Now, since the object is perfectly conducting, E_x has the most rapid variation with x at the z-location of the object itself; as we move away, this variation progressively decreases. Equation (7.4) then implies a rapidly increasing amplitude of E_z as we approach the object. This can actually be seen in Figure 7.18, which shows the variation of E_z in the x–z plane. Furthermore, Figure 7.16 shows that at the location of the object, as we move from the bottom to the top plate, E_x first drops

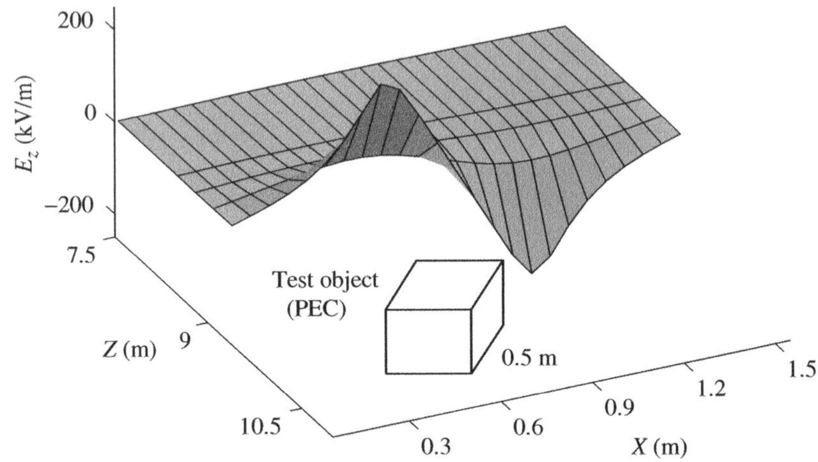

Figure 7.18 Variation of E_z in the *xz*-plane in the vicinity of the test object. For each *z*-location, reference time is chosen corresponding to peak electric field at the center of *xy*-plane. The *y*-value is taken at the center of the simulator.

from a finite value to zero, stays at zero throughout the object, and then rises from zero to a finite value at the top plate. This means that $\partial E_x/\partial x$ exhibits a change of sign. $\partial E_z/\partial z$ must, therefore, exhibit the same change of sign, which is consistent with the structures of higher order TM modes.

Figure 7.19 shows the *x*-variation of E_z at different *z*-values. All E_z values at a given *z*-location are normalized to the peak value at that *z*. This figure clearly shows the excitation of higher order modes.

7.4.2 Quantitative Analysis Using SVD of E_x Data

We can now proceed with a quantitative modal analysis. As stated earlier, the discontinuity in $E_x(x)$, produced by the perfectly conducting object, can only be resolved by using an infinite number of modes. Since this is not meaningful, we restrict our attention to a range of *z*-values that approaches, but does not overlap with, the *z*-range spanned by the object.

Figure 7.20a–c shows the variation of S_0, S_1, and S_2, respectively, between the feed and the termination. The location of the test object is indicated in Figure 7.20a.

A comparison with Figure 7.12 shows no significant change in the nature of the dominant mode inside the tapered section, and also in the parallel-plate section far

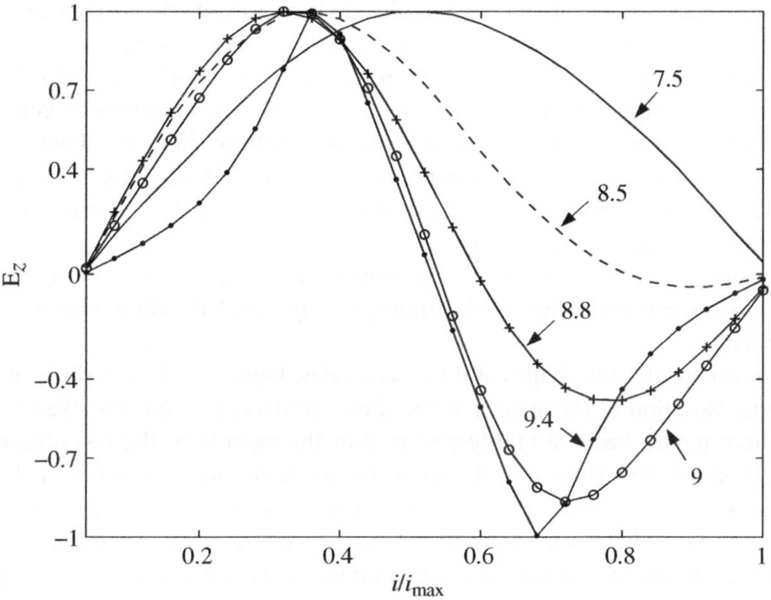

Figure 7.19 Variation of normalized E_z in the *xz*-plane in the vicinity of the test object. For each *z*-location, reference time is chosen corresponding to peak electric field at the center of the *xy*-plane. The *y*-value is taken at the center of the simulator.

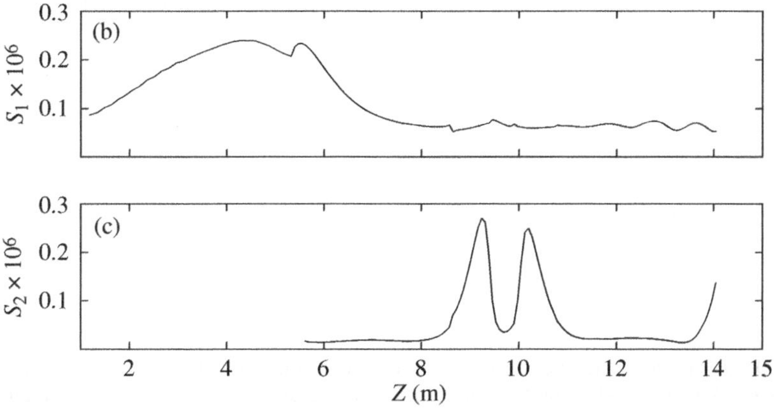

Figure 7.20 Longitudinal (*z*-) variation of singular values of important modes in the presence of a perfectly conducting solid cube of side 0.5 m. Cases (a–c) correspond to the three most important modes.

from the test-object. However, there is a significant increase in S_0 in the immediate vicinity of the test object. This can be explained as follows. The presence of the perfectly conducting object acts like an electrical short-circuit, reducing the effective gap between the parallel plates. This means that the electric field must become stronger in the gap, so as to maintain a reasonable voltage difference between the plates. Now, the SVD analysis is done on the entire E_x data. Since S_0 represents the dominant mode, and E_x is the dominant electric field component, the enhancement in E_x also appears in S_0.

Another physical interpretation for this increase in the electric field in the gap can be given in terms of the flow of electromagnetic power. This will be presented in Section 7.5.

Let us now consider the shapes of the eigenmodes. Figures 7.21–7.24 show the longitudinal variation of the shapes of the three strongest eigenmodes. We find that all three modes become highly distorted in the vicinity of the test object. Figure 7.25 shows the shapes of the same modes at a z-location lying at the center of the object, where the distortion reaches a maximum. The distortion in the modes arises because, in the SVD method used here, no constraint has been imposed upon the form of the eigenmodes. This distortion was, in any

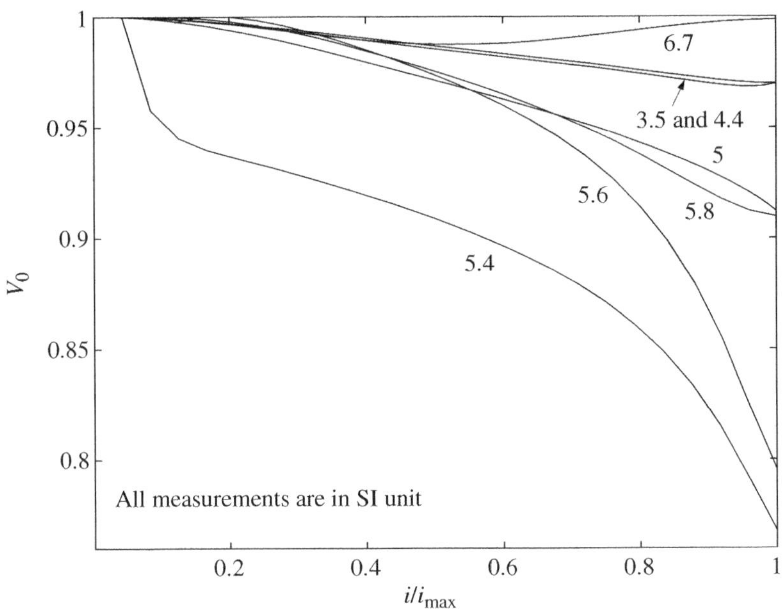

Figure 7.21 Longitudinal (z-) variation of $m = 0$ eigenmode inside the simulator, over the range $z = 3.5$–6.7 m. The numbers next to the curves denote the z-location in meters. The structure in this region is largely unaffected by the presence of the test object.

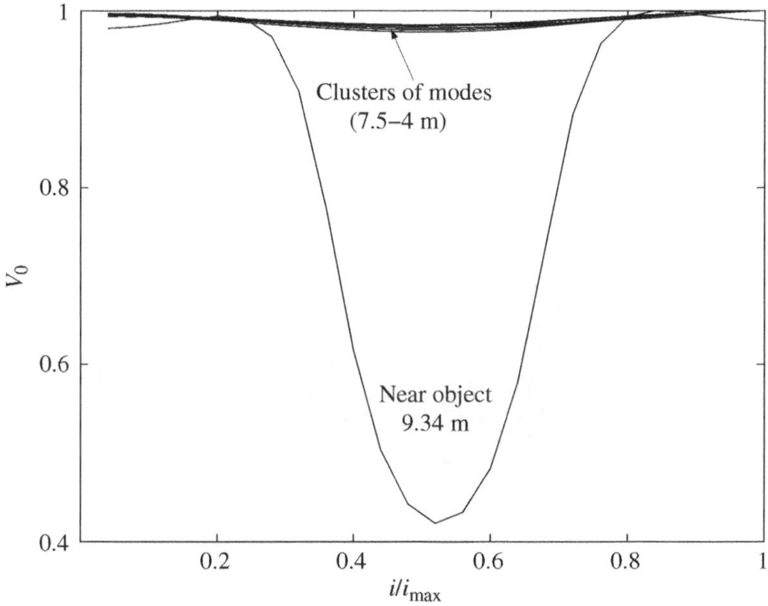

Figure 7.22 Longitudinal (z-) variation of $m = 0$ eigenmode inside the simulator, over the range $z = 7.5$–14 m, in the presence of a test object. The numbers next to the curves denote the z-location in meters. Note the large distortion near the object.

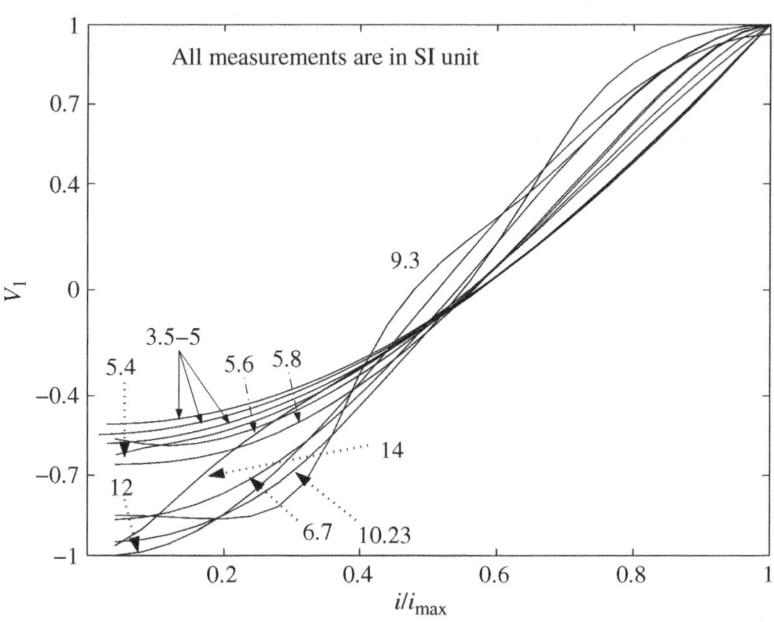

Figure 7.23 Longitudinal (z-) variation of $m = 1$ eigenmode inside the simulator, in the presence of a test object. The numbers next to the curves denote the z-location in meters.

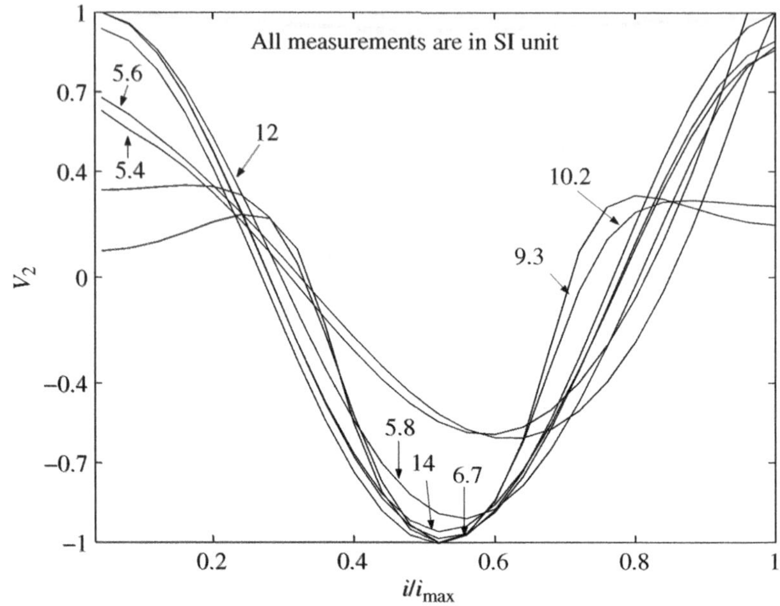

Figure 7.24 Longitudinal (z-) variation of $m = 2$ eigenmode inside the simulator, in the presence of a test object. The numbers next to the curves denote the z-location in meters.

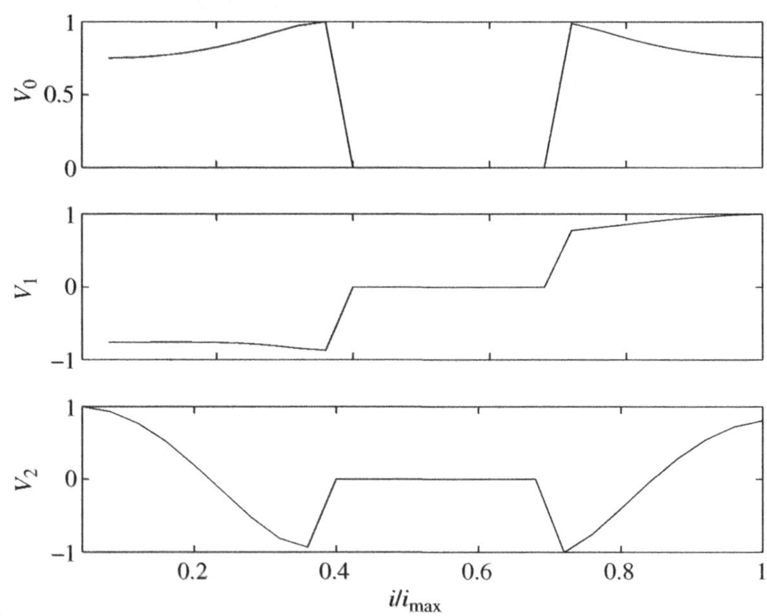

Figure 7.25 Shapes of the three strongest eigenmodes at a z-location corresponding to the center of the perfectly conducting test object. Note the large distortion, particularly inside the object, where the electric field falls to zero.

case, expected from the qualitative discussion given earlier, which indicated the presence of higher order modes.

The significance of this distortion is that the singular values shown in Figure 7.20 accurately reflect the variation in amplitude of various "m"-numbered modes only as long as we are at least ~0.5 m away from the object. Closer to the object, the s-values only represent the strengths of the distorted eigenmodes.

From a practical point of view, we only need to know the extent to which the electric field, seen by the object, deviates from the ideal planar waveform. Since one of the eigenmodes dominates by far, this deviation is perfectly well quantified by the amount of distortion of the dominant eigenmode. This distortion has already been obtained. Hence, strictly speaking, it is not necessary to further resolve these eigenmodes into modes represented by integral m-values. However, it would be more satisfying to see if the distortion could be further resolved into cosinusoidal modes represented by integral "m" values. An attempt to achieve this resolution is discussed below.

In the qualitative discussion earlier in this section, we have explained why we expect progressively higher order modes to appear as we approach the object. In the above analysis, for the parallel-plate section, we have used 25 channels in the x-direction. It is reasonable to think that the distorted eigenmodes probably arise because the spatial resolution of our data is not enough to resolve high-order modes near the object. The solution is to use FDTD data that is more closely spaced in the x-direction. This technique need only be applied in the vicinity of the object, where SVD has, so far, yielded distorted modes.

We have, therefore, repeated the above analysis with 171 channels instead of 25. Higher order modes, e.g. $m = 6$, are clearly resolved, but their singular values turn out to be rather small. The dominant mode still shows a eigenmode shape that is highly distorted, closely resembling the shape that was obtained with 25 channels. This shows that the spatial resolution is adequate – eigenmode distortion must be due to some other effect.

7.4.3 Quantitative Analysis Using SVD of E_z Data

The SVD technique, as it is used here, does not impose any constraint upon the shapes of the eigenmodes. The $E_x(x)$ data, on which SVD is performed, also does not demonstrate any specific boundary condition. This is because the parallel plates of the test section force only E_z to zero, but impose no constraint on E_x. Hence, it is not surprising that, in regions of large field distortion, the "best-fit" eigenmodes are not necessarily cosinusoidal.

An obvious solution is to use E_z instead of E_x data. This has the limitation that the $m = 0$ mode cannot be resolved, since E_z falls to zero at both ends of the x-range. However, the analysis will at least allow us to compare the relative strengths of all modes other than $m = 0$. We can also hope that the eigenmodes, even in the vicinity of the object, will follow a sinusoidal variation more closely than they did with E_x data.

Figure 7.26 shows the variation with z of the ratios of singular values of different modes. As the object is approached, the rapid increase in S_2 leads to a progressive decrease in all ratios. However, S_3 also increases to some extent, so that S_3/S_2 does not decrease to the same extent as S_1/S_2. The increase in S_4 is almost as dramatic as in S_2, so that S_4/S_2 is almost constant. This analysis clearly shows the increase in amplitude of higher order modes.

Figures 7.27–7.30 show the variation with longitudinal position of the shapes of these eigenmodes. The $m = 1$ mode remains fairly close to sinusoidal until we get very close to the object. The $m = 2$ mode retains its basic shape even at $z = 9.27$ m, a location at which SVD analysis of E_x data yielded a highly distorted mode. Both $m = 3$ and $m = 4$ retain their basic shapes, although the negative and positive peaks no longer have the same magnitude. Since the mode shapes are "reasonable," the singular values can, indeed, be said to represent their respective m-numbers.

Two significant conclusions have come out of this analysis. Firstly, SVD analysis of E_z data is better able to resolve sinusoidal modes, corresponding to integral

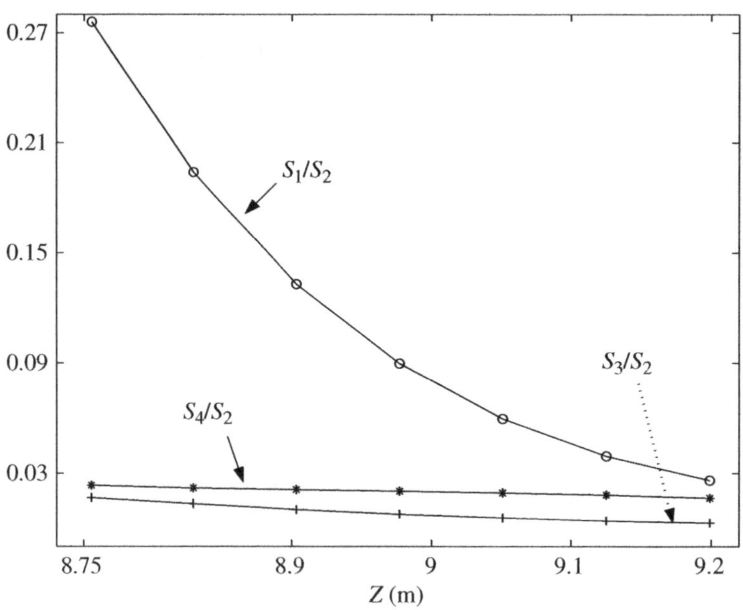

Figure 7.26 Variation of relative strengths of higher order modes with respect to $m = 2$ in the immediate vicinity of the object. SVD analysis has been done on the E_z data.

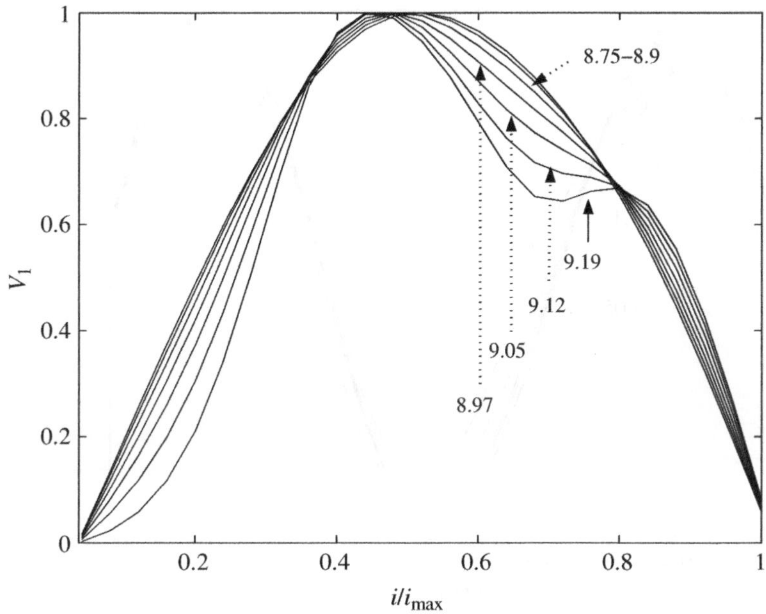

Figure 7.27 Evolution of $m = 1$ eigenmode in the immediate vicinity of the object. SVD analysis has been done on the E_z data. Numbers on the figure indicate longitudinal (z) position in meters.

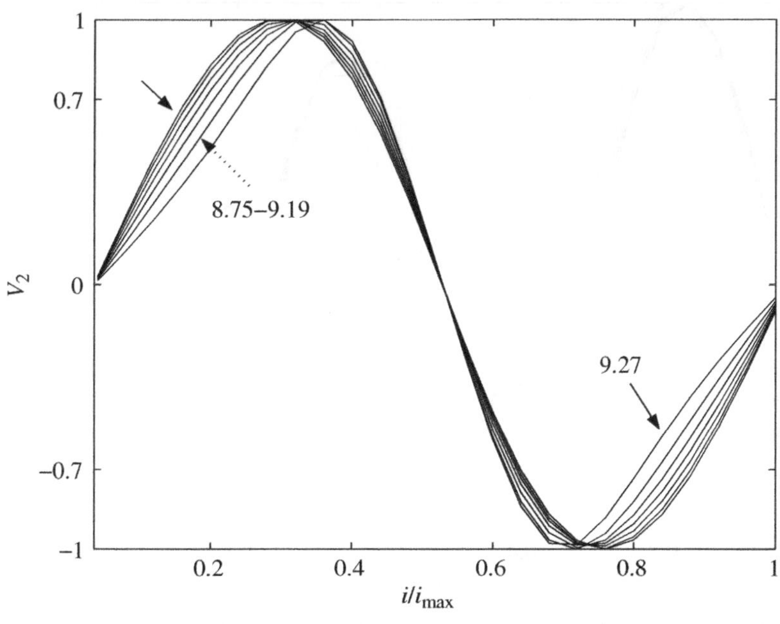

Figure 7.28 Evolution of $m = 2$ mode in the immediate vicinity of the object. SVD analysis has been done on the E_z data. Numbers on the figure indicate longitudinal (z) position in meters.

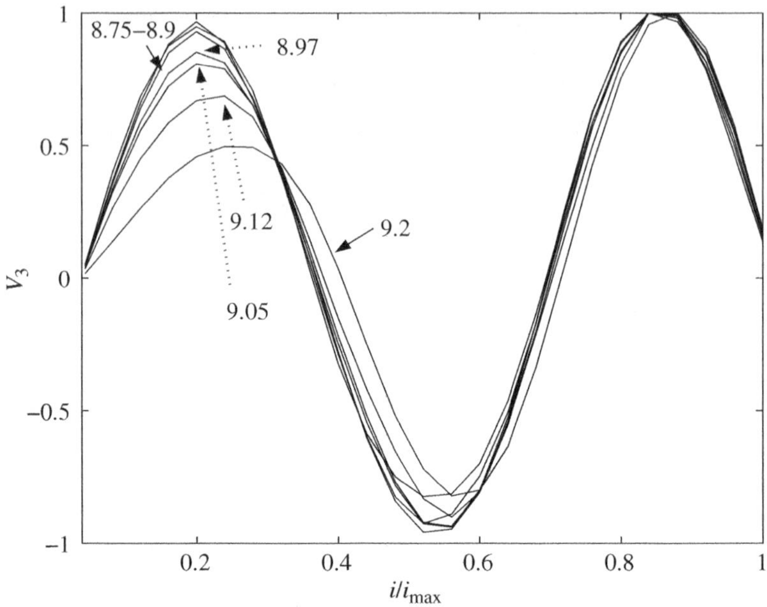

Figure 7.29 Evolution of $m = 3$ mode in the immediate vicinity of the object. SVD analysis has been done on the E_z data. Numbers on the figure indicate longitudinal (z) position in meters.

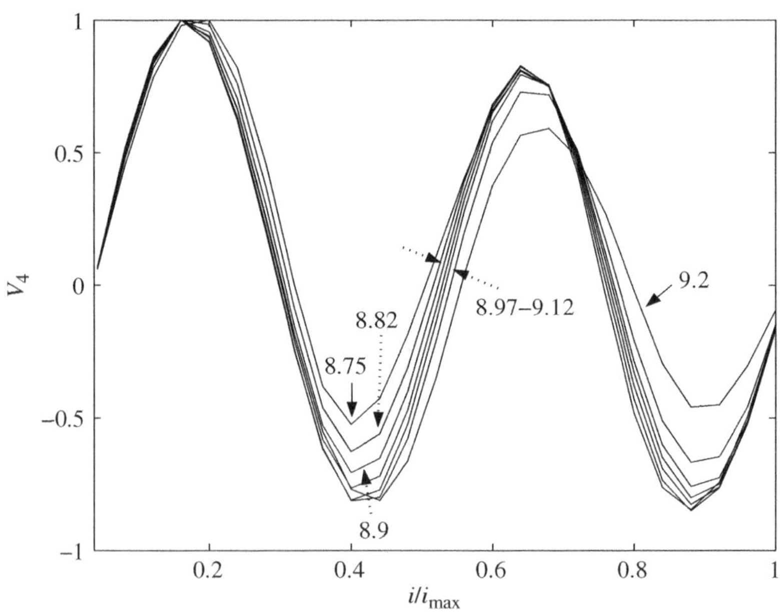

Figure 7.30 Evolution of $m = 4$ mode in the immediate vicinity of the object. SVD analysis has been done on the E_z data. Numbers on the figure indicate longitudinal (z) position in meters.

m-values, as compared to analysis of E_x data. The difference probably lies in the boundary conditions, which force E_z to go to zero at the two plates. Secondly, this analysis shows a rapid increase in the $m = 2$ and $m = 4$ modes, and also some increase in $m = 3$, in the vicinity of the test object.

Hence, we conclude that TEM mode "purity" is considerably reduced by the presence of a reasonably sized test object. The main point of practical interest is the emergence of strong TM modes in the immediate vicinity of the object, which is a major deviation from the situation encountered during free-space illumination.

7.5 Physical Interpretation for Electric Field Increase

In the last subsection, we discussed the increase in electric field in the gap between the test object and the top and bottom plates. That increase can also be explained in terms of the flow of electromagnetic energy through the bounded volume.

FDTD simulations provide a detailed 3-D, time dependent picture of the evolution of electromagnetic fields. This makes it possible to calculate the Poynting flux distribution in time and space, and thereby gain insight into the flow of energy. This flow is studied by calculating the instantaneous distribution of the Poynting flux through each x–y cross section of the test volume. Once the z-component of the Poynting vector is known as a function of x and y, we can calculate the instantaneous z-directed power P_z at different z-locations through the integral $\int_S (\vec{E} \times \vec{H})_z \mathrm{d}S$. Here, d$S$ is an infinitesimal area in the xy-plane.

Figure 7.31 shows P_z as a function of longitudinal (z) position. The simulator has a perfectly conducting, cube-shaped test object of side 0.5 m, mounted at the center of the test volume. Instantaneous electromagnetic field values at each z-location have been taken at a time when the first peak of the electric field reaches that location. The first and last points of the plot shown in Figure 7.31 lie outside the z-directed faces of the object, while the three intermediate points lie inside the test object. We observe that the instantaneous power flow varies within ~0.25%, which is negligible. This confirms that the presence of the object does not change the total power flow in the z-direction.

Clearly, if the perfectly conducting object blocks power flow, electromagnetic power will flow around, rather than through, the object. That requires the Poynting flux to becomes higher in the gap between the object and the simulator plates. Higher Poynting flux corresponds to higher electric field levels in the gap, which is just what we have observed.

Figure 7.32a–c shows the distribution of Poynting flux in an x–z cross section of the test volume, in the immediate vicinity of a perfectly conducting (PEC) solid cube of side 0.5 m. The plots are shown at the mid-point in the y-direction. The three sub-plots (a–c) show snapshots taken at $t = 8.32\tau$, 8.44τ, and 8.5τ

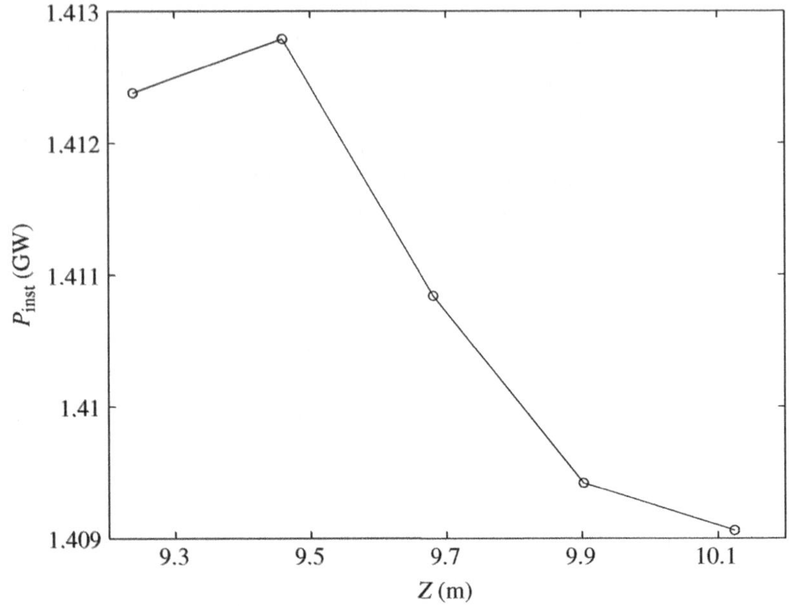

Figure 7.31 Instantaneous z-directed power flow as a function of longitudinal position z. The simulator has a perfectly conducting, cube-shaped test object of side 0.5 m, mounted at the center of the test volume. Instantaneous electromagnetic field values at each z-location have been taken at a time when the first peak of the electric field reaches that location.

respectively, with $\tau = 19.7$ ns. These time values have the following significance. The first is before the main E-field peak reaches the front surface of the object. The second is when the pulse reaches the center of the object and the third corresponds to the time when the pulse crosses the rear surface of the object. As expected, well "upstream" of the object, the Poynting vectors are mainly z-directed. As the object is approached, the flux tends to diverge, such that the energy flows through the upper and lower gaps between the object and the simulator plates. The two flow channels again come together downstream of the object.

At first sight, this flow of electromagnetic energy around the test object appears similar to the flow of a fluid around an obstacle. However, there turns out to be a major difference. A fluid would tend to flow symmetrically all around an obstacle. The Poynting flux, on the other hand, tends to focus only into the gaps between the x-directed faces of the object and the simulator plates, with little enhancement around the y-directed faces of the object. This difference is discussed below.

Figure 7.33a–c shows the variation of Poynting flux in an x–y cross section of the test volume in the immediate vicinity of the test object. The three sub-plots

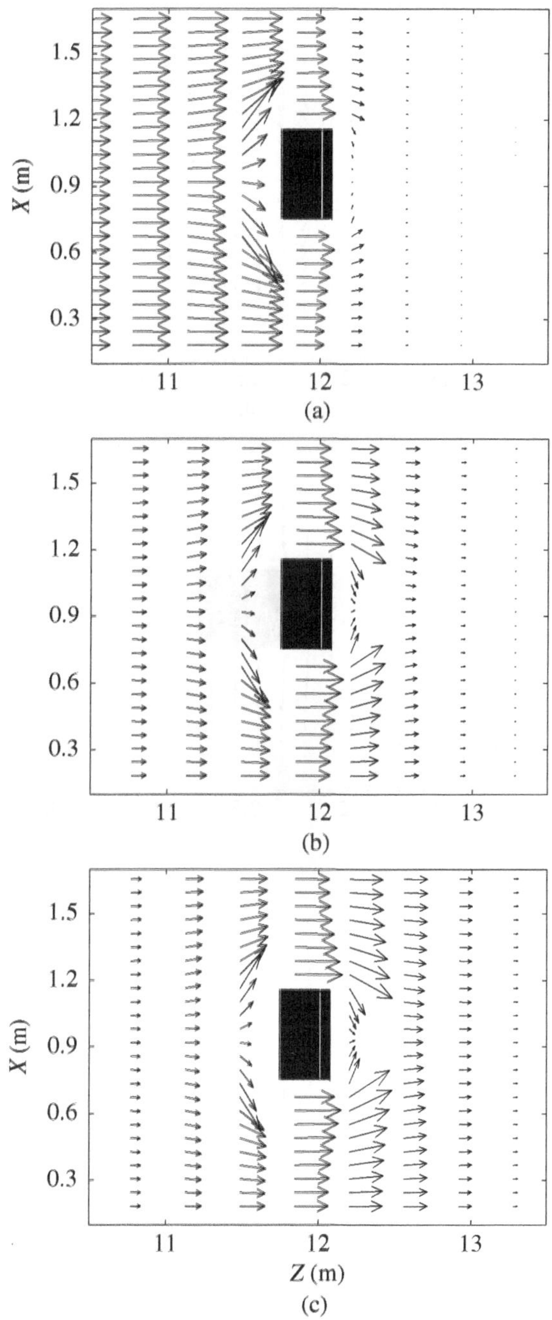

Figure 7.32 Snapshot of Poynting flux distribution in the *xz*-plane in the immediate vicinity of a solid PEC cubic object of side length 0.5 m inside the test volume. The plots are shown at the mid-point in the *y*-direction. Plots (a)–(c) correspond to $t = 8.32\tau$, 8.44τ, and 8.5τ respectively with $\tau = 19.7$ ns. "x" and "z" correspond to height and length of simulator, as shown in Figure 7.1.

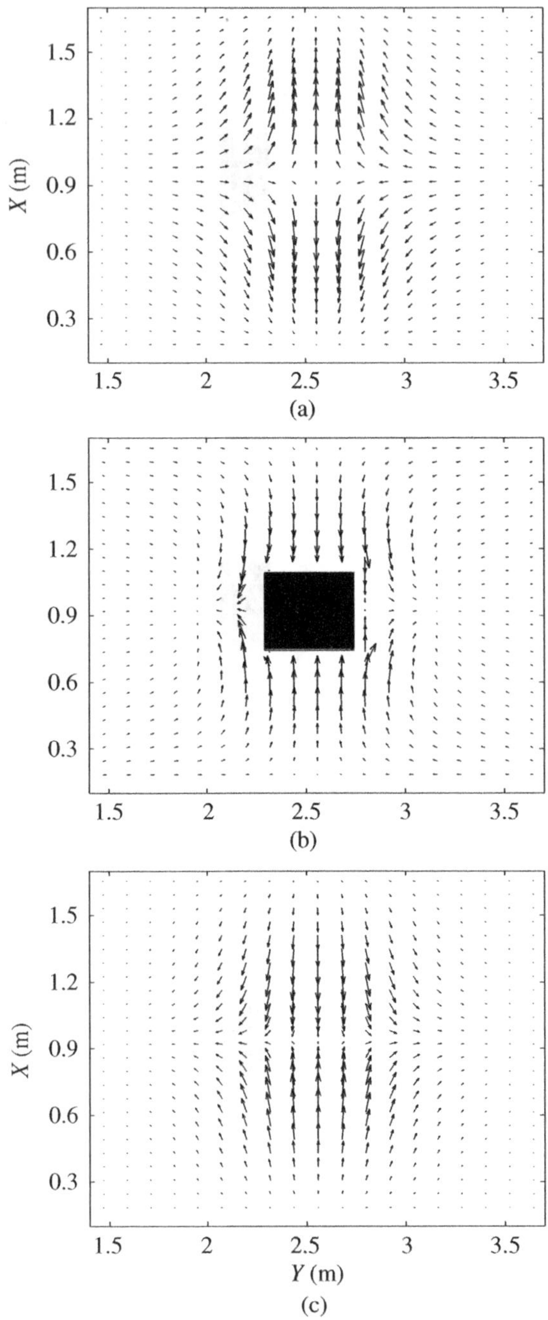

Figure 7.33 Snapshot of Poynting flux distribution in the *xy*-plane in the immediate vicinity of a solid PEC cubic object of side length 0.5 m inside the test volume. Plots (a)–(c) correspond to $t = 8.32\tau$, 8.44τ, and 8.5τ respectively with $\tau = 19.7$ ns. "x" and "y" correspond to the height and width of simulator, as shown in Figure 7.1. Each plot is shown at a *z*-location where electric field reaches its peak at the specified time.

correspond to the same time values as those in Figure 7.32. Each plot is shown at a z-location where the electric field reaches its peak at the specified time.

Now, P_x is produced by a combination of E_z and H_y. The latter is the dominant magnetic field component, while E_z, associated with the TM modes, becomes significant in the vicinity of the test object. Hence P_x also becomes significant in this region. On the other hand, P_y would have to be produced either by a combination of E_x and H_z, or E_z and H_x. Both these magnetic field components are rather small. Hence, P_y is generally rather small as compared to P_x. This explains why the flow of electromagnetic energy tends to concentrate above and below the object, but not on its sides.

The foregoing discussion emphasizes the power of FDTD analyses to yield in-depth physical understanding of pulsed electromagnetic systems.

7.6 Summary

We have performed an analysis of the electromagnetic mode structure inside a bounded-wave EMP simulator, both with and without a test object. This has been done by the application of the SVD method to time-domain data generated by self-consistent, three-dimensional FDTD simulations. This combination of two powerful techniques yields a wealth of information about the internal mode structure which cannot be otherwise obtained. This is particularly important, given that a real-life simulator, with a realistic test object, has a complex geometry and consists of a range of materials from good conductors to dielectrics; and this is further complicated by the existence of broad-band excitation. To our knowledge, this is the first time such a comprehensive study has been done.

E_x and H_y are the dominant field components in the absence of the object. We also find a significant E_z in the vicinity of the test object or the resistive termination. This shows that TEM and TM modes exist, while TE modes are not present to any significant extent.

The conclusions given below are for the case with a switch closure time of 19.7 ns, corresponding to an electric field risetime of 4.3 ns. However, the analysis can easily be repeated for any other value of the switching time.

In the absence of the object, the TEM mode is dominant throughout the simulator length. TM_1 dominates over other TM modes over most of the length. Within the tapered section, in particular, TM_2 and higher modes do not exist, since their cutoff frequencies are far too high for the frequency spectrum excited by the above-mentioned switching time. Close to the termination, however, the TEM mode weakens marginally, while higher order TM modes become stronger. The enhancement of TM_2 near the termination has been explained in terms of the induced current distribution in the resistive sheet and the resulting magnetic field in the vicinity of the termination. The weakening of TEM near the termination has been explained in terms of reflections from the termination.

Placement of a reasonably-sized, perfectly-conducting test object in the parallel-plate section produces major changes in the mode structure in the vicinity of the object. Higher order TM modes become stronger, at the cost of the $m = 0$ mode. This finding has considerable practical significance, since it implies that the object would be subjected to electromagnetic fields that deviate significantly from the desired $m = 0$ form. Also, the object is entirely to be "blamed" for this deviation. We have provided a physical explanation for the deviation in terms of induced currents on the object and the resulting electromagnetic fields in its vicinity.

Another interesting observation is that the eigenmodes computed using SVD analysis of E_x data get considerably distorted near the object, such that the dominant eigenmode consists of a superposition of $m = 0$ and several higher order modes. We have ruled out inadequate spatial resolution as the cause of this distortion. SVD analysis of E_z data is better able to resolve sinusoidal modes, corresponding to integral m-values, as compared to analysis of E_x data. The difference probably lies in the boundary conditions, which force E_z to go to zero at the two plates. Secondly, this analysis shows a rapid increase in the $m = 2$ and $m = 4$ modes, and also some increase in $m = 3$, in the vicinity of the test object.

Finally, we have explained the enhancement of electromagnetic fields near the top and bottom faces of the object in terms of the flow of electromagnetic energy, as represented by the Poynting flux.

Exercises

7.1 Modal analysis of an EMP simulator having a test object setup in the test-volume is a realistic approach to understand the energy transfer in an event of EMP interaction with a test object. Compare the modal analysis using SVD method with the other approaches such as Hilbert Space.

7.2 The test object is assumed to be made of perfect electrical conductor. A realistic test object could consist of multiple materials, including both conductors and lossy dielectrics, which would significantly affect the mode structure. Extend the modal analysis to examine such effects. Comment on the results by comparing with the cuboid shaped PEC.

7.3 Some of the important features and limitations of the SVD method are discussed in [48]. One important limitation observed here is that in regions where mode strength changes rapidly, the eigenmodes can get considerably distorted. In such regions, it is necessary to check for distortion. If distortion is considerable, some other method may be more suitable, e.g. a combination of SVD with fast fourier transform (FFT).

8

Parametric Study of Radiation Leakage from a Bounded-Wave Simulator

8.1 Introduction

In Chapter 7, we presented the general features of the electromagnetic modes inside a bounded-wave electromagnetic pulse (EMP) simulator. In this chapter, we will address radiation leakage out of the bounded volume. Given the high electromagnetic (EM) power levels, it is important to minimize radiation leakage out of the bounded volume, which could cause electromagnetic interference (EMI) with surrounding equipment.

Several researchers have examined different ideas for confining EM energy inside such simulators. Yang and Lee have reported on the role of parallel-plate extensions [20]. They have used the conformal mapping transformation method for the calculation of charge distribution over the plate and the field distribution in and around the plate region.

King and Blejer [17] showed a safe frequency limit of operation for a bounded-wave simulator, which can efficiently guide only those waves for which the lowest wavelength is at least of the order of aperture height. Wavelengths shorter than the aperture height are radiated into the surrounding space. Their work is, however, limited only to single frequencies of excitation.

The general leakage behavior of a simulator is complex due the existence of a wide-band pulse and complex geometry of the simulator and test object. It is further complicated by the possible presence of different materials, both dielectrics and conductors, within the test object. Furthermore, neither of the above methods gives complete information, like the directional distribution of the power leakage and the far-field pattern, which are important from the point of view of EMI.

The finite-difference time-domain (FDTD) method is a powerful tool for analyzing problems involving three-dimensional objects with complex geometries and

Electromagnetic Pulse Simulations Using Finite-Difference Time-Domain Method, First Edition.
Shahid Ahmed.

multiple materials [32]. Hence, it is ideally suited for analysis of radiation leakage from bounded-wave simulators.

This chapter is organized as follows: Section 8.2 presents details of the computational model. The effect of different design features on radiation leakage is examined in subsequent sections. The sensitivity to different lengths of the parallel-plate extension is discussed in Section 8.3, while Section 8.4 assesses the role of changing the angle between the simulator plates. The effect of using different kinds of resistive terminations has been examined in Section 8.5. Section 8.6 examines the sensitivity to closure times of the switch, while the effect of different sizes of the test object is studied in Section 8.7. A physical interpretation of some of the results is presented in Section 8.8. Specific remarks regarding this study are summarized in Section 8.9.

8.2 Details of Computational Model

The experimental device described in [9] is driven by a ~82 pF capacitor charged to an initial voltage of ~1 MV. In our simulation, these numbers have been altered to some extent, so that the total stored energy is 35.7 joules (J).

The standard FDTD computational domain consists of $119 \times 60 \times 155$ cells in the x, y, and z directions, with uniform cell sizes Δy and Δz of 7.4 cm and Δx of 1.85 cm, respectively. This corresponds to a time-step of 58 ps according to the Courant criterion [32].

In each case, the electromagnetic field structure computed using FDTD is used to calculate the net power P_{net} flowing out of the FDTD domain. One way of estimating leakage would be to calculate the far-field from the structure and integrate over a bounding surface. We have opted for a simpler method, as follows. The net power outflow from the FDTD domain, P_{net}, is calculated by taking the surface integral of the Poynting flux, $\vec{E} \times \vec{B}$, over a bounding surface enclosing the entire simulator. It must be noted that the instantaneous power outflow consists of both radiative and reactive terms. The reactive power refers to changes in energy stored in the near field, and can be either positive or negative. The radiative power refers to power entering the far field. Hence, a negative value of P_{net} means that the inward-directed reactive component exceeds the radiated power at that moment. The integral of P_{net} with respect to time, carried out until the integrand falls to a negligible value yields the net energy outflow.

The appropriate bounding surface for determining the surface integral of the Poynting flux must be specified. We have found that P_{net} changes little with the choice of this surface, as long as it is located a few computational cells inside the FDTD domain boundary.

8.3 Sensitivity to Length of Parallel-plate Extension

Robertson and Morgan [23] have stated that the parallel-plate extension to a transverse electromagnetic (TEM) horn structure plays an important role in reducing reflections, especially for low-frequency signals. Yang and Lee have reported that the parallel-plate extension plays an important role in the confinement of EM energy inside the bounded volume [20]. They have used the conformal mapping transformation method for the calculation of charge distribution over the plate and the field distribution in and around the plate region.

In this section, we study the effect of using different sizes of the parallel plate extension on the confinement of EM energy. Three different plate lengths have been considered, viz. 1.5, 3.0, and 4.5 m. These lengths have been chosen to correspond to integral multiples P_{ext} of the separation "h_a" between the plates, i.e. ~1.5 m. In all cases, we have set up a resistive sheet termination.

Figure 8.1 shows the temporal evolution of P_{net} for three values of P_{ext}, viz. 1, 2, 3. The method used for calculating power outflow is as described in Section 8.2. The simulations are run until P_{net} falls to very low levels. The area under the curve, which yields the total outflow energy, is 8.80 J, 10.78 J, and 12.03 J for $P_{ext} = 1.0$, 2.0, and 3.0, respectively. This means that ~24–34% of the initial stored energy in the capacitor, i.e. 35.7 J, is lost by leakage.

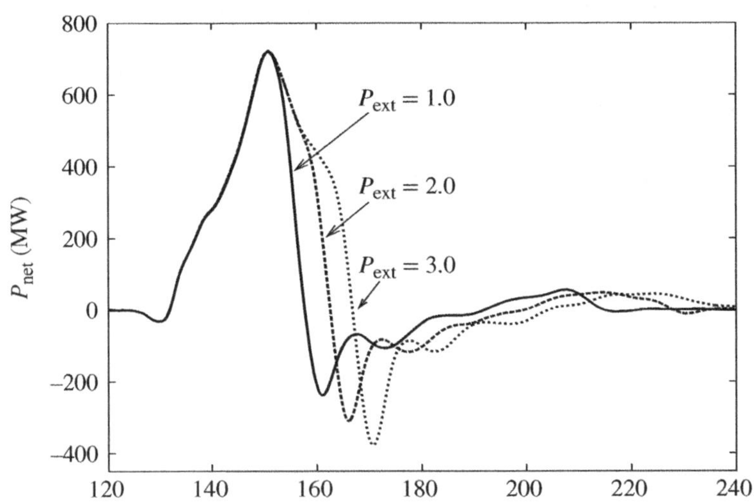

Figure 8.1 Net outward power flow from FDTD domain as a function of parallel-plate length. The simulator has a matched resistive sheet termination.

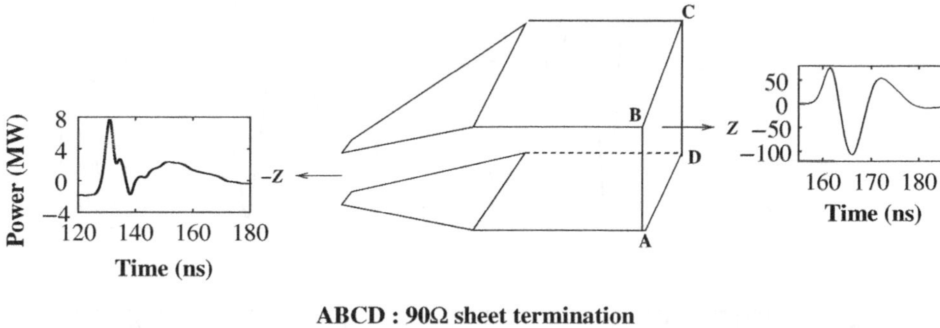

Figure 8.2 Net power outflow in the forward and backward directions, for $P_{ext} = 2$. The simulator is terminated by a matched sheet resistance.

This result has the following practical significance. The length of the test volume is normally dictated by the size of the test object. The above result shows that leakage is a major concern and increases significantly with the length of the test volume. For a given simulator, therefore, longer test objects imply higher EMI problems in the vicinity.

Figure 8.2 shows the net power flow out of the "front" and "back" sections of the bounding surface. These sections correspond to the planes having their outward normals pointing in the positive and negative z-directions, respectively. We see that the net radiation in the forward direction far exceeds that in the reverse direction. This means that there is little power flow back toward the pulse-power source ("pulser"), a highly desirable feature from the point of view of protecting the pulser.

Figure 8.3a,b shows the radiated far-field at two locations – one along the boresight ($\theta = 90$, $\phi = 0$) and the other along side-opening ($\theta = 90$, $\phi = 90$) for two different lengths of parallel-plate extension, i.e. $P_{ext} = 1.0$ and 3.0, respectively. These far-field values are normalized to a distance of 1 m. We see that the side lobes are much weaker than the forward lobes. We also see that increasing the length of the parallel-plate section enhances the forward lobe without significantly affecting the side lobe. This means that the increase in radiation leakage with P_{ext} mainly increases EMI levels along the boresight. Such localization of EMI would make it somewhat easier to provide EMI shielding.

8.4 Sensitivity to Angle Between Tapered Plates

Baum [21] has qualitatively shown that the EM energy leakage from a bounded-wave simulator can be reduced by using a smaller angle β between the tapered plates. Sancer and Varvatsis [22] have numerically determined the

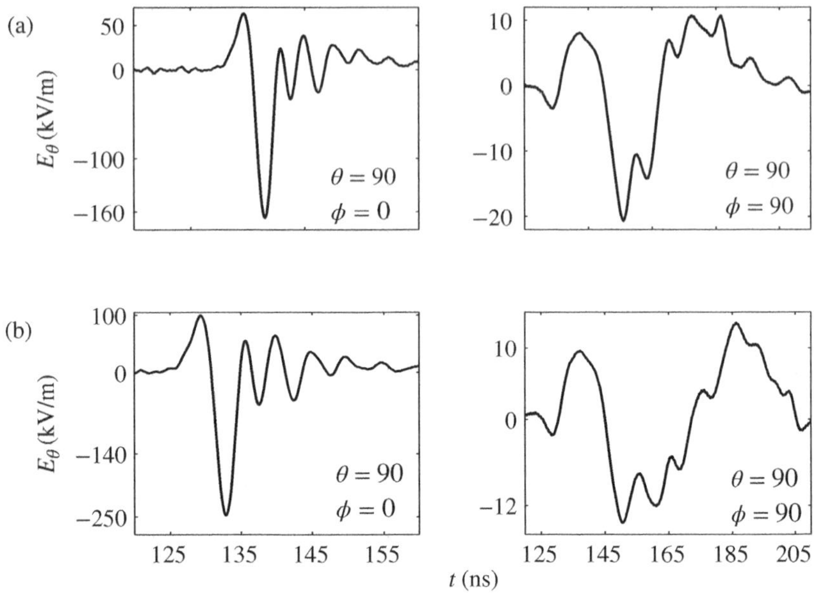

Figure 8.3 Radiated far-field at two angular locations, normalized to a distance of 1 m. Cases (a) and (b) correspond to two different lengths of plate extension, viz. $P_{ext} = 1$ and 3, respectively. The simulator has a matched resistive sheet termination.

surface current density in the parallel-plate section and used that for estimating the EM energy radiated out of the working volume of the simulator. Their results support those of Baum [21].

This observation can first be explained qualitatively in terms of waveguide and antenna theory. The simulator acts like a transmission line for wavelengths much greater than the aperture height "h_a" and behaves like an antenna for wavelengths smaller than, or of the order of "h_a" [17–20]. Reducing the angle β reduces "h_a." Now, the pulsed signal has an ultra-wideband frequency spectrum, where the highest frequency having significant energy content is determined by the switch closure time. Hence, as β is reduced, a progressively smaller part of the spectrum can be radiated, reducing leakage.

Figure 8.4 shows the net power outflow from a simulator for three different values of the angle β viz. 10°, 15°, and 20° and a switching time τ_s of 19.7 ns. In all cases, the simulator has a parallel-plate test section that is 9 m long, terminated by a matched resistive sheet. We find a net leakage of 9.4 J, 12.5 J, and 15.2 J for $\beta = 10°$, 15°, and 20°, respectively. There is a clear trend of reduction in both the peak radiated power and the total radiated energy with reducing β.

A detailed explanation of this trend in terms of the electromagnetic mode structure inside the bounded volume will be presented in Chapter 9.

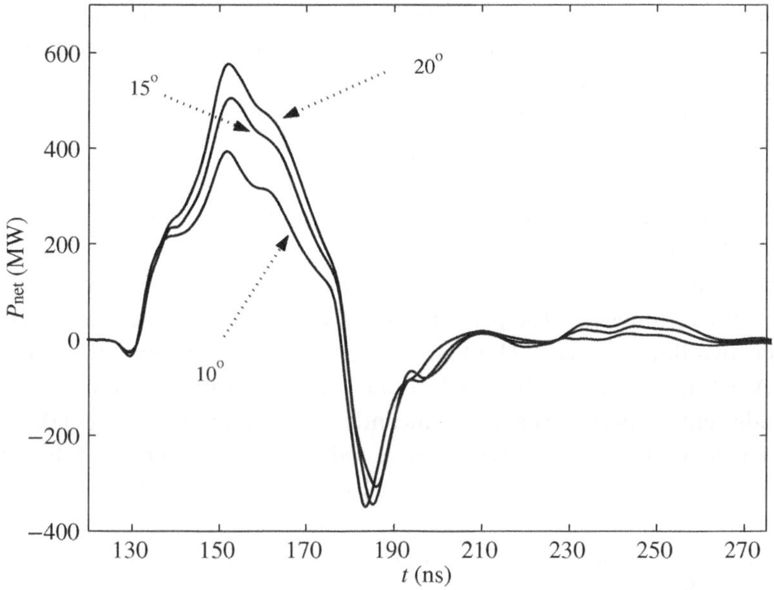

Figure 8.4 Net power outflow for three different values of angle between the simulator plates. The simulator has a 9-m long parallel-plate extension terminated by a matching resistive sheet and a switching time τ_s of 19.7 ns.

8.5 Effect of Type of Termination

An open-ended EMP simulator, i.e. one without a matching termination at the aperture, offers a very high impedance for the low frequency part of the pulse. This is because there is a large discontinuity between the characteristic impedance of the horn and free-space impedance (377 Ω). As a result, a major portion of the energy input at low frequencies will be reflected back toward the pulsed power source, which could damage the source.

Matched resistive terminations are commonly used for reducing reflections in the case of transmission lines. Several alternatives to this option have also been examined in the literature, some of which are summarized below.

Kanda [24] have resistively loaded a TEM horn from the apex to the aperture, such that the resistance varies continuously from a low value at the apex to a high value at the aperture. The functional form of this variation was similar to that proposed by Wu and King for reducing reflections from the open ends of a dipole antenna [25]. The small impedance at the feed was used for matching to the feeding transmission line and the high impedance at the aperture for matching to free-space. Baum [26] has proposed the use of an admittance sheet for terminating broadband signals. The resistive part effectively terminates lower frequencies, while the reactive part is useful at high frequencies. Hence, the termination is effective over a wide frequency range.

The physical shape of the resistive termination is also important, as it significantly affects low-frequency performance [27].

In this section, we limit ourselves to the option of a resistive termination. Two possible variations are examined. The first is related to the geometry and the second to the value of the termination resistance.

TEM-horn type bounded-wave simulators often use a number of parallel chains of resistors as a termination. The performance of such an arrangement would lie in-between two limiting cases. The first is a resistive sheet covering the entire aperture ABCD in Figure 6.1. This idealization would be applicable if the spacing between adjacent chains were made significantly smaller than the wavelengths of interest. The second consists of two resistive rods, connecting corners A-B and C-D in Figure 6.1. In the latter case, the rods would be electrically in parallel, yielding a net termination resistance equal to half the resistance of each rod. The performance of these limiting cases, from the point of view of radiation leakage, is examined in this section.

The electromagnetic field structure inside the simulator, and hence the leakage, would also depend upon the value of termination resistance. The normal procedure is to take a termination resistance equal to the characteristic impedance of the TEM structure. On the one hand, it is true that an unmatched termination would increase reflections toward the pulser. On the other hand, by altering the electromagnetic field structure, there is also the possibility that it could reduce leakage and hence the EMI level. Hence, we have also studied three different sheet resistances, viz. 30, 90, and 200 Ω. This study has been performed for a simulator without a test volume, i.e. the termination is placed at the end of the TEM horn structure.

Let us first consider the effect of changing the termination on reflections. Figure 8.5 shows the vertical component of the electric field E_x at the feed of the TEM structure, for four different terminations, viz. 30 Ω, 90 Ω, and 200 Ω resistive sheets and a 90 Ω rod, respectively. The rod termination consists of two parallel rods, each of 180 Ω resistance.

As expected, the primary peak remains the same for all kind of terminations. However, there are major differences in the secondary peaks. This can be understood in terms of the expression for the reflection coefficient from a transmission line with a resistive termination [41]:

$$R = \frac{Z_L - Z_c}{Z_L + Z_c} \tag{8.1}$$

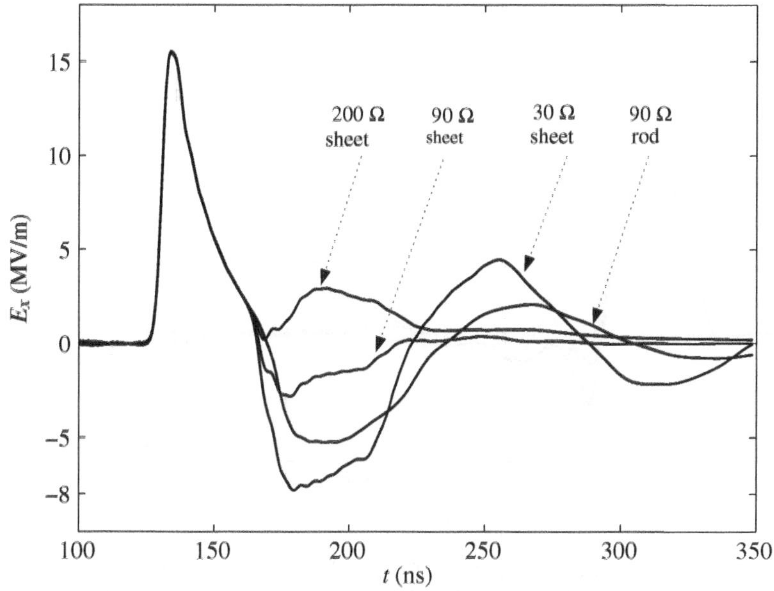

Figure 8.5 Vertical component of electric field E_x at feed of the TEM structure, with four different resistive terminations. The structure does not have a parallel-plate test section.

Here, R is the reflection coefficient, i.e. the ratio of the reflected signal to the incident signal, Z_L is the impedance of the termination and Z_c is the characteristic impedance (~90–95 Ω) of the TEM structure.

The following points are noteworthy. Firstly, the formula is exact only in the case where Z_L is a constant. For the wide-band problem considered here, Z_L varies significantly as a function of frequency, due to variation in the amplitude of current along the length of the termination. This means that R varies significantly with frequency. Hence, we should not expect an exact match between this formula and the reflection observed in the simulator. Secondly, R should be negative for all values of $Z_L < Z_c$ and positive otherwise. This can indeed be seen from Figure 8.5. In particular, the 200 Ω termination yields a positive peak, while the 30 Ω termination has a strong negative peak. The 90 Ω case is marginal, as one would expect from the formula.

We now determine the performance of these terminations from the point of view of energy leakage. Figure 8.6 shows the temporal evolution of P_{net} for three different values of sheet impedance viz. 30, 90, and 200 Ω, respectively, and a matched rod (90 Ω) termination. For a sheet termination, we see that the peak power level increases monotonically with the sheet resistance, the increase being especially marked between 90 and 200 Ω. The peak power level is particularly high with a rod termination. This is what we expect on physical grounds, since a sheet would

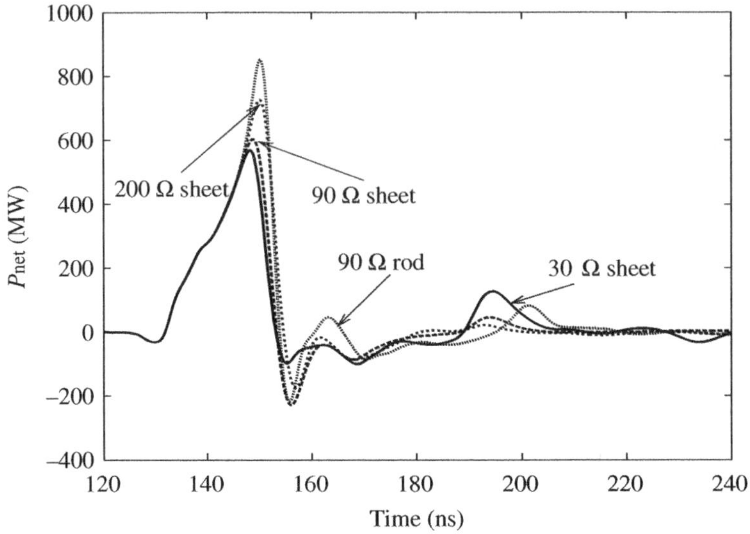

Figure 8.6 Net outward power flow from FDTD domain, using sheet terminations with different resistances, and also a pair of rods with a combined resistance of 90 Ω. The simulator does not have a parallel-plate extension.

tend to "block" the escape of radiation, while a pair of rods permits essentially free escape.

The area under the curve, which yields the total outflow energy, is 6.52, 6.06, 7.30, and 7.94 J for the four cases, respectively. The difference in these levels is significant, considering that the total stored energy in the driving capacitor is 35.7 J.

We would next like to understand the differences in the power flow pattern that produce this change in P_{net}. This has been done by determining the far-field patterns for the case of matched sheet and rod terminations. Figure 8.7 shows the temporal evolution of the far-zone electric field, E_θ, produced by the simulator in the horizontal plane ($\theta = 90$) as a function of angle ϕ. At time $t \sim 133$ ns, there is a strong negative peak at $\phi = 0$. In both cases, the side lobes of E_θ ($\phi = 90$) are much weaker than in the boresight. There are significant differences between the two kinds of terminations – the rod termination yields higher peak values of E_θ, and the far-field also takes longer to decay to low levels.

At the negative peak for $\phi = 0$, the electric field with a rod termination is ~17% higher than that with a sheet termination. Recalling that the Poynting flux is related to E^2, we get a ratio of peak radiated powers of 1.36, reasonably close to the observed ratio of P_{net} values.

From this study, it is clear that a matched resistive sheet is the best termination, since it minimizes reflections as well as energy leakage from the bounded volume.

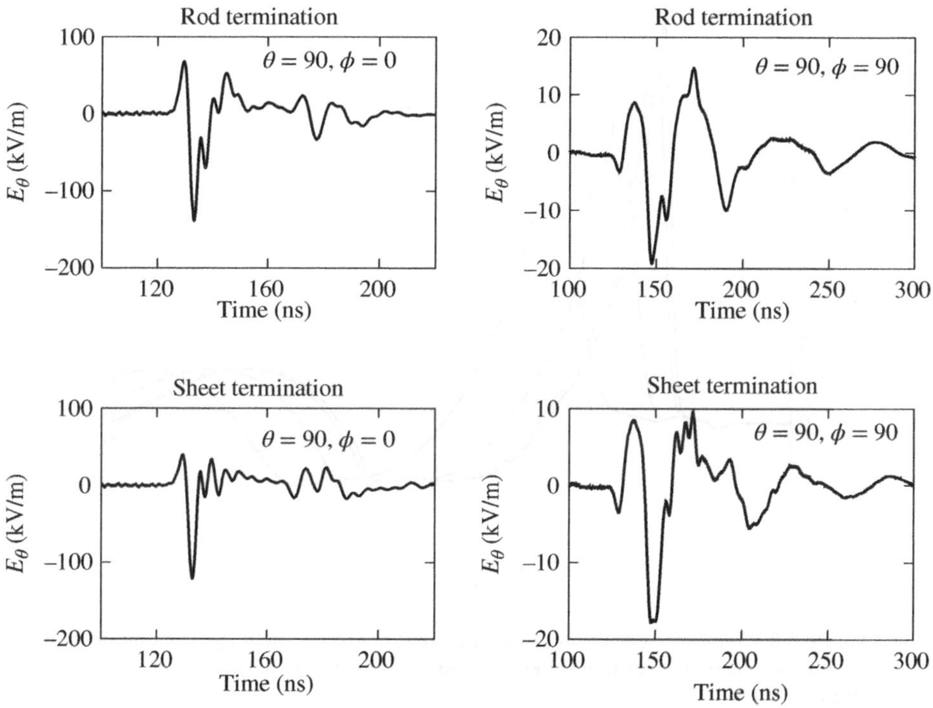

Figure 8.7 Far-zone electric field, normalized to unit distance. The simulator has matched sheet and rod terminations and the switching time $\tau_s = 19.7$ ns.

8.6 Sensitivity to Closure Time of Switch

In Section 6.6, we have already explained the need to vary the switch closure time in a given simulator. In that discussion, it was pointed out that while shorter closure times permit testing at higher frequencies, they also have a deleterious side effect in terms of radiation leakage. This is discussed in this section.

King and Blejer [17] have reported a condition for the confinement of electromagnetic energy inside the closed volume of a bounded-wave EMP simulator, viz.

$$kh_a \ll 1$$

Here "k" is the wavenumber ($= 2\pi/\lambda$), and "h_a" is the height of the aperture of the TEM structure, i.e. the separation between the parallel plates shown in Figure 6.1. The wavelength λ corresponds to the frequency of interest. According to the above relation, electromagnetic waves with wavelengths higher than or comparable with "h_a" are confined, while those smaller than "h_a" are radiated. This means that the

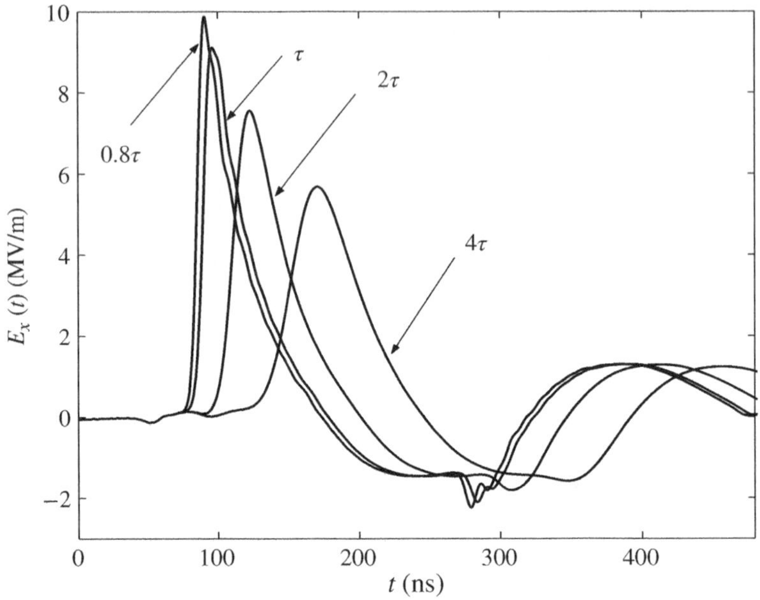

Figure 8.8 Temporal evolution of x-component of electric-field $E_x(t)$ at simulator feed. Results are presented for four different switching times $\tau_s = 0.8\tau$, τ, 2τ, and 4τ, respectively, with $\tau = 19.7$ ns.

upgradation of a given simulator to handle higher frequencies may enhance the leakage.

We have studied the leakage resulting from the use of four different values of the switching time τ_s, viz. 0.8τ, τ, 2τ, and 4τ, where $\tau = 19.7$ ns.

Figure 8.8 shows the temporal evolution of E_x at the simulator feed for these switching times. As expected, the waveform progressively gets compressed with decrease in τ_s. Furthermore, the peak value of E_x progressively increases with decreasing τ_s – the physical reason is as follows.

Decrease in τ_s reduces the time over which energy flows to the TEM structure. This corresponds to a higher average value of the z-component of the Poynting flux at any given z-location. This, in turn, implies generally higher values of E_x at a given z-location, as observed from Figure 8.8.

We next describe the effect of switching time on the net power outflow. Figure 8.9 shows the temporal evolution of P_{net} for the four cases. The area under the curve, which yields the total outflow energy, is 6.5 J, 6 J, 4.8 J, and 3.3 J, respectively, for $\tau_s = 0.8\tau$, τ, 2τ and 4τ, respectively. As expected, pulse compression enhances the radiation leakage significantly.

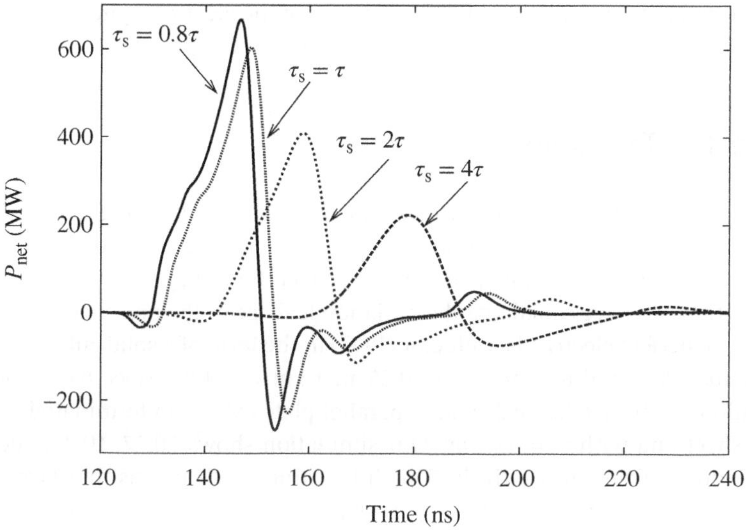

Figure 8.9 Net power outflow for different values of switch closure time. The simulator has a matching resistive sheet termination and is without a parallel-plate extension.

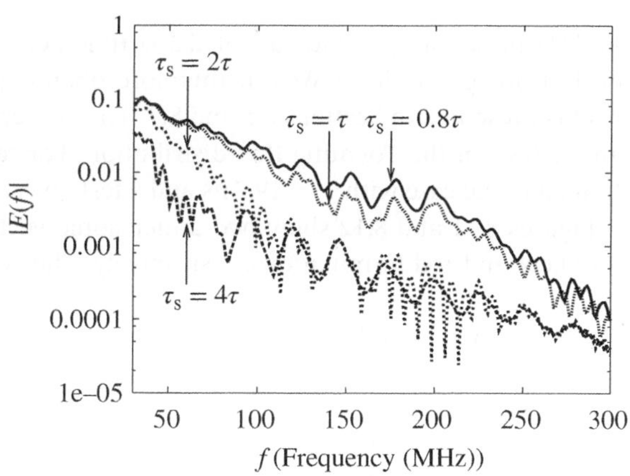

Figure 8.10 Frequency spectrum of $|E_x(f)|$ at simulator feed for different switch closure times. The simulator has a 90 Ω resistive sheet termination and is without an extension.

This result can be physically understood in terms of the frequency spectrum appearing at the feed point of the simulator. Figure 8.10 shows the frequency spectrum $|E_x(f)|$ for the above values of τ_s. The smallest and highest values of τ_s show significant values of $|E_x(f)|$ up to 300 and 150 MHz, respectively. The corresponding wavelengths are 1 m ($<h_a$) and 2 m ($>h_a$). We therefore get better energy confinement for larger τ_s values.

A physical explanation for this variation, in terms of the electromagnetic mode structure inside the bounded volume, will be presented in Chapter 9.

8.7 Effect of Test Object

A variety of test objects, made out of different materials, and having different geometries, are tested in EMP simulators. The test object is expected to distort the electromagnetic fields in its vicinity, and thereby affect the leakage.

For illustration, we have calculated the radiation leakage in the presence of test objects made of perfect electrical conductor (PEC) in the form of a solid cube. The side of the cube takes on three values, viz. 0.25 m, 0.5 m, and 1 m respectively. For this calculation, we have taken a 3-m long parallel-plate extension terminated by a matched sheet and with $\tau_s = 19.7$ ns. Our simulation shows 10.77, 10.73, and 9.8 J of radiation leakage, respectively. This interesting result means that larger test objects reduce leakage. Of course, larger objects are also likely to distort the electromagnetic fields in their vicinity.

8.8 Physical Interpretation

FDTD simulations provide a detailed 3-D, time-dependent picture of the evolution of electromagnetic fields. We can, therefore, obtain a physical understanding of the results presented in Sections 8.5 and 8.7. This understanding is obtained through an analysis of the Poynting flux distribution. For reasons of space, this study is limited to the case with $\tau_s = 19.7$ ns and a test object with side 1 m.

Figures 8.11 and 8.12 show two-dimensional vector plots of the Poynting flux for sheet and rod terminations, respectively. The vectors, comprising the y- and z-components P_y and P_z, are represented in the yz-plane exactly between the top and bottom plates of the test volume shown in Figure 6.1. In order to illustrate the role of the terminations, the vectors are shown over a region slightly bigger than the test volume.

Sub-plots (a–d) of Figures 8.11 and 8.12 correspond to four different time snapshots, viz. $t = 7.8$–8.4τ in steps of 0.2τ, with $\tau = 19.7$ ns. These time values have the following significance. The first is the time just before the main E-field peak reaches the front surface of the test object. The second is when the pulse just reaches the back surface of the object. Sub-plot (c) shows the distribution shortly after the pulse crosses the termination, and (d) comes after another 4 ns. Some points of interest are discussed below.

Firstly, with both kinds of termination, the P_y components are small on the y-directed boundaries, corresponding to weak side-lobes, consistent with the

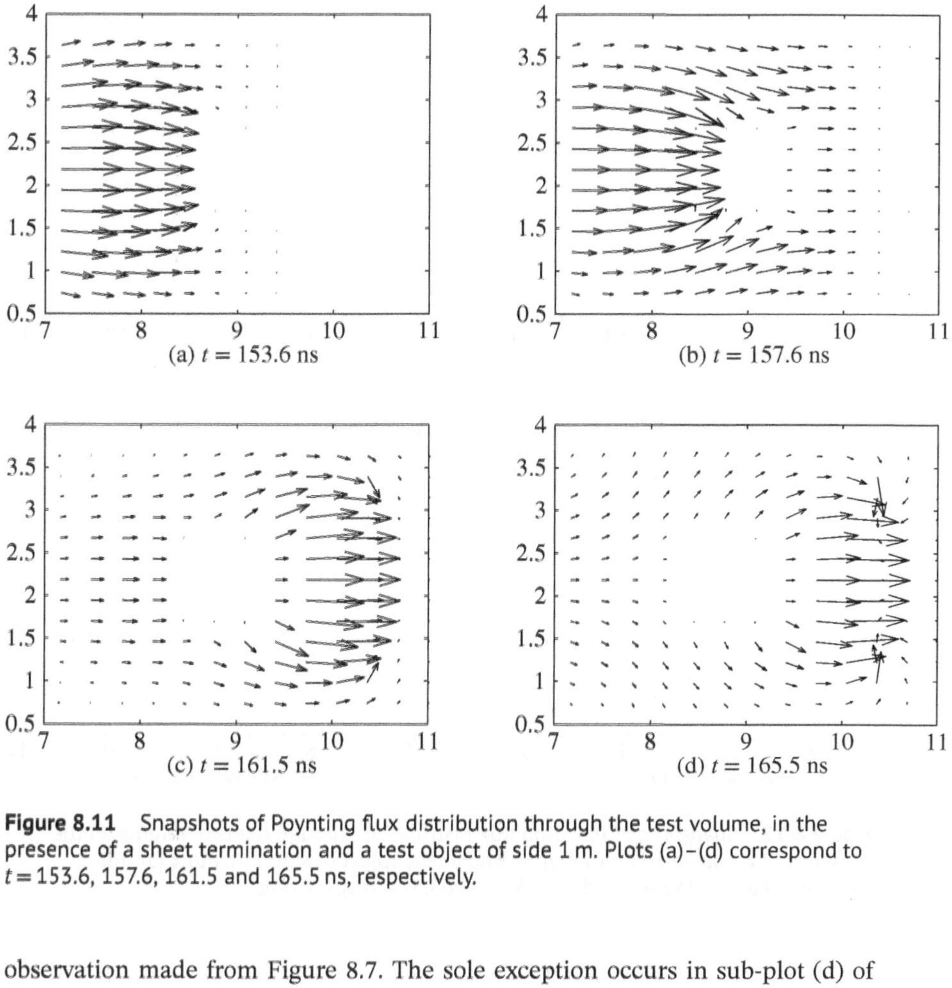

Figure 8.11 Snapshots of Poynting flux distribution through the test volume, in the presence of a sheet termination and a test object of side 1 m. Plots (a)–(d) correspond to t = 153.6, 157.6, 161.5 and 165.5 ns, respectively.

observation made from Figure 8.7. The sole exception occurs in sub-plot (d) of Figure 8.12 – this is discussed later.

Secondly, sub-plots (a) and (b) are similar in both figures. This is to be expected, considering that the main positive peak has not yet reached the termination. However, a portion of the pulse preceding the main peak would have reached the termination, and this accounts for minor differences of detail between the two sub-plots (b).

Thirdly, in both sub-plots (b), we note that the incoming flux lines tend to bend toward the test object. This can be understood as follows. The applied E_x would induce transient x-directed currents on both y-directed faces of the object. Idealizing these as sheet currents in the x-direction, we would get equal and opposite magnetic field components H_z outside the two faces. It is easily seen that

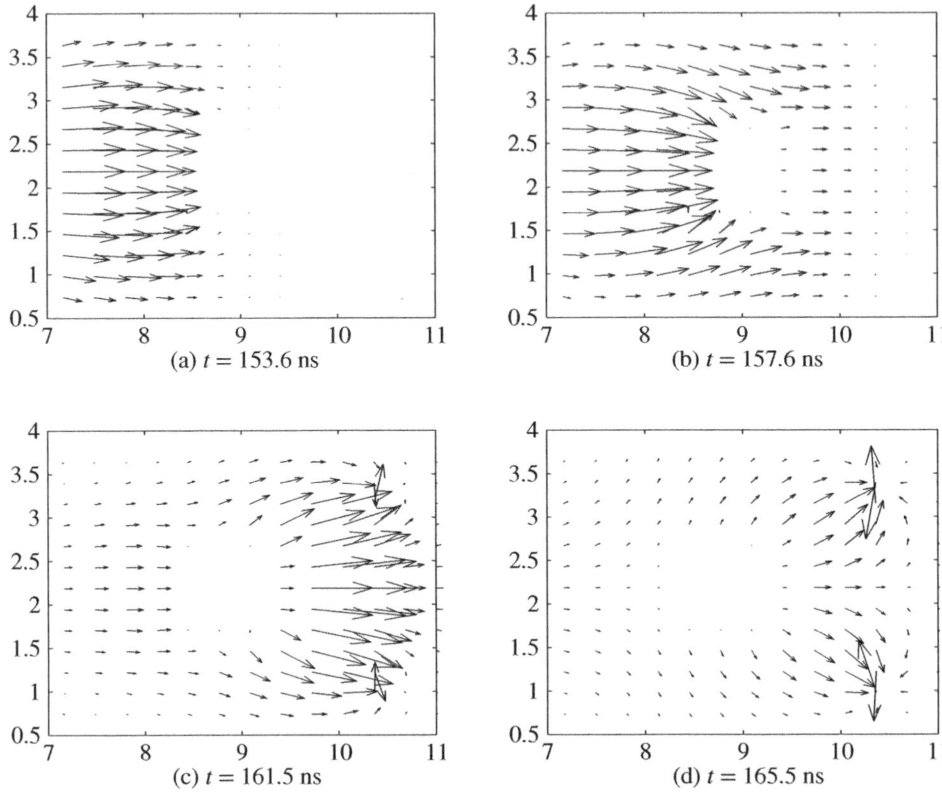

Figure 8.12 Snapshots of Poynting flux distribution through the test volume, in the presence of a double-rod termination and a test object of side 1 m. Plots (a)–(d) correspond to $t = 153.6$, 157.6, 161.5 and 165.5 ns, respectively.

the resulting P_y would point toward the object on both faces, giving rise to the observed "inward bending" of the power flow.

An interesting difference can be seen between the two sub-plots (c), corresponding to a time shortly after the main pulse crosses the termination. Near the sheet termination (Figure 8.11c), the Poynting flux has primarily a z-component, while a rod termination results in substantial y-directed divergence of the power "flow." This can be interpreted in terms of the magnetic field produced by the x-directed current through the terminations. A sheet termination would produce primarily H_y inside the bounded volume. This, in combination with E_x, would yield primarily P_z. A rod, on the other hand, would produce an azimuthal magnetic field around itself, containing both H_y and H_z. Inside the bounded volume, the resulting

H_z would have equal and opposite values at the two limiting values of y, giving rise to the observed divergence.

Finally, we see from Figure 8.11c,d that there is a significant reduction in the magnitude of the Poynting flux just outside the sheet termination, as compared to its value inside. With a rod termination, the decrease is not so marked. This explains the observed reduction in total energy leakage when a sheet termination is used.

8.9 Summary

Leakage of electromagnetic energy from a bounded-wave EMP simulator is an important concern, since it causes EMI with equipment in the vicinity. We have performed self-consistent, three-dimensional, time-domain calculations for calculation of radiation leakage from such a simulator. To our knowledge, this is the first time such a study has been done.

Large bounded-wave structures can be expensive and take time to set up. Given a simulator that was originally set up for lower frequencies, it is often attractive to use pulsers with shorter risetimes to allow testing at higher frequencies [9]. However, we find that this comes at the cost of increased radiation leakage. This trade-off must be borne in mind.

Given a test object with fixed dimensions in two directions, and a variable length in the z-direction, we need to increase the test section length to accommodate longer objects. Increasing the length of the parallel-plate extension shows significant enhancement of forward radiation in comparison with flow toward the pulser. This is advantageous from the point of view of protecting the pulser. However, it is accompanied by an increase in total energy leakage.

The use of a resistive sheet as a matched termination significantly reduces reflections as compared to the case of an unmatched sheet termination. If the termination resistance is constrained to match the characteristic impedance of the TEM structure, the use of a resistive sheet geometry significantly reduces reflections as compared to the case of two parallel resistive rods.

We find that a matched sheet resistance provides better confinement of electromagnetic energy as compared to rod terminations. This has also been explained in terms of the Poynting flux distribution. Our study has been limited to these two simple options. In a more detailed study, it would be desirable to examine the possibility of further reducing leakage by using more complex terminations, e.g. a sheet admittance having both resistance and inductance, as described in [26].

The size of the test object has been found to affect the leakage, with larger objects reducing leakage. This effect is also likely to vary with the shape and electrical properties of the object. We have studied the behavior of the Poynting flux inside the bounded volume, and found that there is a tendency for the flux to get focused toward the object. This has been explained in terms of currents induced in the conducting object.

These results show that a detailed study, using the methodology described here, is necessary for each combination of a simulator and a test object.

Exercises

8.1 In the illustration of simulation presented, a cubic object made of PEC has been considered as an equipment-under-test (EUT); however, for the purpose of practical interest, a test-object may involve more complex shape made of multiple materials. These should be examined in a detailed study.

8.2 A variety of terminations may be considered, the study illustrated determines the leakage in the case of resistive terminations connecting the upper and lower plates. Develop a computer code to simulate resistive loading on the simulator plates, and admittance sheet at the final end of the test-volume.

9

Modal Perspective of Radiation Leakage from a Bounded-Wave Simulator

9.1 Introduction

In Chapter 7, we studied the mode structure inside a bounded-wave electromagnetic pulse (EMP) simulator. A parametric study of radiation leakage from the bounded wave EMP simulator has been explored in Chapter 8. Apart from affecting the purity of the transverse electromagnetic (TEM) mode, transverse magnetic (TM) modes can also enhance radiation leakage from the simulator. This is undesirable for two reasons. Firstly, it results in electromagnetic interference (EMI) with surrounding equipment. Secondly, it reduces the amount of energy available for coupling to the test object.

Various researchers have reported that higher order TE and TM modes tend to cause energy leakage [28, 29]. However, these studies have not accurately quantified the relative contribution of different modes to radiation leakage. Here, we will learn how to establish the correlation between higher order modes and the radiated far-field in a realistic EMP simulator using the full-wave time-domain simulations.

This chapter is organized as follows: Section 9.2 presents details of the computational model for calculating the cross-correlation. The effect of different design features on the modal perspective of radiation leakage is examined in subsequent sections. Section 9.3 assesses the role of angle between the simulator plates in the tapered section. The effects of pulse compression have been illustrated in Section 9.4. Finally, the specific remarks regarding this study are summarized in Section 9.5.

Electromagnetic Pulse Simulations Using Finite-Difference Time-Domain Method, First Edition.
Shahid Ahmed.

9.2 Calculation Procedure

In an EMP simulator of the type considered here, radiation in the forward direction ($\theta = 90, \phi = 0$) is significantly stronger than sideways radiation ($\theta = 90, \phi = \pm 90$) [39]. The angles θ and ϕ are defined in Figure 6.1. As shown earlier, the TE-mode is negligibly small, hence the major contribution to radiation must come from TM and/or TEM modes. Thus, we only need to estimate the correlation between the radiated far-field in the forward direction and the temporal variation of the dominant modes close to the termination.

The degree of correlation between two time-dependent signals can be quantified by the cross-correlation, defined as

$$C(\tau) = \int_{-\infty}^{\infty} u(t)\, E_\theta(t - \tau)\mathrm{d}t \tag{9.1}$$

where $E_\theta(t)$ is the radiated far-field, $u(t)$ is the principal component of the mode at a z-location close to the termination and τ is the temporal delay between measurement of the two variables. This τ bears no relation to the switching time mentioned earlier. If the two fluctuations are measured at different locations, the value of τ for which the cross-correlation is significant provides information about the propagation speed, direction and spatial coherence of the fluctuations [53].

The auto-correlation is a measure of the temporal coherence of a time dependent physical quantity like $E_\theta(t)$, which is defined as

$$A(\tau) = \int_{-\infty}^{\infty} E_\theta(t)E_\theta(t - \tau)\mathrm{d}t \tag{9.2}$$

Since $u(t)$ values for different modes vary by more than an order of magnitude, we have normalized each variable to its standard deviation. For example, the cross-correlation $C(\tau)$ between $u_0(t)$ and $E_\theta(t)$ is given by

$$C(\tau) = \int y_1(t)y_2(t - \tau)\mathrm{d}t \tag{9.3}$$

where $y_1(t) = u_0(t)/\sigma_{u_0}$ and $y_2(t) = E_\theta(t)/\sigma_{E_\theta}$. The standard deviation of $u_0(t)$ is given by

$$\sigma_{u_0} = \sqrt{\frac{1}{N-1}\sum_{j=1}^{N}\left(u_0(t_j) - \overline{u_0(t)}\right)^2} \tag{9.4}$$

where N is the total number of time points. A similar definition is used for σ_{E_θ}.

9.3 Effect of Angle of Inclination Between Tapered Plates

Baum [21] has qualitatively shown that the EM energy leakage from a bounded-wave simulator can be reduced by using a smaller angle β between the tapered plates. This angle has been defined in Figure 7.1. Sancer and Varvatsis [22] have numerically determined the surface current density in the parallel-plate section and used that for estimating the EM energy radiated out of the working volume of the simulator. Their results support those of Baum [21].

This observation can first be explained qualitatively in terms of waveguide and antenna theory. The simulator acts like a transmission line for wavelengths much greater than the aperture height "h" and behaves like an antenna for wavelengths smaller than, or of the order of "h" [17], [20]. Reducing the angle β reduces "h." Now, the pulsed signal has an ultra-wideband frequency spectrum, where the highest frequency having significant energy content is determined by the switch closure time. Hence, as β is reduced, a progressively smaller part of the spectrum can be radiated, reducing leakage.

9.3.1 Correlation Study

Figure 9.1a–c shows the variation of $\log(S_1/S_0)$ and $\log(S_2/S_0)$ inside the simulator for different values of the slope angle $\beta = 10°, 15°$ and $20°$ respectively. In all cases, the switching time τ_s is held constant at 19.7 ns. An increase in β clearly increases the relative strength of both TM_1 and TM_2 with respect to the TEM mode.

Figure 9.2a–c shows the corresponding temporal waveforms of normalized E_θ, u_0, u_1, and u_2. These are, respectively, the vertical component of the radiated

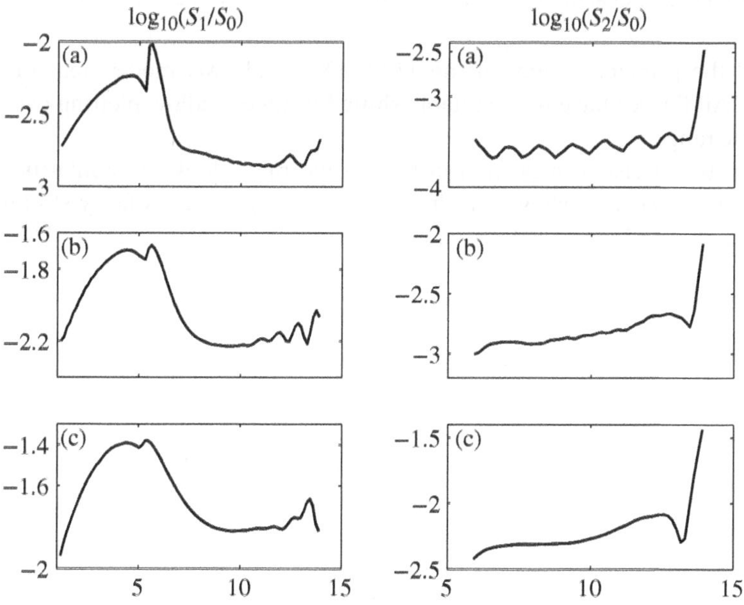

Figure 9.1 Variation of $\log(S_1/S_0)$ and $\log(S_2/S_0)$ along the simulator length. Cases (a)–(c) correspond to $\beta = 10°, 15°$ and $20°$ respectively. The switching time $\tau_s = 19.7$ ns.

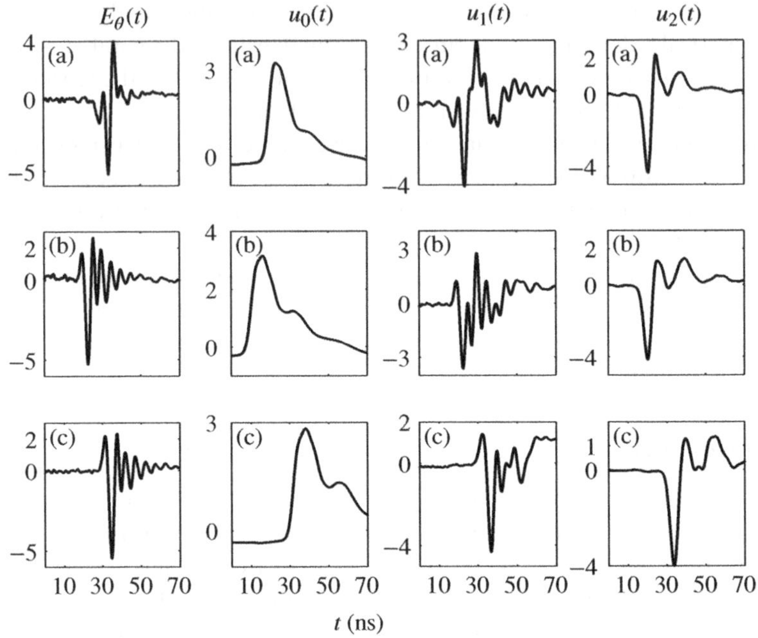

Figure 9.2 Temporal evolution of normalized E_θ, u_0, u_1 and u_2. Cases (a)–(c) correspond to three different values of angle β, viz. 10°, 15° and 20°, respectively. Switching time τ_s is 19.7 ns.

far-field and the principal values of the TEM, TM_1, and TM_2 modes near the termination. All figures have been suitably shifted in time to allow plotting over the same time range.

The first step is a visual inspection of the waveforms to look for similarities between E_θ and $u_i(t)$. This shows that for all values of β, there is fairly strong similarity of E_θ with TM_1 and TM_2. There is only limited similarity with the TEM waveform.

The second step involves calculation of the cross-correlations. Figure 9.3a–c shows the normalized cross-correlations, $R_{E_\theta u_0}(\tau)$, $R_{E_\theta u_1}(\tau)$ and $R_{E_\theta u_2}(\tau)$, as functions of delay τ, for the three values of β listed earlier. In each plot, we have used the normalization factor $\sqrt{R_{EE}(0)R_{uu}(0)}$, where $R_{EE}(0)$ and $R_{uu}(0)$ are the peak autocorrelation values of E_θ and u_i, measured at $\tau = 0$, respectively. This normalization is required because $E_\theta(t)$ and $u_i(t)$ have different levels of fluctuation, as shown in Figure 9.2. We have suitably shifted the figures to allow plotting over the same time range. This means that no physical significance should be attached to the actual values of time-delay τ shown in the figures.

In all cases, TM_1 shows strong as well as long-time correlation with the far-field, which is consistent with the similarity between their temporal waveforms.

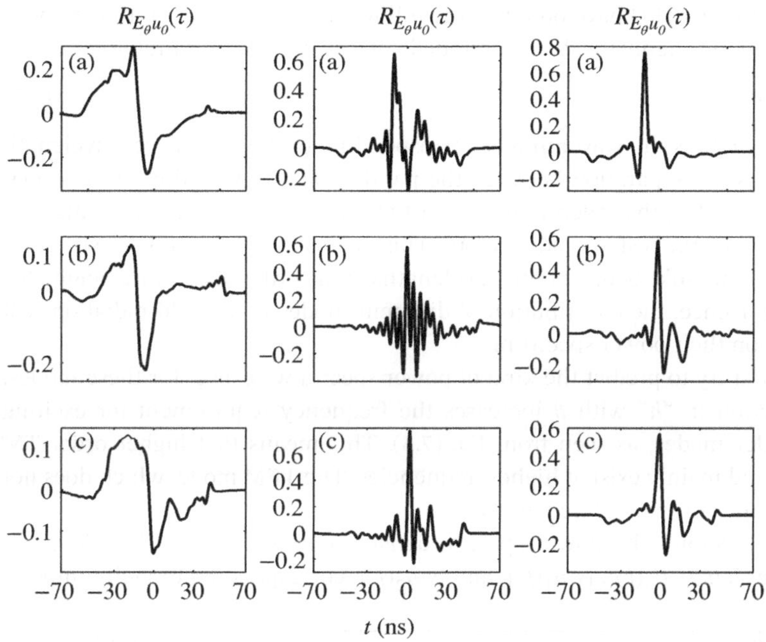

Figure 9.3 $R_{E_\theta u_0}(\tau)$, $R_{E_\theta u_1}(\tau)$, and $R_{E_\theta u_2}(\tau)$ showing the normalized cross-correlation between the vertical component of far-field E_θ and the TEM, TM_1 and TM_2 modes, respectively. The cross-correlation is shown as a function of delay τ. Cases (a)–(c) correspond to three different values of angle β, viz. 10°, 15° and 20°, respectively. In all cases, the switching time is taken as $\tau_s = 19.7$ ns.

The TM_2 mode also shows strong correlation, but this lasts only for small values of τ. The TEM mode, on the other hand, is only weakly correlated. This implies that if all three modes had the same strength inside the bounded volume, TM_1 would contribute most to the far-field, followed by TM_2, with TEM coming a weak third. However, given that TEM dominates, by far, near the termination, it could still be contributing significantly to the far-field. The cross-correlation should only be used as a measure of the effectiveness of a particular mode in radiating, rather than as an absolute measure of its contribution to the far-field.

A physical interpretation of these results is given in Section 9.3.2.

9.3.2 Physical Interpretation

We have already seen that the TM_1 and TM_2 modes show strong correlation, while the TEM mode is only weakly correlated with the far-field. It is necessary to understand this in physical terms.

King and Blejer [17] have obtained the following condition for confinement of electromagnetic energy inside the bounded volume of an EMP simulator:

$$kh \ll 1 \tag{9.5}$$

where $k = 2\pi/\lambda$ is the wavenumber corresponding to the free-space wavelength λ and "h" is the separation between the parallel plates. According to this relation, only wavelengths much longer than the separation between the plates are confined, while the rest are radiated out. This means that the simulator structure exhibits antenna-like behavior for wavelengths of the order of aperture height "h" or smaller. Hence, the contribution of different modes to far-field radiation will depend upon their power spectrum.

Let us first try to predict the kind of power spectra we expect for these modes. The reduction in "h" with β increases the frequency requirement for exciting higher order modes, as seen from Eq. (7.3). This means that higher order TM modes would mainly exist at higher frequencies. The TEM mode, which does not have a cutoff, can span all frequencies.

Figure 9.4 shows the power spectral density amplitudes of TEM, TM_1, and TM_2 modes, i.e. $|s_0 u_0(f)|$, $|s_1 u_1(f)|$, and $|s_2 u_2(f)|$, corresponding to their temporal

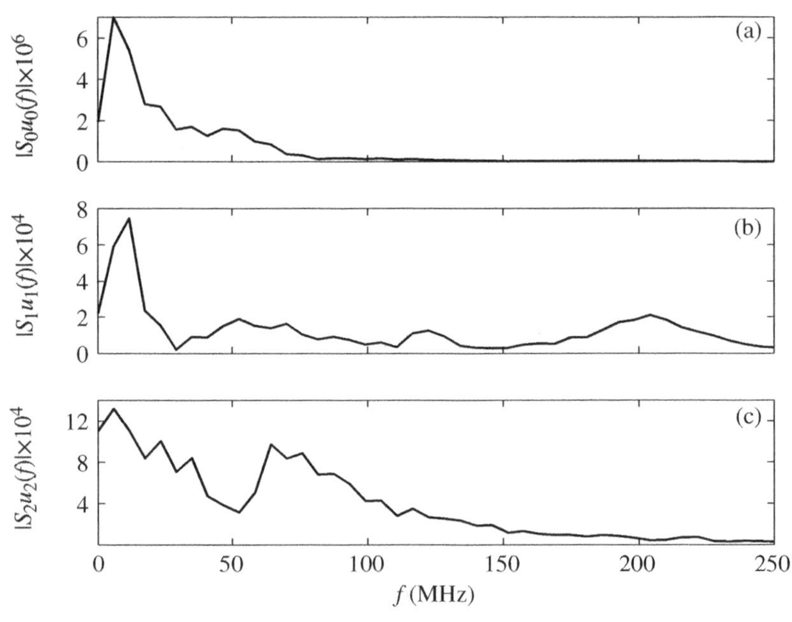

Figure 9.4 Power spectral density (PSD) amplitudes $|s_0 u_0(f)|$, $|s_1 u_1(f)|$ and $|s_2 u_2(f)|$ of TEM, TM_1, and TM_2 modes respectively, calculated near the termination. The amplitudes are the square roots of the actual PSD. The switching time $\tau_s = 19.7$ ns and angle between plates $\beta = 15°$.

amplitude evolution $s_0 u_0(t)$, $s_1 u_1(t)$, and $s_2 u_2(t)$, respectively. These time series data are obtained just next to the termination, for a switching time τ_s of 19.7 ns and for an angle $\beta = 15°$. The TEM spectrum has significant power content upto 50 MHz, which corresponds to wavelengths of 6 m and higher, much greater than the aperture height "h" (~1.5 m). King's condition then implies that most of the TEM energy will be confined, in accordance with the cross-correlation study. By contrast, the TM_1 and TM_2 modes have significant spectral content up to ~225 MHz, which corresponds to wavelengths ~1.3 m, less than the aperture height "h" (~1.5 m). This implies that a significant portion of their energy can be radiated out, again in agreement with the cross-correlation.

This trend has been confirmed with the calculation of the radiated far-field and the total energy outflow, as shown below.

Figure 9.4 shows an unexpected feature – the TM_1 and TM_2 plots also show significant power at low frequencies. Now, if each mode has some lower cutoff frequency, it should not, ideally speaking, exist below that frequency. The explanation is that in a waveguide of finite length, even a mode that is evanescent can exist, although its amplitude drops exponentially along the length of the guide [50].

9.3.3 Variation of Leakage with Plate Angle

The variation of EM energy leakage with the angle between the plates has already been discussed in Chapter 8. A short discussion is also presented here for the sake of completeness.

Figure 9.5a–c shows the temporal variation of the vertical component (E_θ) of the radiated far-field, normalized to unit distance, for three different values of angle β, viz. 10°, 15° and 20°, respectively. The values are shown on the horizontal plane ($\theta = 90$), for two different values of the azimuthal angle ϕ. These angles are defined in Figure 6.1. Here, $\phi = 0$ corresponds to the forward direction (boresight). For all cases, the side lobes ($\theta = 90, \phi = 90$) are weak compared to the forward lobes ($\theta = 90, \phi = 0$), which is consistent with the characteristics of a TEM-cell. As expected, the overall level of the radiated far-field increases with β. This is consistent with the singular values of the higher order modes, shown in Figure 9.1.

It is also of interest to estimate the total energy outflow from the boundaries of the finite-difference time-domain (FDTD). This has been calculated from the net power outflow, $P_{net}(t)$, which is the surface integral of the Poynting flux $\vec{E} \times \vec{B}$ over a surface enclosing the entire simulator. Details of this method are given in Chapter 8. The time integral of $P_{net}(t)$, up to the time where its value falls to very low levels, yields the total energy outflow. We find a net leakage of 9.4 J, 12.5 J, and 15.2 J for $\beta = 10°$, 15° and 20°, respectively, out of a total initial energy of 35.4 J stored in the capacitor.

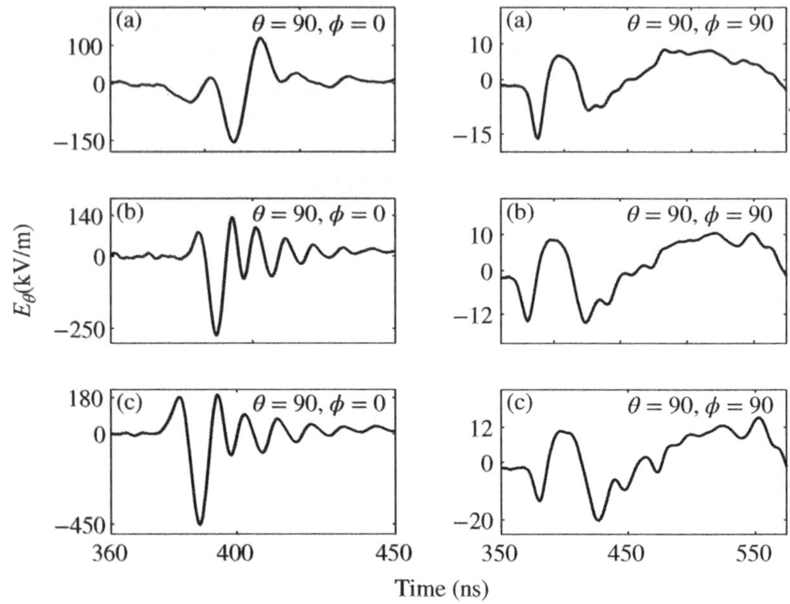

Figure 9.5 Vertical component (E_θ) of radiated far-field normalized to unit distance. Cases (a)-(c) correspond to angle $\beta = 10°, 15°$ and $20°$, respectively.

Hence, the trend predicted by mode analysis is confirmed, both for the radiated far-field and for the net energy leakage.

9.4 Effect of Pulse Compression

In Section 6.6, we have already explained the need for using switches with different closure times on a given EMP simulator. Pulsers with different risetimes, viz. ≤ 7 ns and 1 ns, have been used by Schilling et al. [9] for driving a 90 Ω TEM-cell.

Pulse compression increases the high-frequency content of the electromagnetic field. As reported in Section 9.3, higher frequencies tend to be radiated by the simulator to a greater extent. In this section, we study the effect of the switching time upon the radiated far-field and the total energy outflow. This is then explained in terms of the cross-correlation between the far-field and the dominant modes.

9.4.1 Effect on Radiation Leakage

Figure 9.6a–d shows the vertical component E_θ of the radiated far-field in the horizontal plane ($\theta = 90$), for two values of ϕ, for four different switching times, viz.

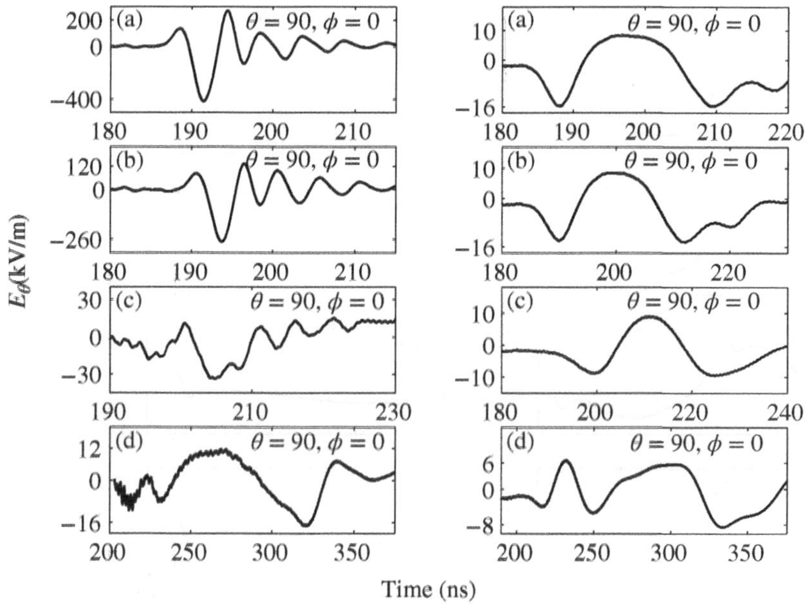

Figure 9.6 Vertical component (E_θ) of radiated far-field normalized to unit distance. Cases (a)–(d) are for four different values of switching time $\tau_s = 0.8\tau$, τ, 2τ and 4τ, respectively, and $\beta = 15°$.

$\tau_s = 0.8\tau$, τ, 2τ and 4τ, respectively, where $\tau = 19.7$ ns. In these plots, E_θ has been normalized to unit distance, and the angle β is taken as 15°. For all switching times, we see that the side lobes (θ =90, ϕ = 90) are very small as compared to the forward lobes (θ = 90, ϕ = 0). This is consistent with the known characteristic of a TEM-cell. We also see that forward radiation increases significantly with pulse compression. The total energy outflow from the bounded volume of the simulator, for $\tau_s = 0.8\tau, \tau, 2\tau$ and 4τ, is 13.0 J, 12.5 J, 10.9 J, and 8.4 J, respectively, out of a total initial energy budget of 35.4 J. This is consistent with the radiated far-field level.

9.4.2 Explanation in Terms of Mode Structure

It is possible to explain this variation in terms of the mode structure. Figure 9.7a–d shows the variation of $\log(S_1/S_0)$ and $\log(S_2/S_0)$ through the simulator length, from the feed to the aperture, for the same switching times as above. Clearly, pulse compression increases the relative strength of both TM_1 and TM_2 with respect to the TEM mode. We have seen earlier that there is strong cross-correlation of the radiated far-field with the TM_1 and TM_2 mode. This implies that pulse compression increases leakage through an enhancement in the levels of higher order modes.

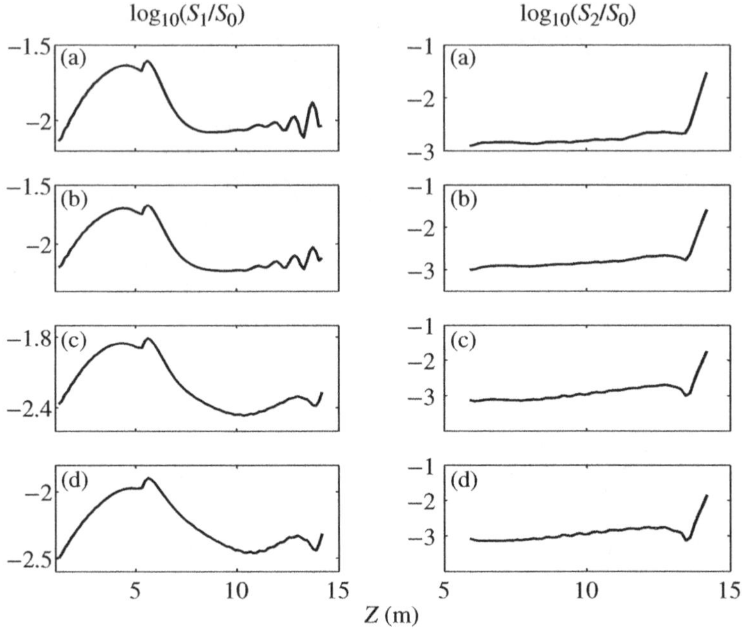

Figure 9.7 Variation of $\log(S_1/S_0)$ and $\log(S_2/S_0)$ throughout the simulator length. Cases (a)–(d) correspond to switching times $\tau_s = 0.8\tau$, τ, 2τ, and 4τ, respectively, where $\tau =$ 19.7 ns and $\beta = 15°$.

The cross-correlations were obtained in Section 9.3 using $\tau_s = 19.7$ ns. Now, pulse compression alters the frequency spectrum inside the simulator, and hence the power spectrum of the TM modes. It is possible, therefore, that the cross-correlation of these modes with the far-field may change as well. Hence, it is necessary to determine the cross-correlations for each of the switching times mentioned above.

Figure 9.8a–d shows temporal waveforms of normalized E_θ, u_0, u_1, and u_2, for the four values of τ_s listed earlier. In all cases, the TM_1 and TM_2 waveforms have fairly good similarity with the far-field E_θ. This is consistent with our earlier observation based on Figure 9.2a–c.

Figure 9.9a–d shows the normalized cross-correlations $R_{E_\theta u_0}(\tau)$, $R_{E_\theta u_1}(\tau)$ and $R_{E_\theta u_2}(\tau)$. For short switching times, the result is the same as earlier, i.e. TM_1 shows strong as well as long-time correlation with the far-field, TM_2 shows strong but only short-time correlation, while the TEM mode shows only weak correlation. However, two important changes becomes apparent for longer switching times. Firstly, the cross-correlation of the TEM mode has peak values that are comparable with those for the TM modes. Secondly, for $\tau_s = 0.8\tau$ and τ,

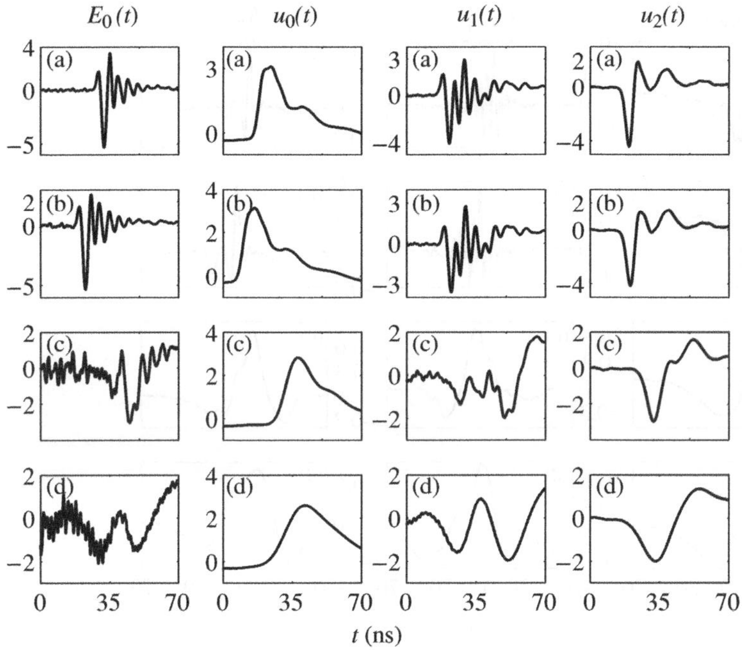

Figure 9.8 Temporal evolution of normalized E_θ, u_0, u_1 and u_2. Cases (a)-(d) correspond to $\tau_s = 0.8\tau$, τ, 2τ and 4τ, respectively, where $\tau = 19.7$ ns and the angle between the plates $\beta = 15°$.

there are significant values of the cross-correlation at a large number of delay values, which are clustered together. As τ_s increases, the number of such values becomes smaller, and also more widely spaced. Hence, we conclude that pulse compression increases the relative amplitude of higher order modes, and also their correlation with the far-field, and thereby enhances radiation leakage from the simulator.

9.5 Summary

The TEM cell radiates mainly in the forward direction, with much weaker side lobes. We have calculated the temporal cross-correlation between the modes inside the simulator and in the forward-radiated far-field. The principal TEM mode turns out to be poorly correlated with the far-field. The dominant TM modes, viz. TM_1 and TM_2, exhibit fairly strong correlation with the far-field. TM_1 exhibits long-time correlation with the far-field, while TM_2 only shows short-time correlation. This shows that, for a given mode amplitude, higher order TM modes

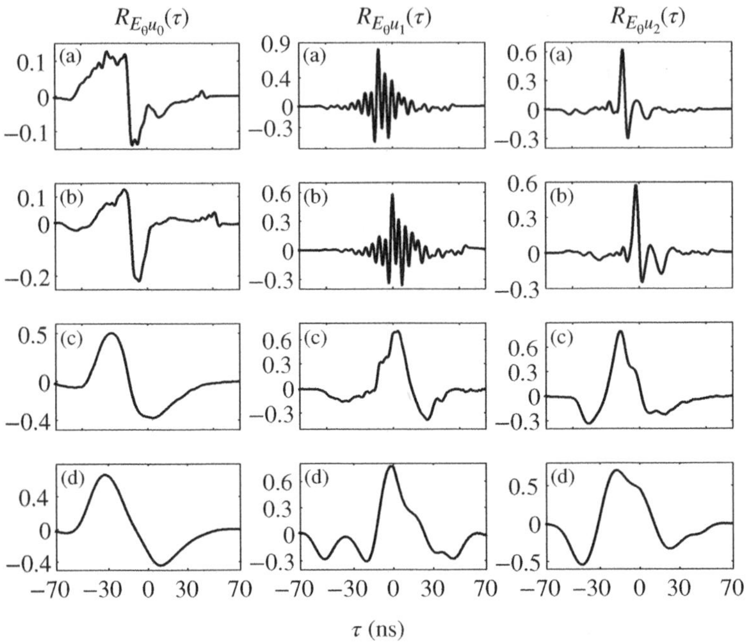

Figure 9.9 Variation of normalized cross-correlation $R_{E_\theta u_0}(\tau)$, $R_{E_\theta u_1}(\tau)$ and $R_{E_\theta u_2}(\tau)$ with time-delay. Other parameters are the same as in Figure 9.8.

are more effective in producing radiation leakage from a bounded-wave EMP simulator as compared to the TEM mode. However, given that the TEM mode is far stronger than the TM modes near the termination, in absolute terms it could still be contributing a significant amount to the far-field. Increasing the angle between plates results in excitation of more higher order modes and increases their strength, which in turn enhances radiation leakage. For a given simulator geometry, pulse-compression increases the strength of higher order modes in relation to the TEM mode. The TM_1 and TM_2 modes show fairly strong correlation with the far-field for all switching times considered here. An interesting development is that for larger values of the switching time, the TEM mode also exhibits peak values of the cross-correlation that are comparable with those for the TM modes. The observations listed above open the way to a new method of understanding EMI with surrounding equipment.

Exercises

9.1 Examine the radiation leakage due to an EMP interaction with an object-under-test made of anisotropic materials. How the radiated far-fields are correlated with the modes confined in the simulator?

9.2 In the present illustration, the cross-correlation has been established between time series data. It is worth extending these studies for spatial variation of modes excited inside the EMP simulator and the electromagnetic fields leaking out of the confined volume.

10

Spatial Mode Filter for Reducing Radiation Leakage

10.1 Introduction

In Chapter 9, we have studied the modal perspective of the radiation leakage from a bounded wave electromagnetic pulse (EMP) simulator and identified the modes responsible for radiation leakage. It is desirable to find some way of suppressing these modes. Giri et al. [29] have suggested the use of electrical conductors called spatial mode filters (SMF) for suppressing certain modes. However, that analysis has been done under several simplifying assumptions. For example the calculation of the angle of inclination of the SMF with respect to the plates of the horn is based upon the high frequency assumption, i.e. frequencies for which the wavelengths are small as compared to the aperture height of the transverse electromagnetic (TEM) horn. It is important to examine the utility of SMFs in a self-consistent full-wave numerical simulations involving real-life geometry to estimate the effectiveness of a mode suppression technique.

This chapter is organized as follows: Section 10.2 provides the comprehensive picture of a specific type of SMF design together with the full-wave self-consistent simulations illustrating the effectiveness of SMF in suppressing the unwanted modes without affecting the desired TEM mode. Finally, the closing remarks are discussed in Section 10.3.

10.2 Suppression of Higher Order Modes

Giri et al. [29] have suggested the idea of using an SMF for suppressing higher order modes. An SMF consists of a two-dimensional resistive sheet mounted between the pulser and the test volume. In general, the sheet should offer a

Electromagnetic Pulse Simulations Using Finite-Difference Time-Domain Method, First Edition. Shahid Ahmed.

longitudinal (z-directed) impedance Z_{TM} and a transverse (y-directed) impedance Z_{TE}, given by [29]

$$Z_{\mathrm{TM}} = \frac{Z_0 \sin(\zeta)}{2}, \quad Z_{\mathrm{TE}} = \frac{2}{Z_0 \sin(\zeta)} \tag{10.1}$$

where an electromagnetic signal, travelling through free space with impedance Z_0, is incident upon the SMF at an angle ζ.

TM modes can be suppressed by the magnetic field H_y produced by a longitudinal current I_z driven by E_z on the longitudinal impedance Z_{TM}. In the same way, TE modes can be suppressed by H_y driven by E_y through the transverse impedance Z_{TE}.

One option is to restrict the SMF to the tapered section of the simulator, so that it does not occupy space in the test volume. If necessary, however, it can also be extended into the test volume in order to suppress those higher order modes which only appear in that region.

For the simulator being examined here, E_y is negligibly small, as shown in Figure 7.2. This means that the TE mode is insignificant. Hence, we restrict our attention to the TM mode. Since there is no need to "short-out" E_y, we have used a suppressor design different from the planar sheet proposed by Giri et al. [29].

Figure 10.1 shows a three-dimensional schematic of the simulator with the SMF and the test object. The suppressor consists of a set of parallel rods of cross-sectional area ~13.6 cm^2, elongated in the z-direction and separated by ~7.4 cm gaps in the y-direction. There is no special reason for choosing these dimensions – they have simply been dictated by the finite-difference time-domain (FDTD) mesh. The portion of the SMF within the test section lies in the yz plane. Figure 10.2a,b shows top and side views of the SMF set up in the FDTD mesh. The resistivity of the material is chosen so as to obtain the required longitudinal resistance.

An optimized SMF would reduce the strength of higher order modes, without significantly reducing the strength of the main TEM mode and the peak electric field in the test volume. Having chosen the basic geometry of the SMF, we still

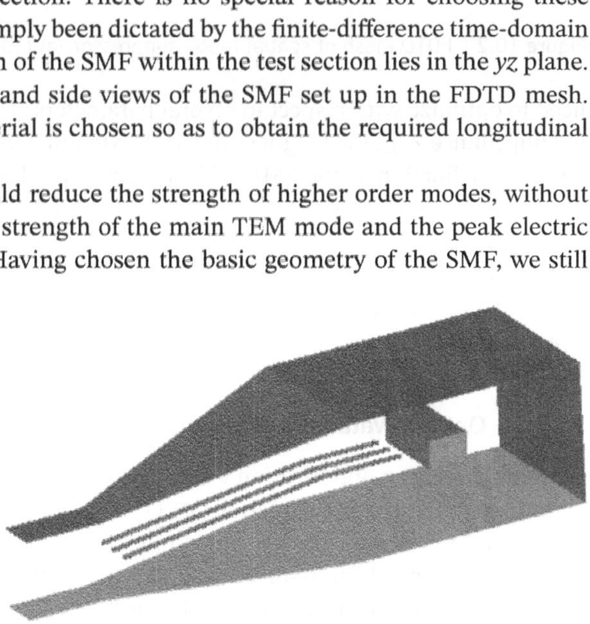

Figure 10.1 3-D view of bounded-wave EMP simulator with SMF and test object.

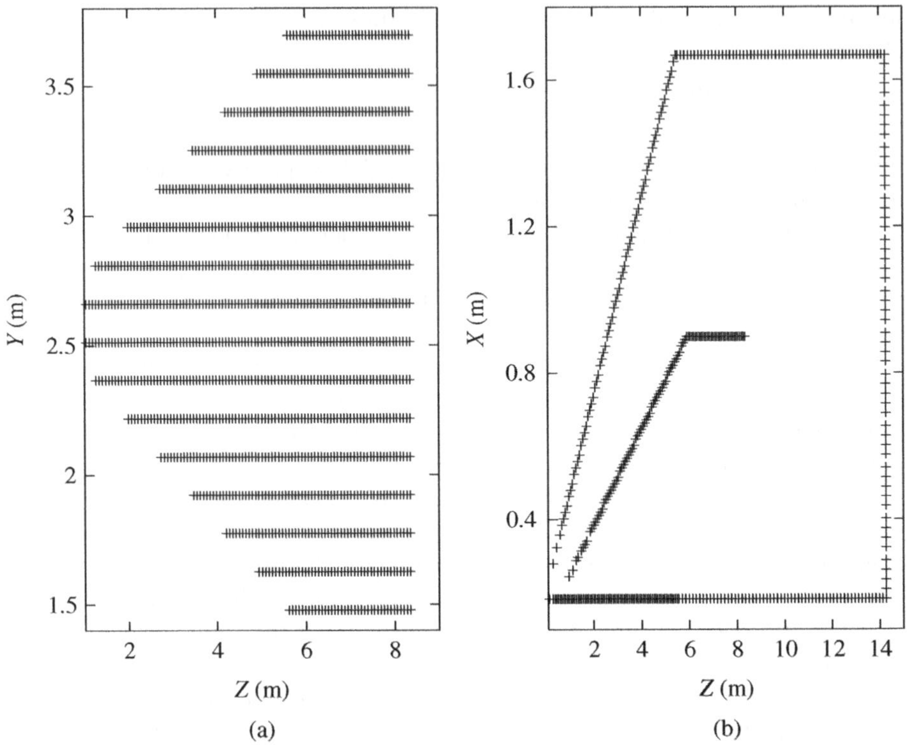

Figure 10.2 FDTD mesh of spatial mode suppressor, (a) top-view (b) side-view.

need to optimize with respect to several parameters. These include the longitudinal impedance Z_{TM}, the length of the SMF extending into the test volume, and the angle of inclination of the SMF with respect to the horizontal plane. In order to reduce the parameter space, we have fixed the angle of inclination of the SMF at 7.5°, which is half the angle between the two plates.

Details of the optimization are given in the following subsections. As a matter of practical interest, we have also studied the effectiveness of the SMF in the presence of a test object.

10.2.1 Optimal Value of Longitudinal Resistance

We first perform an optimization with respect to the longitudinal resistance in the absence of the test object. Equation (10.1) yields a starting point around which the optimization is performed. For our problem, it is reasonable to assume that the incident wave travels more-or-less parallel to the SMF. That implies a very small value of ζ, in turn yielding a small value of Z_{TM}.

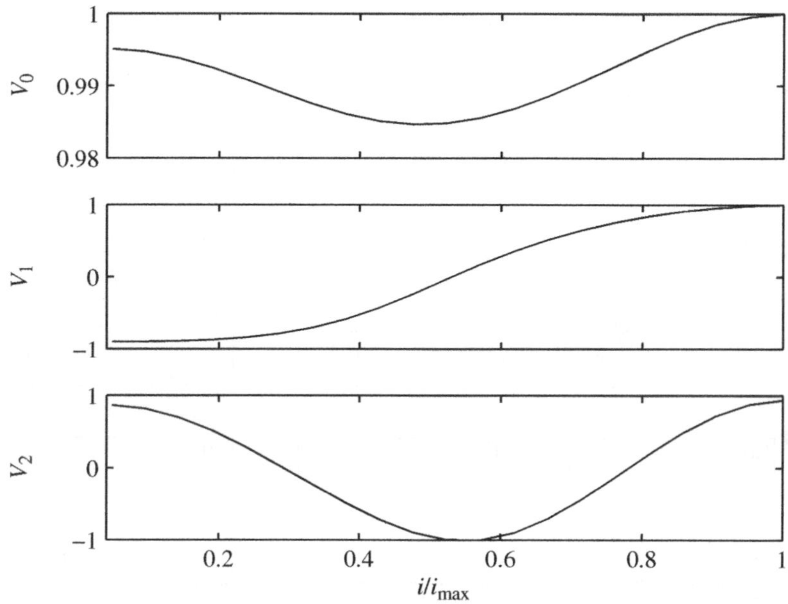

Figure 10.3 Eigenmode structures at $z = 8.9$ m, just beyond the end of the SMF. Results obtained with an SMF impedance of 132 Ω. The SMF projects to a distance of 3 m into the test section and there is no test object.

In the following calculations, the suppressor is assumed to extend to a distance of ~3 m into the test volume, which has a total length of ~9 m. The switching time τ_s is taken as 19.7 ns and the angle β as 15°.

Before we start examining the effect of the SMF on mode strengths, it is necessary to ensure that the mode shapes are acceptable. Figure 10.3 shows the shapes of the important eigenmodes at $z = 8.9$ m, just beyond the end of the SMF. These results have been obtained with an SMF impedance of 132 Ω. The mode shapes match well with the ideal cosinusoidal forms.

Figure 10.4a–c shows singular values of the dominant modes for three different values of SMF impedance. For greater clarity, we have shown the results only for longitudinal locations "z" following the end of the suppressor in the test volume.

Three points are noteworthy. Firstly, we recall that the test object may be placed anywhere within the test volume, hence it is important to reduce TM_1 strength throughout that region. Now, we observe that TM_1 is generally reduced within the test volume for all values of the impedance. The insertion of an SMF halves the effective height of the "waveguide," doubling the cutoff frequencies. Hence, a smaller portion of the input power spectrum remains accessible for TM_1, reducing its amplitude. It is only in case (a) that the modified value

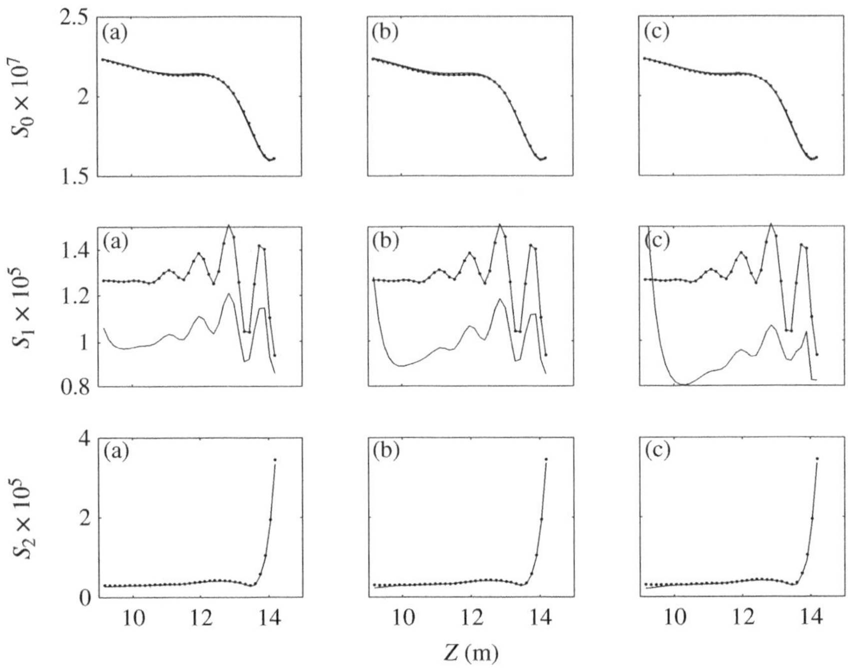

Figure 10.4 Comparison of singular values of TEM, TM_1, and TM_2 modes, as a function of longitudinal position in the test volume. There is no test object. Cases (a)–(c) correspond to SMF impedances of 52, 132, and 528 Ω. Data is shown only for longitudinal locations beyond the SMF. Dotted curves show the result without SMF.

remains below its no-SMF value at all *z*-locations. In case (c), in particular, the value actually becomes higher than the no-SMF value just after the end of the SMF.

Secondly, the TEM mode is not significantly affected by the presence of the SMF, as desired.

Thirdly, the TM_2 mode remains largely unchanged. A comprehensive explanation for this is not yet available. However, a limited explanation is possible in terms of the input power spectrum. We have seen that a propagating TM_2 mode only appears inside the test volume, being evanescent inside most of the tapered section. This mode is at least a factor of 2 weaker than TM_1 over most of the length of the test volume. This is consistent with the input power spectrum, since TM_1 has a cutoff frequency half that of TM_2, and thus has access to a significantly larger fraction of the input power. The doubling of cutoff frequencies due to the SMF means that TM_2 can now propagate only for frequencies >400 MHz. There is negligible energy content at such frequencies, so that TM_2 continues to be very weak in the region of the SMF. Let us now consider the region after the SMF. TM_2 has a full-wave structure in the x-direction, with E_z falling to zero at the two simulator plates and having another zero in-between. In addition, it has one maximum and one minimum in this region. Now, an SMF of the form we are using tends to short out the E_z component. A single SMF placed midway between the two plates would thus be trying to eliminate E_z in a region where it is anyway close to zero. This explains why the TM_2 mode is unaffected.

Based on this logic, it is reasonable to suppose that the insertion of two SMFs, one each at the location of the maximum and minimum, would be more effective in reducing TM_2 than a single SMF of the kind mentioned above. We have examined such a system. The angles of inclination of these suppressors are chosen such that their extensions are separated by a distance of "$h/4$" from the parallel plates inside the test volume. Here, "h" is the height of the test volume in the x-direction. Since TM_2 was evanescent in the SMF region even with one SMF, it remains evanescent in the double-SMF case as well. As in the single-SMF case, it reappears beyond the end of the SMF. There is, however, a major negative aspect, viz. the use of two SMFs adversely affects the TEM mode. Hence, we conclude that two SMFs are not desirable.

This study has only considered three values of the resistance, out of which a 52 Ω longitudinal resistance gives the best results. Hence, we cannot claim to have determined the optimal value, which would require systematic optimization. However, the study clearly shows the sensitivity of mode structure to the choice of resistance, and indicates that systematic optimization would be well worth the effort.

10.2.2 Optimal Length of Suppressor Inside Test Volume

The objective of the SMF is to suppress higher modes in the test volume. Clearly, it must extend throughout the tapered section of the simulator. If the SMF were to stop at the end of the tapered section, it is possible that some modes might appear inside the test volume itself. For example, for a switching time $\tau_s = 19.7$ ns, we have seen that the input pulse has significant energy content only up to ~240 MHz. At this frequency, the waveguide cutoff condition implies that the TM_1 mode first appears somewhere in the middle of the tapered section. However, TM_2 first arises only in the test section. Hence, we must also investigate the possibility of extending the SMF to some distance inside the test volume. There is, of course, an upper limit to this extension, imposed by the location of the test object.

In this subsection, we assume that the SMF resistance is the value determined in case (a) above and determine its optimal length inside the test volume. We have

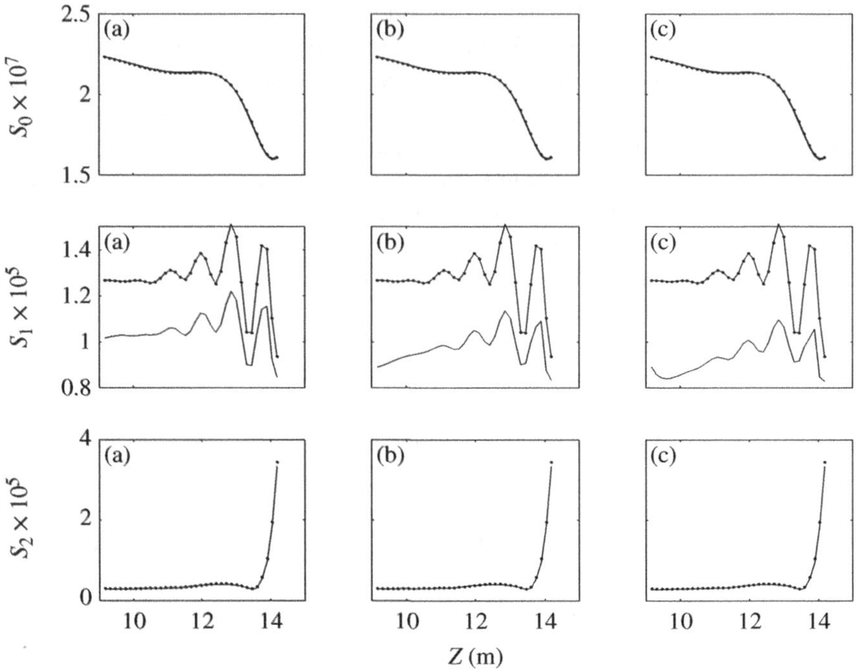

Figure 10.5 Singular values of TEM, TM_1, and TM_2 modes as a function of longitudinal position in the test volume. Cases (a)–(c) correspond to SMF lengths of 0.75, 1.5, and 2.25 m. Dotted curves show the result without SMF. Data is shown only for longitudinal locations beyond the SMF.

examined three different lengths, viz. 0.75 m, 1.5 m, and 2.25 m. Figure 10.5a–c shows the singular values of TEM, TM_1, and TM_2 with and without an SMF, as a function of longitudinal position. The dotted curves show results in the absence of SMF. We find maximum suppression with a 2.25 m length.

10.2.3 Mode Structure with Suppressor in Presence of Test Object

We have already seen the large modifications produced in the mode structure by the presence of a test object. In particular, we have seen that higher order TM modes become significant as we approach the object, so that the object is not truly subjected to the desired "free-space" illumination. It is of practical interest to determine the extent to which an SMF can help reduce the TM content near the object.

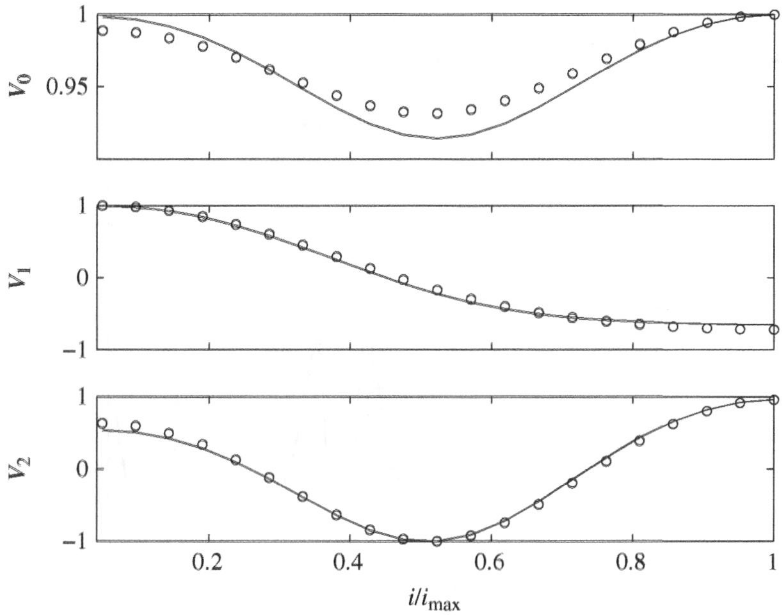

Figure 10.6 Comparison of shapes of important modes at a location ~0.6 m before the object, both with ("circles") and without ("solid line") the SMF.

In this subsection, we determine the mode structure in the presence of both, the "optimized" SMF and a cube-shaped object of side 0.5 m, made out of perfectly conducting material (PEC). Here, the optimized SMF is taken to have a longitudinal resistance of 52 Ω and projects upto 2.25 m within the test section.

Figure 10.6 shows the shapes of different eigenmodes at a distance of ~0.6 m before the object, both with and without an SMF. Clearly, there is no significant distortion in the modes due to insertion of the SMF in the region between the object and the pulser.

Having verified the mode shapes, we now look at their amplitudes. Figure 10.7a,b shows the variation of S_1/S_0 and S_2/S_0 with longitudinal position, with and without an SMF, in the presence of a perfectly conducting, solid cube-shaped object of side 0.5 m. The results are shown beyond the SMF region. The solid and dotted lines correspond to the result with and without an SMF, respectively. Clearly, the SMF has significantly reduced S_1/S_0, while S_2/S_0 remains unaffected. These results are consistent with the earlier analysis without an object, which are discussed in Sections 10.2.1 and 10.2.2.

We conclude that an optimized SMF significantly reduces the level of the TM_1 mode, but does not affect the TM_2 mode, either with or without a test object.

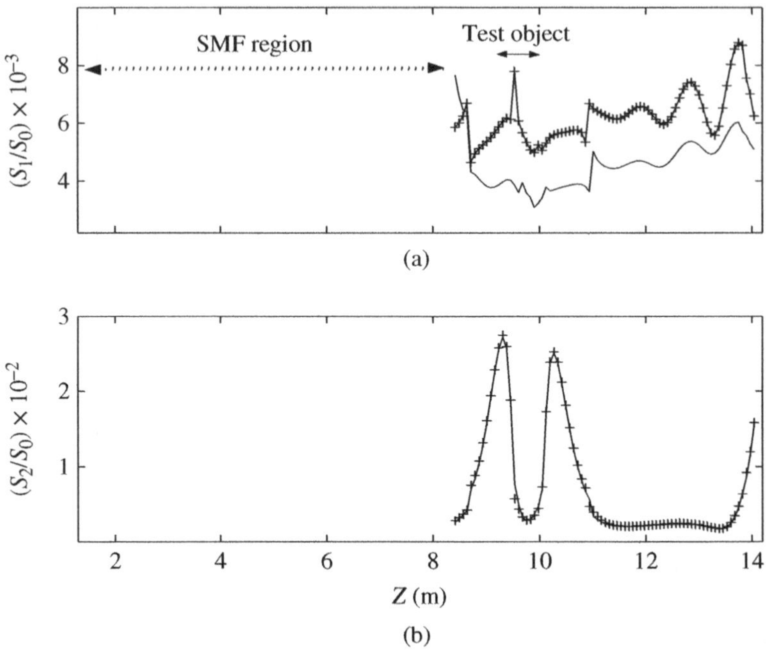

Figure 10.7 Simulator with optimized SMF and test object of side length 0.5 m. Cases (a) and (b) show the relative strengths of TM_1 and TM_2 modes with respect to the TEM mode, respectively. The Solid line and line with cross correspond to results with and without an SMF, respectively. The location of the test object is indicated in sub-plot (a).

10.3 Summary

The TEM cell radiates mainly in the forward direction, with much weaker side lobes. Placing a spatial mode filter in the region between the pulser and the test volume can significantly reduce the TM_1 mode, without significantly modifying the desired TEM mode. However, the TM_2 mode remains largely unchanged. We have provided an explanation in terms of the power spectrum and the cutoff frequencies of TM modes in a parallel-plate waveguide. We find that the reduction in TM_1 mode strength depends sensitively upon the choice of SMF parameters, such as the longitudinal resistance and the length of the SMF within the test section.

The use of two equally spaced mode filters is not advisable, for two reasons. Firstly, it does not yield a significant improvement in suppressing higher order modes. Secondly, it actually reduces the strength of the TEM mode, which is undesirable.

The observations of the studies open the way to a new method of understanding and improving simulator design from the point of view of electromagnetic interference with surrounding equipment.

Exercises

10.1 For the given design of a spatial-mode-filter (SMF) illustrated, the sensitivity of mode suppression to the choice of SMF parameters e.g. conductivity, length etc. needs a systematic optimization study.

10.2 As seen, the higher order TM modes are mainly responsible for the radiation leakage. Design a SMF based on the conventional approach of low-pass filter where the high frequency components can be filtered out. Verify the effectiveness.

11

EMP Interaction with Biological Tissues

11.1 Introduction

Simulations of ultrawide band (UWB) short pulse electromagnetics have been the subject of utmost interest in pulsed power engineering. However, the recent research into biological cells has opened another interesting area for UWB pulses [54]. Studies performed on human cells demonstrate the effects of UWB pulses depending on the duration and the intensity. An interesting application discussed in [54] reveals that intense sub-microsecond pulses can induce programmed cell death in biological cells, which reduces the growth of tumor. Therefore, the exposure of UWB pulses could be a boon for human body or may cause serious damage. There are several physical phenomena occurring in our environment, among them lightening and electrostatic discharges are most common. The corresponding electromagnetic pulses (EMPs) known as lightning electromagnetic pulse (LEMP) and electrostatic discharge-electromagnetic pulse (ESD-EMP) are intense, carrying hundreds of kilovolt per meter (kV/m) electric field for sub-microsecond duration. Therefore, the study of ultra-short pulses is important from the view point of its effect on biological cells.

Such studies are experimentally performed by placing the test-body in a desired EMP environment created inside the test-volume of an EMP simulator and the effects are assessed from the electromagnetic (EM) fields induced. The human body is inherently a complex object; its presence in the test-volume would then modify the spectrum of the incident EM fields. This in turn affects the induced field structures. The EMP is further modified due to reflections and scattering from different sections of the simulator and the ambient objects. These are practical phenomena, which should be automatically accounted in a realistic model. This necessitates the self-consistent full-wave analysis of an EMP simulator with test-body. Several studies on EMP interaction with human body have been reported [54, 55]. However, to author's knowledge, none of these studies are solved self-consistently.

Electromagnetic Pulse Simulations Using Finite-Difference Time-Domain Method, First Edition. Shahid Ahmed.

Chapter 6 is dedicated to the self-consistent analysis of a bounded-wave EMP simulator and the interaction of an EMP with a perfect electric conductor (PEC) solid cube. The pulsed power source used for driving the simulator involved air-gap parallel-plate capacitor and the pressurized section was not modeled. This study is a continuation of Chapter 6 in which the following important improvements are made. First, the air-gap parallel-plate capacitor is replaced by a real dielectric compact size parallel-plate capacitor. Second, the narrow region of the tapered section of the simulator is normally pressurized with SF_6 [9] to avoid breakdown. This has been now modeled by including the dielectric properties of SF_6 with losses. Third, using a fine mesh and running the code in Itanium II 160 processors cluster allows a risetime of 1.4 ns, which is close to the experimental design of 1 ns [9]. Fourth, the interaction of high-power UWB EMP with the human body holding a handset designed at 2 GHz is studied. Therefore, the present analysis is a significant development from the practical design of EMP simulation and its application to a new area of science.

This chapter is organized as follows: Section 11.2 describes the details of setting up the model used in the full-wave finite-difference time-domain (FDTD) simulations. The important analysis of the study, results, and discussions are focused in Section 11.3. This chapter is finally concluded with a brief summary in Section 11.4.

11.2 Model Description

The basic parameters of the EMP simulator, e.g. the capacitance of source, characteristic impedance of the transverse electromagnetic (TEM) cell, etc. have been chosen similar to [9]. Figure 11.1 shows the three dimensional schematic of the bounded-wave EMP simulator together with a human model and a handset mounted in the test volume. The pulsed power source consists of a parallel-plate capacitor loaded with dielectric having relative permittivity $\epsilon_r = 14.5$. Both plates of the capacitor and the simulator are assumed to be made of PEC. The simulation starts with an uncharged capacitor, and charging to $V_0 = 1$ MV is accomplished by the application of a voltage waveform following a half-Gaussian in time. Details of this procedure are given in Chapter 6. The charged capacitor is then discharged into the TEM "cell" through a closing switch shown by two black resistive strips (see Figure 11.1). Each strip having a length of one computational cell, separating the upper and lower capacitor plates from the corresponding plates of the TEM structure as shown in Figure 11.1. The strip material has a time-dependent conductivity σ, which is initially zero during capacitor charging. The mathematical model representing the physics of a closing

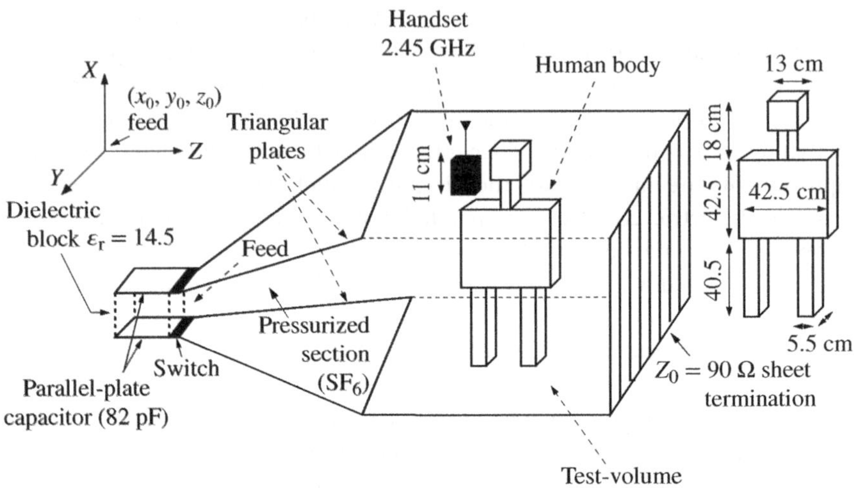

Figure 11.1 3D schematic of Bounded-wave EMP simulator with the human body and a handset designed at 2 GHz. The origin (x_0, y_0, z_0) of the coordinate system is chosen at the vertex of the TEM cell i.e. (0.32, 1.44, 0.45) m.

switch is characterized by the half-Gaussian (details are in Chapter 6)

$$\sigma(t) = \sigma_c \exp[-\alpha_s(t - \tau_s - \tau_0)^2] \qquad \tau_0 \leq t \leq \tau_0 + \tau_s \tag{11.1}$$

where τ_s is a characteristic time for closure, τ_0 is the time when switching starts, $\alpha_s = (4/\tau_s)^2$, and σ_c is the conductivity in the fully closed state. σ_c is chosen such that the switch resistance becomes very small in comparison with the characteristic impedance of the TEM-cell. In the simulations, we have chosen $\tau_s = 7$ ns which generates higher frequencies that can be handled by our FDTD cells. Setting up the capacitor, switch, and simulator in the FDTD domain allows a fully self-consistent analysis of the system, including details of capacitor charging and discharging through the closing switch. Moreover, the volume enclosed between the triangular plates starting from the feed to $z = 3.25$ m is filled with SF_6 having $\epsilon_r = 1.0623$ and $\sigma = 1.0 \times 10^{-7}$. The introduction of SF_6 may affect the transit time which occurs in the real experiment.

The simulations are performed by discretizing the computational domain using $100 \times 156 \times 600$ cells in the x, y, and z directions with uniform cell sizes $\Delta x = \Delta y = \Delta z = 1.85$ cm and running the code in Itanium II 160 processors cluster for 15 hours. The active region, consisting of the capacitor, switch, TEM structure and the test volume, is placed within this domain with a uniform spacing of at least 15 cells to the domain boundaries. The computational domain is truncated by Berenger's perfectly matched layers (PML) [36] occupying eight uniformly spaced edge cells.

For the purposes of illustration, we considered a simplified model of the human body and the dielectric constant is assumed to be homogeneous with $\epsilon_r = 40.0$ and $\sigma = 2.1$. The handset consists of a $\lambda/4$ monopole antenna designed at 2 GHz centered on a box made of PEC having dimension 110 (length) × 55 (thickness) × 90 (width) mm. Subcell model is preferably chosen for accurately handling the antenna.

11.3 Results and Discussion

The results of this study have been presented in two parts. First part will discuss the evolution of the EMP in the bounded-wave simulator. However, the details of EMP interaction with human tissue in the presence of electrical equipment will be demonstrated in the second part.

11.3.1 Pulse Evolution in the TEM Cell

Figure 11.2 shows the evolution of electric field components E_x, E_y, and E_z at $z = 3$ cm inside the tapered section of the simulator. Since E_x is, by far, the

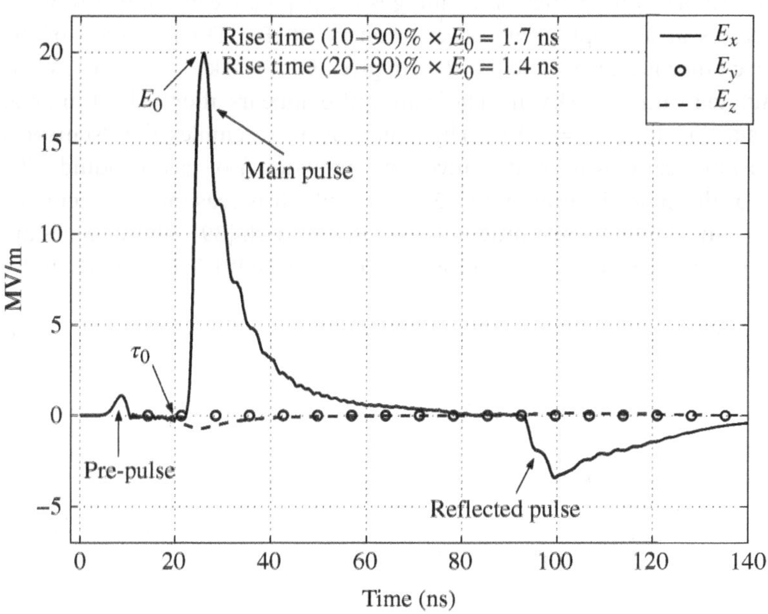

Figure 11.2 Typical level of electric field components at $z = 3$ cm inside the tapered section of the simulator. Distance is measured with respect to the origin defined in Figure 11.1.

dominant component within the capacitor plates, we would expect it to dominate within the bounded volume as well, which is indeed seen. There is a small, but still significant, longitudinal component E_z which arises due to reflection from the termination, while the transverse component E_y is negligibly small. It is desirable to check if the pulse propagating in the tapered section satisfies a basic requirement of a spherical TEM wave, i.e. the radial component of the electric field $E_r = 0$. Now, $RE_r = (x - x_0)E_x + (y - y_0)E_y + (z - z_0)E_z$, where (x, y, z) are the coordinates of the observation point. R is the distance between the origin and the observation point. Now, at $(x, y, z) = (0.46, 1.44, 1.4)$ m and the time corresponding to the main peak in E_x, we have $E_x = 2.02$ MV/m, $E_y \simeq 0$ and $E_z = -0.22$ MV/m, yielding $E_r = 73$ kV/m, which is much smaller than both E_x and E_z. Hence $E_r \sim 0$, for all practical purposes.

The "pre-pulse" appears during the charging state of the capacitor ($t < \tau_0 = 20$ ns) when the switch is fully open. This field develops due to the flow of electric displacement current $I_D(t) = C dV(t)/dt$, where C is the switch capacitance and $V(t)$ is the time dependent charging voltage having waveform represented by the half-Gaussian. The detailed analysis of the "pre-pulse" can be seen in Chapter 6. However, a simple estimate of the "pre-pulse" electric field $E_p = I_D(t)Z_0/g$, where $Z_0 = 90\ \Omega$, the characteristic impedance of the TEM cell which is almost constant over a broad frequency band and g is the gap between the plates of the TEM structures at an appropriate observation point. FDTD calculation of the switch capacitance is 2.25 pF, and $g = 7.4$ cm gives $E_p \approx 940$ kV/m which is close to the simulated value 1100 kV/m. The main pulse appears after a short interval of switch starting time $\tau_0 = 20$ ns. This small delay resembles the "statistical delay" of a practical closing switch involving breakdown of gas or liquid. The risetime t_r of the pulse is around 20–25% of τ_s, which is reasonably acceptable from the avalanche breakdown process of a closing switch [39]. These values are reasonably close to the designed risetime (20–90%) 1 ns of [9]. The negative pulse is a reflection from the termination.

Now, it is important to mention that the wave becomes slow during its propagation inside the SF_6 region which leads to $t_r(10\text{–}90) = 1.6$ ns. However, the initial risetime of 1.4 ns is achieved in the free space.

11.3.2 Interaction of EMP with Human Body

Human tissues are mixtures of different chemical compositions having complex physical properties – electromagnetic (EM) is one of them. In general, the EM properties of a substance are expressed in terms of complex permittivity $\epsilon^* = \epsilon - j\sigma$, where the real part gives the measure of electrical strength known as dielectric constant, however, the imaginary part includes losses. The response of EM impact on a substance is dictated by the "relaxation time" $\tau_r = \epsilon/\sigma$. Therefore, its study

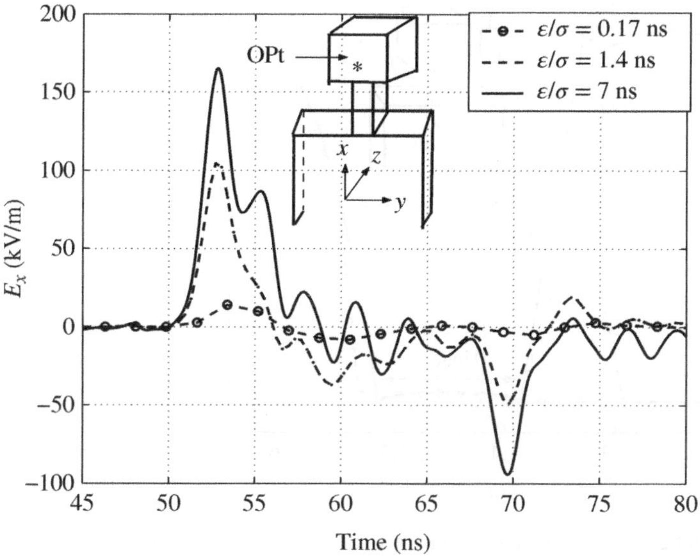

Figure 11.3 Induces electric field inside the human-head located at the centre marked by "★" for $\tau_r = 0.17$, 1.4 and 7 ns, respectively. Incident EMP corresponds to $t_r = 1.4$ ns and $E_0 = 500$ kV/m.

is important in order to abstract in-depth understanding of the biological tissues response to an EM signal. For the purpose of illustration, we have considered a biological tissue for a human model with $\epsilon_r = 40.0$ and $\sigma = 0.05$, 0.25, and 2.1 such that $\tau_r = 0.17$, 1.4, and 7 ns. These numbers are closely related to τ_s and t_r. The first number is approximately 1/10th of t_r, second is equal to t_r, and the third one corresponds to τ_s. The impact of the EMP for these values of τ_r is illustrated in Figure 11.3. The tissue with higher conductivity redistributes charges quickly (τ_r is small) in order to nullify the effects, resulting in the shielding of incident EMP. That is why induced E_x is minimum for $\tau_r = 0.17$ ns; however, it increases with decreasing σ. The brief summary of the study is as follows. The human body is covered with skin whose conductivity increases at higher frequencies (known from experiments [56]), low intensity shorter pulses may not be fatal. However, the intense EMP or continuous exposure may cause injury.

For understanding the interaction of EMP with real life complex problems, we have simulated a human model holding an electrical equipment working at 2.45 GHz. In Figure 11.4, we have scanned the E_x field from $y = 1.15$ to 1.65 m passing through the head in the xz-plane. The cross-sectional cut lies exactly at the center of the human head. The motivation behind this simulation is to understand the effect of electrical/electronic equipment such as a mobile phone to the human brain. In the region of a mobile phone with PEC box, E_x becomes zero

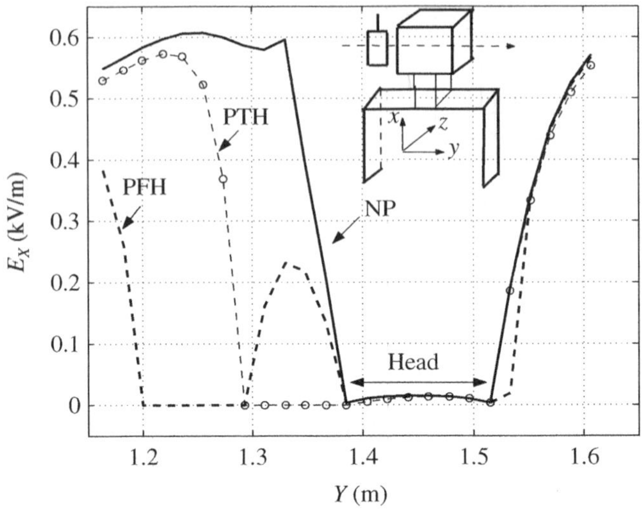

Figure 11.4 Variation of electric field along the y-direction in the xz-plane passing through the head shown by dashed arrow. The acronym "NP," "PTH," and "PFH" correspond to no phone, phone touching head, and phone 92 mm away from the head. Case for human tissue with $\epsilon_r = 40.0$ and $\sigma = 2.1$ ($\tau_r = 0.17$ ns).

which is expected because it is tangential to the walls, however, finite everywhere else. The field inside the human head follows the trend discussed in the earlier paragraph.

Figure 11.5 illustrates the comparative study of current evolution corresponding to the cases "NP," "PTH," and "PFH" discussed earlier. For the purpose of illustration, results are shown at two different locations along the height of the human body marked by the dashed arrows "A" and "B." These currents are computed using the Ampere's circuital law $\oint \vec{H} \cdot \vec{dl}$. We observe that the induced currents are same in all the three cases "NP," "PTH," and "PFH," respectively. However, the currents at "B" is almost double the currents at "A" despite the similarity in their waveforms. This difference in amplitudes at positions "A" and "B" could be estimated from the surface area of the FDTD cells enclosed by the loops which are $S_A = 7 \times 7$ and $S_B = 23 \times 3$, respectively. The ratio $S_A/S_B = 0.71$, directly takes into account the enhancement of 71%; however, the additional 29% could be the effects of cells in the neighborhood of the location "B."

Power absorbed by the entire body cells for the three different cases "NP," "PTH," and "PFH," respectively are illustrated in Figure 11.6. The absorbed power is calculated using $\int_V \sigma E^2 dV$, where the integral is taken over the entire volume of the body. We observe that power absorption in "PTH" case is marginally smaller than that of "NP" and "PFH." This small difference occurs due to the boundary condition.

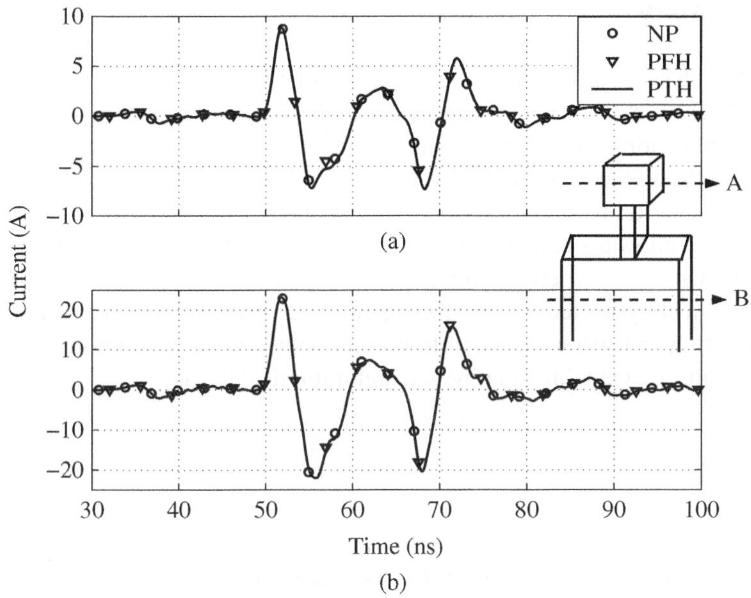

Figure 11.5 Comparison between induced currents for cases "NP," "PTH," and "PFH" discussed in Figure 11.4 at two different locations of human body marked by dashed arrows "A" and "B." The respective plots are shown in subplots (a) and (b).

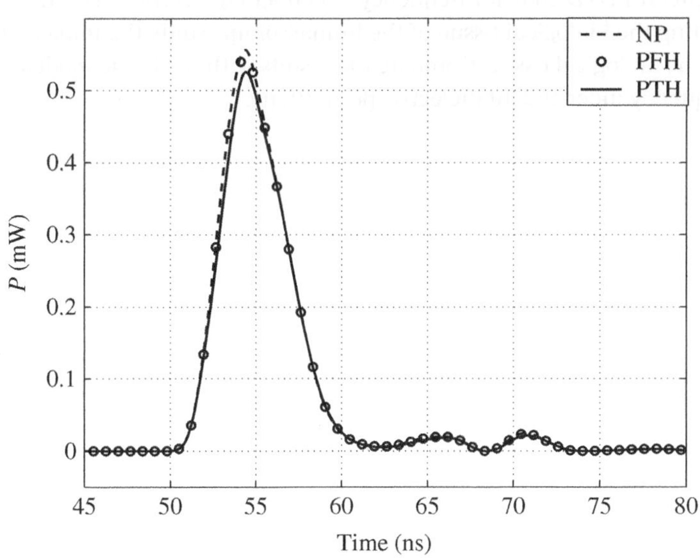

Figure 11.6 Comparison between powers absorbed by the entire human body for cases "NP," "PTH," and "PFH" discussed in Figure 11.4.

11.4 Summary

A 3D FDTD analysis of a bounded-wave EMP simulator from the view point of experimental design has been performed. This study has practical significance in UWB pulse applications such as investigation of ultra-short pulse effect on biological cells and pulsed power. In this study, we have utilized the massive computer resources available. Refinement of mesh and running the code in Itanium II 160 processors cluster allow to obtain risetime 1.4 ns – very close to the practical design 1 ns. The obtained EMP is then subjected to biological cells for understanding its effects on human body. Moreover, for investigating the effects of EMP in a real life environment study is further extended to human body with mobile phone. The simulation results are explained in physical and numerical terms.

Exercises

11.1 Optimize the monopole antenna of the handset corresponding to the peak power spectral density (PSD) of the EMP generated by the EMP simulator. Study the response of the biological tissue and calculate the specific absorption ratio (SAR).

11.2 Implement an FDTD code for frequency dependent dielectric permittivity representing the biological tissue of the human brain. Study the impact of EMP on the biological tissue. Compare the results with the tissue modeled with frequency independent dielectric permittivity.

12

FDTD Computer Program

12.1 Introduction

The computer program "EMPSIM.F" listed here is full-wave three-dimension in space and one dimension in time. This computer code is written in FORTRAN. It utilizes the finite-difference time-domain numerical method for the implementation of Maxwell's equations for simulating an electromagnetic pulse (EMP) environment. The code can handle a lossy dielectric material. The simulation space is truncated by a perfectly matched layer (PML) boundary condition. The radiated far-field simulation has been incorporated in the code.

The simulation setup starts with the charging of an uncharged capacitor bank, which upon switch-closure drives a TEM-cell consisting of a conical TEM section followed by a test-volume terminated through a resistive sheet. The open and closed switch models follow a half-Gaussian waveform.

The nomenclature used for a subroutine is consistent with its functionality, for example, subroutine "ZERO" initializes all the variables and the subroutine "BUILD" construct the geometry of the system of interest.

12.2 Computer Code Details

```
C
C      PROGRAM      : EMPSIM.F
C      PURPOSE      : SETTING UP AN EMP SIMULATOR
C
       PROGRAM EMPSIMA
       INCLUDE 'COMMON.H'

INTEGER MFAR,I,J,K,KK,MXDIM,NN,NN1,NN2,NN3,NBOUND
REAL*8 UXARR(NDIM,MXDIM),UYARR(NDIM,MXDIM),UZARR(NDIM,MXDIM)
         REAL*8 WXARR(NDIM,MXDIM),WYARR(NDIM,MXDIM),WZARR(NDIM,MXDIM)

         REAL*8 EXSURFY(NX,NZ),EXSURF_Y(NX,NZ),EXSURFZ(NX,NY),
     &          EXSURF_Z(NX,NY)
```

Electromagnetic Pulse Simulations Using Finite-Difference Time-Domain Method, First Edition.
Shahid Ahmed.

```
          REAL*8 EYSURFX(NY,NZ),EYSURF_X(NY,NZ),EYSURFZ(NX,NY),
     &           eysurf_z(nx,ny)
          Real*8 ezsurfx(ny,nz),ezsurf_x(ny,nz),ezsurfy(nx,nz),
     &           ezsurf_y(nx,nz)
          Real*8 hxsurfy(nx,nz),hxsurf_y(nx,nz),hxsurfz(nx,ny),
     &           hxsurf_z(nx,ny)
          Real*8 hysurfx(ny,nz),hysurf_x(ny,nz),hysurfz(nx,ny),
     &           hysurf_z(nx,ny)
          Real*8 hzsurfx(ny,nz), hzsurf_x(ny,nz), hzsurfy(nx,nz),
     &           hzsurf_y(nx,nz)
      Real*8 EXS(NX,NY,NZ),EYS(NX,NY,NZ),EZS(NX,NY,NZ)
      Real*8 HXS(NX,NY,NZ),HYS(NX,NY,NZ),HZS(NX,NY,NZ)
      Real*8 fartimx,fartim_x,fartimy,fartim_y
      Real*8 fartimz,fartim_z,xfar(ndim),yfar(ndim),zfar(ndim)
      Real*8 fartimhx,fartimh_x,fartimhy,fartimh_y,fartimhz,fartimh_z

      Real*8 tt,toffset,xc,yc,zc,tc,tstart,forpicdt,xref,yref,zref,
     &           radfar,p,Px,Py,Pz,pxp,pxn,pyp,pyn,pzp,pzn,dum1,dum2,
     &           dum3,dum4,dum5,dum6,areax,areay,areaz,tauc,tau0

      Common/ comswitch/ tauc
      Common/ comdischarg/ tau0


C OPEN DATA FILES.
         OPEN(unit=17,file='diags3d.dat',status='unknown')
         OPEN(unit=10,file='nzout3d.dat',status='unknown')
C Define the Cell Size (meter) :
      DELX = 18.475e-3
      DELy = DELX
      DELZ = DELX

C Assignment for PML SETUP
         ESIGMAM=3.*EPS0*3.E8*ALOG(1./5.E-3)/(2.* NPML *DELX)
         HSIGMAM=ESIGMAM*XMU0/EPS0

C Initialize the electric and magnetic vector potentials array :
        do 1112 nn = 1,ndim
        do 1111 kk = 1,mxdim
          uxarr(nn,kk) = 0.0d0
          uyarr(nn,kk) = 0.0d0
          uzarr(nn,kk) = 0.0d0
          wxarr(nn,kk) = 0.0d0
          wyarr(nn,kk) = 0.0d0
          wzarr(nn,kk) = 0.0d0
1111      continue
1112      continue

          CALL ZERO
          CALL BUILD
          CALL SETUP

C**********************
C  PML SUBROUTINES     *
C**********************
      CALL PMLCUBE
           CALL PMLSETUP
```

```
C End of initialization block :

tstart = 0.0d0

C Time for the signal to reach the far-zone point :
C tfar = distance of far zone point from the reference / 3.0d8 (units of dt)
C toffset, is the time for the signal to reach the far field point from the
C source which is the radiator in this case.

radfar  = 1.0d3
toffset  = radfar / c
forpicdt = 4.0d0 * pi * c * dt

C NUMBER OF CELLS LEFT FROM THE BOUNDARIES IN ALL THE DIRECTIONS FOR
C INTEGRATION OVER THE SURFACE SURROUNDING THE OBJECT FOR FAR-FIELD:
nbound = 12

C Assign the reference point of the domain to measure the far field:
xref = dfloat(iminf + imaxf) * 0.5d0 * delx
yref = dfloat(jminf + jmaxf) * 0.5d0 * dely
zref = dfloat(kfeed) * delz

C ASSIGN THE CO-ORDINATE OF THE FAR-FIELD POINTS :
C We are here scanning the angular locations in the YZ-plane, X = 0:
C Here X of the far zone point is always zero.
C theta = -90, -60, -30, 0,30,60,90,120,180

C theta = -90, phi = -90; y<y0
xfar(1) = xref
yfar(1) = yref - radfar
zfar(1) = zref

C theta = -60, phi = -90; y<y0
xfar(2) = xref
yfar(2) = yref - radfar * sin(pi/3.0d0)
zfar(2) = zref + radfar * cos(pi/3.0d0)

C theta = -30, phi = -90; y<y0
xfar(3) = xref
yfar(3) = yref - radfar * sin(pi/6.0d0)
zfar(3) = zref + radfar * cos(pi/6.0d0)

C theta = 0, phi = 90; y=y0
xfar(4) = xref
yfar(4) = yref
zfar(4) = zref + radfar

C theta = 30, phi = 90; y>y0
xfar(5) = xref
yfar(5) = yref + radfar * sin(pi/6.0d0)
zfar(5) = zref + radfar * cos(pi/6.0d0)

C theta = 60, phi = 90; y>y0
xfar(6) = xref
yfar(6) = yref + radfar * sin(pi/3.0d0)
zfar(6) = zref + radfar * cos(pi/3.0d0)
```

```
C theta = 90, phi = 90; y>y0
xfar(7) = xref
yfar(7) = yref + radfar * sin(pi/2.0d0)
zfar(7) = zref + radfar * cos(pi/2.0d0)

C theta = 120, phi = 90; y>y0
xfar(8) = xref
yfar(8) = yref + radfar * sin(2.0d0 * pi/ 3.0d0)
zfar(8) = zref + radfar * cos(2.0d0 * pi/ 3.0d0)

C theta = 150, phi = 90; y>y0
xfar(9) = xref
yfar(9) = yref + radfar * sin(5.0d0 * pi/ 6.0d0)
zfar(9) = zref + radfar * cos(5.0d0 * pi/ 6.0d0)

C theta = 180, phi = 90; y>y0
xfar(10) = xref
yfar(10) = yref + radfar * sin(pi)
zfar(10) = zref + radfar * cos(pi)

C ******************************************************************
C MAIN LOOP FOR FIELD COMPUTATIONS AND DATA SAVING
C ******************************************************************

      T = tstart

      DO 100 N=1,NSTOP

      WRITE (*,*) N

C ADVANCE SCATTERED ELECTRIC FIELD

   CALL EXFLD
        CALL EYFLD
        CALL EZFLD

C Assign fields :

do i = 1,NX
     do j = 1,NY
        do k = 1,NZ
     EXS(i,j,k) = EX(i,j,k)
     EYS(i,j,k) = EY(i,j,k)
     EZS(i,j,k) = EZ(i,j,k)
        enddo
           enddo
enddo

C Positive  X-directed face:
C******************************************************************

do 191 nn1 = 1,ndim

areax = dely * delz C area of the x-directed face
xc = dfloat(nx-nbound) * delx
```

```
do 121 j = nbound+1, ny-nbound
        yc = dfloat(j) * dely
do 121 k = nbound+1, nz-nbound
       zc = dfloat(k) * delz

eysurfx(j,k) = 0.5d0*(eys(nx-nbound,j,k)+eys(nx-nbound,j,k+1))
ezsurfx(j,k) = 0.5d0*(ezs(nx-nbound,j,k)+ezs(nx-nbound,j+1,k))

fartimx = dsqrt((xfar(nn1)-xc)**2+(yfar(nn1)-yc)**2+
     &    (zfar(nn1)-zc)**2)/c
tc = dfloat(n) * dt + fartimx - toffset
mfar = nint(tc / dt + 0.5d0) CConvert the real into nearest integer
if(mfar.le.nbound .or. mfar.ge.mxdim)then

go to 121

else

uyarr(nn1,mfar-1) =  areax * ezsurfx(j,k) * (0.5d0 - tc/dt
     &          + dfloat(mfar)) / forpicdt + uyarr(nn1,mfar-1)

uyarr(nn1,mfar) = areax * ezsurfx(j,k) * 2.0d0 * (tc/dt
     &           - dfloat(mfar)) / forpicdt  +  uyarr(nn1,mfar)

uyarr(nn1,mfar+1) = - areax * ezsurfx(j,k) * (0.5d0 + tc/dt
     &           - dfloat(mfar)) / forpicdt + uyarr(nn1,mfar+1)


uzarr(nn1,mfar-1) = - areax * eysurfx(j,k) * (0.5d0 - tc/dt
     &           + dfloat(mfar)) / forpicdt + uzarr(nn1,mfar-1)

uzarr(nn1,mfar) = - areax * eysurfx(j,k) * 2.0d0 * (tc/dt
     &           - dfloat(mfar)) / forpicdt + uzarr(nn1,mfar)

uzarr(nn1,mfar+1) = areax * eysurfx(j,k) * (0.5d0 + tc/dt
     &           - dfloat(mfar)) / forpicdt + uzarr(nn1,mfar+1)
endif
121  continue

C Negative X-direction :

xc = dfloat(nbound + 1) * delx

do 122 j = nbound + 1, ny-nbound
        yc = dfloat(j) * dely
do 122 k = nbound + 1, nz-nbound
        zc = dfloat(k) * delz

eysurf_x(j,k) =0.5d0 * (eys(nbound+1,j,k) + eys(nbound+1,j,k+1))
ezsurf_x(j,k) =0.5d0 * (ezs(nbound+1,j,k) + ezs(nbound+1,j+1,k))

fartim_x = dsqrt((xfar(nn1)-xc)**2+(yfar(nn1)-yc)**2+
     &    (zfar(nn1)-zc)**2)/c
tc =  dfloat(n) * dt + fartim_x - toffset
mfar = nint(tc/dt + 0.5d0)
```

```
if(mfar .le. nbound .or. mfar .ge. mxdim)then

go to 122

else

uyarr(nn1,mfar-1) = - areax * ezsurf_x(j,k) * (0.5d0 - tc/dt
     &          + dfloat(mfar)) / forpicdt + uyarr(nn1,mfar-1)

uyarr(nn1,mfar) = - areax * ezsurf_x(j,k) *  2.0d0 * (tc/dt
     &         - dfloat(mfar)) / forpicdt + uyarr(nn1,mfar)

uyarr(nn1,mfar+1) =  areax * ezsurf_x(j,k) * (0.5d0 + tc/dt
     &         - dfloat(mfar)) / forpicdt + uyarr(nn1,mfar+1)


uzarr(nn1,mfar-1) =  areax * eysurf_x(j,k) * (0.5d0 - tc/dt
     &           + dfloat(mfar)) / forpicdt + uzarr(nn1,mfar-1)

uzarr(nn1,mfar) = areax * eysurf_x(j,k) * 2.0d0 * (tc/dt
     &         - dfloat(mfar)) / forpicdt + uzarr(nn1,mfar)

uzarr(nn1,mfar+1) = - areax * eysurf_x(j,k) * (0.5d0 + tc/dt
     &           - dfloat(mfar)) / forpicdt + uzarr(nn1,mfar+1)
endif
122  continue

C Positive  Y-directed face :

areay = delx * delz C area of y-directed face
yc = dfloat(ny-nbound) * dely

do 123 i = nbound+1, nx-nbound
        xc = dfloat(i) * delx
do 123 k = nbound+1, nz-nbound
        zc = dfloat(k) * delz

exsurfy(i,k)= 0.5d0*(exs(i,ny-nbound,k)+exs(i,ny-nbound,k+1))
ezsurfy(i,k)= 0.5d0*(ezs(i,ny-nbound,k)+ezs(i+1,ny-nbound,k))

fartimy = dsqrt((xfar(nn1)-xc)**2+(yfar(nn1)-yc)**2+
     & (zfar(nn1)-zc)**2)/c
tc = dfloat(n) * dt + fartimy - toffset
mfar = nint(tc / dt + 0.5d0)

if(mfar .le. nbound .or. mfar .ge. mxdim)then

go to 123

else

uzarr(nn1,mfar-1) = areay * exsurfy(i,k) * (0.5d0 - tc/dt
     &    + dfloat(mfar)) / forpicdt + uzarr(nn1,mfar-1)

uzarr(nn1,mfar) = areay * exsurfy(i,k) * 2.0d0 * (tc / dt
     &    - dfloat(mfar)) / forpicdt + uzarr(nn1,mfar)
```

```
uzarr(nn1,mfar+1) = - areay * exsurfy(i,k) * (0.5d0 + tc/dt
     &     - dfloat(mfar)) / forpicdt + uzarr(nn1,mfar+1)


uxarr(nn1,mfar-1) = - areay * ezsurfy(i,k) * (0.5d0 - tc/dt
     &           + dfloat(mfar)) / forpicdt + uxarr(nn1,mfar-1)

uxarr(nn1,mfar) = - areay * ezsurfy(i,k) * 2.0d0 * (tc/dt
     &           - dfloat(mfar)) / forpicdt + uxarr(nn1,mfar)

uxarr(nn1,mfar+1) = areay * ezsurfy(i,k) * (0.5d0 + tc/dt
     &           - dfloat(mfar)) / forpicdt + uxarr(nn1,mfar+1)
endif
123  continue

C Negative Y- direction:

yc = dfloat(nbound + 1) * dely
do 212 i = nbound+1, nx-nbound
        xc = dfloat(i) * delx
do 212 k = nbound+1, nz-nbound
        zc = dfloat(k) * delz

exsurf_y(i,k) = 0.5d0 * (exs(i,nbound+1,k)+exs(i,nbound+1,k+1))
ezsurf_y(i,k) = 0.5d0 * (ezs(i,nbound+1,k)+ezs(i+1,nbound+1,k))

fartim_y = dsqrt((xfar(nn1)-xc)**2+(yfar(nn1)-yc)**2+
       &    (zfar(nn1)-zc)**2)/c
tc =  dfloat(n) * dt + fartim_y - toffset
mfar = nint(tc/dt + 0.5d0)

if(mfar .le. nbound .or. mfar .ge. mxdim)then

go to 212

else

uzarr(nn1,mfar-1) = - areay * exsurf_y(i,k) * (0.5d0 - tc/dt
       &           + dfloat(mfar)) / forpicdt + uzarr(nn1,mfar-1)

uzarr(nn1,mfar) = - areay * exsurf_y(i,k) * 2.0d0 * (tc/dt
       &           - dfloat(mfar)) / forpicdt + uzarr(nn1,mfar)

uzarr(nn1,mfar+1) = areay * exsurf_y(i,k) * (0.5d0 + tc/dt
       &           - dfloat(mfar)) / forpicdt + uzarr(nn1,mfar+1)

uxarr(nn1,mfar-1) =  areay * ezsurf_y(i,k) * (0.5d0 - tc/dt
       &           + dfloat(mfar)) / forpicdt + uxarr(nn1,mfar-1)

uxarr(nn1,mfar) = areay * ezsurf_y(i,k) * 2.0d0 * (tc/dt
       &           - dfloat(mfar)) / forpicdt + uxarr(nn1,mfar)

uxarr(nn1,mfar+1) = - areay * ezsurf_y(i,k) * (0.5d0 + tc/dt
       &           - dfloat(mfar)) / forpicdt + uxarr(nn1,mfar+1)

endif
```

```
212  continue

C Positive Z-directed face :

areaz = delx * dely C area of z-directed face
zc = dfloat(nz-nbound) * delz

do 127 i = nbound+1, nx-nbound
        xc = dfloat(i) * delx
do 127 j = nbound+1, ny-nbound
        yc = dfloat(j) * dely

exsurfz(i,j) = 0.5d0*(exs(i,j,nz-nbound)+exs(i,j+1,nz-nbound))
eysurfz(i,j) = 0.5d0*(eys(i,j,nz-nbound)+eys(i+1,j,nz-nbound))

fartimz = dsqrt((xfar(nn1)-xc)**2+(yfar(nn1)-yc)**2+
     &    (zfar(nn1)-zc)**2)/c
tc = dfloat(n) * dt + fartimz - toffset
mfar = nint(tc / dt + 0.5d0)

if(mfar .le. nbound .or. mfar .ge. mxdim)then
go to 127
else

uxarr(nn1,mfar-1) = areaz * eysurfz(i,j) * (0.5d0 - tc/dt
     &           + dfloat(mfar)) / forpicdt + uxarr(nn1,mfar-1)

uxarr(nn1,mfar) = areaz * eysurfz(i,j) * 2.0d0*(tc/dt
     &           - dfloat(mfar)) / forpicdt + uxarr(nn1,mfar)

uxarr(nn1,mfar+1) = - areaz * eysurfz(i,j) * (0.5d0 + tc/dt
     &           - dfloat(mfar)) / forpicdt + uxarr(nn1,mfar+1)

uyarr(nn1,mfar-1) = - areaz * exsurfz(i,j) * (0.5d0 - tc/dt
     &           + dfloat(mfar)) / forpicdt  + uyarr(nn1,mfar-1)

uyarr(nn1,mfar) = - areaz * exsurfz(i,j) * 2.0d0 * (tc / dt
     &           - dfloat(mfar)) / forpicdt + uyarr(nn1,mfar)

uyarr(nn1,mfar+1) = areaz * exsurfz(i,j) * (0.5d0 + tc/dt
     &           - dfloat(mfar)) / forpicdt + uyarr(nn1,mfar+1)

endif
127  continue

C Negative Z-directed face :

zc = dfloat(nbound + 1) * delz

do 128 i = nbound + 1, nx - nbound
        xc = dfloat(i) * delx
do 128 j = nbound + 1 , ny - nbound
        yc = dfloat(j) * dely

exsurf_z(i,j) = 0.5d0*(exs(i,j,nbound+1)+exs(i,j+1,nbound+1))
eysurf_z(i,j) = 0.5d0*(eys(i,j,nbound+1)+eys(i+1,j,nbound+1))
```

```
fartim_z = dsqrt((xfar(nn1)-xc)**2+(yfar(nn1)-yc)**2+
       &     (zfar(nn1)-zc)**2)/c
tc = dfloat(n) * dt + fartim_z - toffset
mfar = nint(tc / dt + 0.5d0)

if(mfar .le. nbound .or. mfar .ge. mxdim)then

go to 128

else

uxarr(nn1,mfar-1) = - areaz * eysurf_z(i,j) * (0.5d0 - tc/dt
       &            + dfloat(mfar)) / forpicdt + uxarr(nn1,mfar-1)

uxarr(nn1,mfar) = - areaz * eysurf_z(i,j) * 2.0d0 * (tc / dt
       &            - dfloat(mfar)) / forpicdt + uxarr(nn1,mfar)

uxarr(nn1,mfar+1) = areaz * eysurf_z(i,j) * (0.5d0 + tc/dt
       &            - dfloat(mfar)) / forpicdt + uxarr(nn1,mfar+1)

uyarr(nn1,mfar-1) = areaz * exsurf_z(i,j) * (0.5d0 - tc/dt
       &            + dfloat(mfar)) / forpicdt + uyarr(nn1,mfar-1)

uyarr(nn1,mfar) = areaz * exsurf_z(i,j) * 2.0d0 * (tc/dt
       &            - dfloat(mfar)) / forpicdt  + uyarr(nn1,mfar)

uyarr(nn1,mfar+1) = - areaz * exsurf_z(i,j) * (0.5d0 + tc/dt
       &            - dfloat(mfar)) / forpicdt + uyarr(nn1,mfar+1)

endif

128  continue
191  continue


C ADVANCE TIME BY 1/2 TIME STEP
T = T + DT/2.0d0

C ADVANCE MAGNETIC FIELD
        CALL HXFLD
        CALL HYFLD
        CALL HZFLD

C Assign magnetic field :

do i = 1,NX
   do j = 1,NY
        do k = 1,NZ
     HXS(i,j,k) = HX(i,j,k)
     HYS(i,j,k) = HY(i,j,k)
     HZS(i,j,k) = HZ(i,j,k)
        enddo
           enddo
        enddo
```

```
C Calculate corresponding to electrical surface current :

C X-directed face  magnetic field

do 192 nn2 = 1,ndim

areax = dely * delz C Area of x-directed face
xc = dfloat(nx-nbound) * delx

do 129 j = nbound+1, ny-nbound
      yc = dfloat(j)*dely
do 129 k = nbound+1, nz-nbound
      zc = dfloat(k)*delz

C positive x ->
          hysurfx(j,k) = 0.25d0 * (hys(nx-nbound,j,k) +
     & hys(nx1-nbound,j,k) + hys(nx-nbound,j+1,k) +
     & hys(nx1-nbound,j+1,k))

        hzsurfx(j,k) = 0.25d0 * (hzs(nx-nbound,j,k) +
     & hzs(nx1-nbound,j,k) + hzs(nx-nbound,j,k+1) +
     & hzs(nx1-nbound,j,k+1))

fartimhx = dsqrt((xfar(nn2)-xc)**2+(yfar(nn2)-yc)**2+
     & (zfar(nn2)-zc)**2)/c
tc = dfloat(n) * dt + fartimhx - toffset
mfar = nint(tc / dt + 1.0d0)

if(mfar .le. nbound .or. mfar .ge. mxdim)then

go to 129

else

wzarr(nn2,mfar-1) = areax * hysurfx(j,k) * (0.5d0 - tc/dt
     &           + dfloat(mfar)) / forpicdt + wzarr(nn2,mfar-1)

wzarr(nn2,mfar) =  areax * hysurfx(j,k) * 2.0d0 * (tc / dt
     &           - dfloat(mfar)) / forpicdt + wzarr(nn2,mfar)

wzarr(nn2,mfar+1) = - areax * hysurfx(j,k) * (0.5d0 + tc / dt
     &           - dfloat(mfar)) / forpicdt + wzarr(nn2,mfar+1)


wyarr(nn2,mfar-1) = - areax * hzsurfx(j,k) * (0.5d0 - tc / dt
     &           + dfloat(mfar)) / forpicdt + wyarr(nn2,mfar-1)

wyarr(nn2,mfar) = - areax * hzsurfx(j,k) * 2.0d0 * (tc / dt
     &           - dfloat(mfar)) / forpicdt + wyarr(nn2,mfar)

wyarr(nn2,mfar+1) = areax * hzsurfx(j,k) * (0.5d0 + tc / dt
     &           - dfloat(mfar)) / forpicdt + wyarr(nn2,mfar+1)

endif
129  continue
```

```
Ccalculation for negative x-directed face:
xc = dfloat(nbound+1) * delx
do 119 j = nbound+1, ny-nbound
      yc = dfloat(j) * dely
do 119 k = nbound+1, nz-nbound
      zc = dfloat(k) * delz

C    negative x ->
        hysurf_x(j,k) = 0.25d0*(hys(nbound+1,j,k)+hys(nbound+2,j,k)+
     &              hys(nbound+1,j+1,k) + hys(nbound+2,j+1,k))

        hzsurf_x(j,k) = 0.25d0*(hzs(nbound+1,j,k)+hzs(nbound+2,j,k)+
     &             hzs(nbound+1,j,k+1) + hzs(nbound+2,j,k+1))
fartimh_x = dsqrt((xfar(nn2)-xc)**2+(yfar(nn2)-yc)**2+
     &    (zfar(nn2)-zc)**2)/c
tc = dfloat(n) * dt + fartimh_x - toffset
mfar = nint(tc /dt + 1.0d0)


if(mfar .le. nbound .or. mfar .ge. mxdim)then

go to 119

else

wzarr(nn2,mfar-1) = - areax * hysurf_x(j,k) * (0.5d0 - tc / dt
     &           + dfloat(mfar)) / forpicdt + wzarr(nn2,mfar-1)

wzarr(nn2,mfar) = - areax * hysurf_x(j,k) * 2.0d0 * (tc / dt
     &           - dfloat(mfar)) / forpicdt + wzarr(nn2,mfar)

wzarr(nn2,mfar+1) = areax * hysurf_x(j,k) * (0.5d0 + tc / dt
     &           - dfloat(mfar)) / forpicdt + wzarr(nn2,mfar+1)

wyarr(nn2,mfar-1) = areax * hzsurf_x(j,k) * (0.5d0 - tc/dt
     &           + dfloat(mfar)) / forpicdt + wyarr(nn2,mfar-1)

wyarr(nn2,mfar) = areax * hzsurf_x(j,k) * 2.0d0 * (tc / dt
     &           - dfloat(mfar)) / forpicdt + wyarr(nn2,mfar)

wyarr(nn2,mfar+1) = - areax * hzsurf_x(j,k) * (0.5d0 + tc / dt
     &           - dfloat(mfar)) / forpicdt + wyarr(nn2,mfar+1)

endif

119  continue


C Y-directed face  magnetic field :

areay = delx * delz C Area of y-directed face
yc = dfloat(ny-nbound) * dely

do 124 i = nbound+1, nx-nbound
      xc = dfloat(i) * delx
do 124 k = nbound+1, nz-nbound
```

```
          zc = dfloat(k) * delz

           hxsurfy(i,k) = 0.25d0*(hxs(i,ny-nbound,k)+hxs(i,ny1-nbound,k)+
     &                hxs(i+1,ny-nbound,k) + hxs(i+1,ny1-nbound,k))

           hzsurfy(i,k) = 0.25d0*(hzs(i,ny-nbound,k)+hzs(i,ny1-nbound,k)+
     &                hzs(i,ny-nbound,k+1) + hzs(i,ny1-nbound,k+1))

      fartimhy = dsqrt((xfar(nn2)-xc)**2+(yfar(nn2)-yc)**2+
     & (zfar(nn2)-zc)**2)/c
      tc = dfloat(n) * dt + fartimhy - toffset
      mfar = nint(tc / dt + 1.0d0)


      if(mfar .le. nbound .or. mfar .ge. mxdim)then

      go to 124

      else

      wxarr(nn2,mfar-1) = areay * hzsurfy(i,k) * (0.5d0 - tc/dt
     &           + dfloat(mfar)) / forpicdt + wxarr(nn2,mfar-1)

      wxarr(nn2,mfar) = areay * hzsurfy(i,k) * 2.0d0 * (tc / dt
     &           - dfloat(mfar)) / forpicdt + wxarr(nn2,mfar)

      wxarr(nn2,mfar+1) = - areay * hzsurfy(i,k) * (0.5d0 + tc/dt
     &           - dfloat(mfar)) / forpicdt + wxarr(nn2,mfar+1)


      wzarr(nn2,mfar-1) = - areay * hxsurfy(i,k) * (0.5d0 - tc / dt
     &           + dfloat(mfar)) / forpicdt + wzarr(nn2,mfar-1)

      wzarr(nn2,mfar) = - areay * hxsurfy(i,k) * 2.0d0 * (tc / dt
     &           - dfloat(mfar)) / forpicdt + wzarr(nn2,mfar)

      wzarr(nn2,mfar+1) = areay * hxsurfy(i,k) * (0.5d0 + tc/dt
     &           - dfloat(mfar)) / forpicdt + wzarr(nn2,mfar+1)

      endif
      124  continue

      C  FOR NEGATIVE Y-DIRECTED FACE :

      yc = dfloat(nbound+1) * dely

      do 126 i = nbound+1, nx-nbound
            xc = dfloat(i) * delx
      do 126 k = nbound+1, nz-nbound
             zc = dfloat(k) * delz

              hxsurf_y(i,k) = 0.25d0*(hxs(i,nbound+1,k)+hxs(i,nbound+2,k)+
     &              hxs(i+1,nbound+1,k) + hxs(i+1,nbound+2,k))

              hzsurf_y(i,k) = 0.25d0*(hzs(i,nbound+1,k)+hzs(i,nbound+2,k+1)+
     &                hzs(i,nbound+1,k+1) +  hzs(i,nbound+2,k+1))
```

```
fartimh_y = dsqrt((xfar(nn2)-xc)**2+(yfar(nn2)-yc)**2+
     &  (zfar(nn2)-zc)**2)/c
tc = dfloat(n) * dt + fartimh_y - toffset
mfar = nint(tc / dt + 1.0d0)

if(mfar .le. nbound .or. mfar .ge. mxdim)then

go to 126

else

wxarr(nn2,mfar-1) = - areay * hzsurf_y(i,k) * (0.5d0 - tc / dt
     &           + dfloat(mfar)) / forpicdt + wxarr(nn2,mfar-1)

wxarr(nn2,mfar) = - areay * hzsurf_y(i,k) * 2.0d0 * (tc/dt
     &           - dfloat(mfar)) / forpicdt + wxarr(nn2,mfar)

wxarr(nn2,mfar+1) = areay * hzsurf_y(i,k) * (0.5d0 + tc / dt
     &           - dfloat(mfar)) / forpicdt + wxarr(nn2,mfar+1)


wzarr(nn2,mfar-1) = areay * hxsurf_y(i,k) * (0.5d0 - tc / dt
     &           + dfloat(mfar)) / forpicdt + wzarr(nn2,mfar-1)

wzarr(nn2,mfar) = areay * hxsurf_y(i,k) * 2.0d0 * (tc / dt
     &           - dfloat(mfar)) / forpicdt + wzarr(nn2,mfar)

wzarr(nn2,mfar+1) = - areay * hxsurf_y(i,k) * (0.5d0 + tc/dt
     &           - dfloat(mfar)) / forpicdt + wzarr(nn2,mfar+1)

endif

126  continue


C Z-directed face :

areaz = delx * dely CArea of the xz-plane

zc = dfloat(nz-nbound) * delz

do 125 i = nbound+1, nx-nbound
      xc = dfloat(i) * delx
do 125 j = nbound+1, ny-nbound
      yc = dfloat(j) * dely

C    positive z -->
        hxsurfz(i,j) = 0.25d0*(hxs(i,j,nz-nbound)+hxs(i,j,nz1-nbound)+
     &            hxs(i+1,j,nz-nbound) + hxs(i+1,j,nz1-nbound))

        hysurfz(i,j) = 0.25d0*(hys(i,j,nz-nbound)+hys(i,j,nz1-nbound)+
     &            hys(i,j+1,nz-nbound) + hys(i,j+1,nz1-nbound))

fartimhz = dsqrt((xfar(nn2)-xc)**2+(yfar(nn2)-yc)**2+
     & (zfar(nn2)-zc)**2)/c
tc = dfloat(n) * dt + fartimhz - toffset
mfar = nint(tc / dt + 1.0d0)
```

```
if(mfar .le. nbound .or. mfar .ge. mxdim)then

go to 125

else

wyarr(nn2,mfar-1) = areaz * hxsurfz(i,j) * (0.5d0 - tc/dt
     &           + dfloat(mfar)) / forpicdt + wyarr(nn2,mfar-1)

wyarr(nn2,mfar) = areaz * hxsurfz(i,j) * 2.0d0 * (tc / dt
     &           - dfloat(mfar)) / forpicdt + wyarr(nn2,mfar)

wyarr(nn2,mfar+1) = - areaz * hxsurfz(i,j) * (0.5d0 + tc / dt
     &           - dfloat(mfar)) / forpicdt + wyarr(nn2,mfar+1)

wxarr(nn2,mfar-1) = - areaz * hysurfz(i,j) * (0.5d0 - tc / dt
     &           + dfloat(mfar)) / forpicdt + wxarr(nn2,mfar-1)

wxarr(nn2,mfar) = - areaz * hysurfz(i,j) * 2.0d0 * (tc / dt
     &           - dfloat(mfar)) / forpicdt + wxarr(nn2,mfar)

wxarr(nn2,mfar+1) =  areaz * hysurfz(i,j) * (0.5d0 + tc/dt
     &           - dfloat(mfar)) / forpicdt + wxarr(nn2,mfar+1)
endif

125  continue

C  FOR NEGATIVE Z-DIRECTED FACE:

zc = dfloat(nbound+1) * delz
do 109 i = nbound+1, nx-nbound
        xc = dfloat(i) * delx
do 109 j = nbound+1, ny-nbound
         yc = dfloat(j) * dely

C    negative z -->
        hxsurf_z(i,j) = 0.25d0*(hxs(i,j,nbound+1)+hxs(i,j,nbound+2)+
     &            hxs(i+1,j,nbound+1)+hxs(i+1,j,nbound+2))

       hysurf_z(i,j) = 0.25d0*(hys(i,j,nbound+1)+hys(i,j,nbound+2)+
     &            hys(i,j+1,nbound+1)+hys(i,j+1,nbound+2))
   fartimh_z=dsqrt((xfar(nn2) - xc)**2+(yfar(nn2) - yc)**2+
     & (zfar(nn2) - zc)**2)/c
tc = dfloat(n) * dt + fartimh_z - toffset
mfar = nint(tc / dt + 1.0d0)

if(mfar .le. nbound .or. mfar .ge. mxdim)then

go to 109

else

wyarr(nn2,mfar-1) = - areaz * hxsurf_z(i,j) * (0.5d0 - tc/dt
     &           + dfloat(mfar)) / forpicdt + wyarr(nn2,mfar-1)
```

```
wyarr(nn2,mfar) = - areaz * hxsurf_z(i,j) * 2.0d0 * (tc / dt
     &          - dfloat(mfar)) / forpicdt + wyarr(nn2,mfar)

wyarr(nn2,mfar+1) = areaz * hxsurf_z(i,j) * (0.5d0 + tc / dt
     &          - dfloat(mfar)) / forpicdt + wyarr(nn2,mfar+1)

wxarr(nn2,mfar-1) = areaz * hysurf_z(i,j) * (0.5d0 - tc / dt
     &          + dfloat(mfar)) / forpicdt + wxarr(nn2,mfar-1)

wxarr(nn2,mfar) = areaz * hysurf_z(i,j) * 2.0d0 * (tc / dt
     &          - dfloat(mfar)) / forpicdt + wxarr(nn2,mfar)

wxarr(nn2,mfar+1) = - areaz * hysurf_z(i,j) * (0.5d0 + tc / dt
     &          - dfloat(mfar)) / forpicdt + wxarr(nn2,mfar+1)

endif

109  continue
192   continue

C ADVANCE TIME ANOTHER 1/2 STEP

         T = T + DT / 2.0d0

C CALCULATION OF POWER RADIATED OUT FROM THE FDTD-DOMAIN AT AN INSTANT OF
C TIME USING POYNTING FLUX:

C CALCULATION OF POYNTING FLUX :
    Px  = 0.0d0 C Net power from x-face
    Py  = 0.0d0 C Net power from y-face
    Pz  = 0.0d0 C Net power from z-face
    pxp = 0.0d0 Cpower along the positive directed x-face
    pxn = 0.0d0 Cpower along the negative directed x-face
    pyp = 0.0d0 Cpower along the positive directed y-face
    pyn = 0.0d0 Cpower along the negative directed y-face
    pzp = 0.0d0 Cpower along the positive directed z-face
    pzn = 0.0d0 Cpower along the negative directed z-face

C  X-DIRECTED FACE -
    do 147 j = nbound+1,ny-nbound
       do 147 k = nbound+1,nz-nbound

C positive x --->
dum1 = eysurfx(j,k) * hzsurfx(j,k) -
     & ezsurfx(j,k) * hysurfx(j,k)
pxp = pxp +  dum1 * dely * delz

C  negative x --->
dum2 = eysurf_x(j,k) * hzsurf_x(j,k) -
     & ezsurf_x(j,k) * hysurf_x(j,k)
pxn = pxn +  dum2 * dely * delz

C Net power-density in +x-direction (W/m2)
px = px + dum1 - dum2

147       continue
```

```
C Net power in x-direction (W)
       px = px * dely * delz

C  Y-DIRECTED FACE -
do 148 i = nbound+1,nx-nbound
     do 148 k = nbound+1,nz-nbound

C  positive y --->
      dum3 = ezsurfy(i,k) * hxsurfy(i,k) -
     &  exsurfy(i,k) * hzsurfy(i,k)
pyp = pyp +  dum3 * delx * delz

C  negative y --->
       dum4 = ezsurf_y(i,k) * hxsurf_y(i,k) -
     & exsurf_y(i,k) * hzsurf_y(i,k)
pyn = pyn +  dum4 * delx * delz

C Power-density in +y-direction (W/m2)
      py = py + dum3 - dum4
 148    continue

C Net power in + y-direction (W)
py = py * delx * delz

C  Z-DIRECTED FACE -
do 149 i = nbound+1,nx-nbound
   do 149 j = nbound+1,ny-nbound

C  positive z --->
dum5 = exsurfz(i,j) * hysurfz(i,j) -
     & eysurfz(i,j) * hxsurfz(i,j)
pzp = pzp +  dum5 * delx * dely

C  negative z --->
dum6 = exsurf_z(i,j) * hysurf_z(i,j) -
     & eysurf_z(i,j) * hxsurf_z(i,j)
pzn = pzn +  dum6 * delx * dely

C net power density in +z-direction (W/m2)
pz = pz + dum5 - dum6

 149  continue


C Net power in z-direction (W)
pz = pz * delx * dely

C Net power flow out of the domain:
p = Px + Py + Pz

write(31,50) t,pxp,pxn,pyp,pyn,pzp,pzn,px,py,pz,p
50  format(1x,11(d12.6,1x))

C SAMPLE FIELDS IN SPACE AND WRITE TO DISK

         CALL DATSAV
```

```
C For discharge the capacitor bank, make the switch on :
tau0 = 2.0d0 * tauc
        if (t .ge. tau0) call setup

 100  CONTINUE
      T = NSTOP*DT
      WRITE (17,200) T,NSTOP
 200  FORMAT(T2,'EXIT TIME= ',E14.7,' SECONDS,  AT TIME STEP',I6,/)

C SAVE THE DATA OF POTENTIAL ARRAY OF DIFFERENT POSITIONS IN DISK :
open(unit = 1, file = 'fort1', status = 'unknown')
open(unit = 2, file = 'fort2', status = 'unknown')
open(unit = 3, file = 'fort3', status = 'unknown')
open(unit = 4, file = 'fort4', status = 'unknown')
open(unit = 5, file = 'fort5', status = 'unknown')
open(unit = 6, file = 'fort6', status = 'unknown')
open(unit = 7, file = 'fort7', status = 'unknown')
open(unit = 8, file = 'fort8', status = 'unknown')
open(unit = 9, file = 'fort9', status = 'unknown')
open(unit = 10, file = 'fort10', status = 'unknown')
open(unit = 11, file = 'fort11', status = 'unknown')

do 193 nn3 = 1, ndim
    do 194 j = 1,mxdim

           tt = tstart + dfloat(j-1) * dt
   write(nn3,1235) tt,uxarr(nn3,j),uyarr(nn3,j),
     & uzarr(nn3,j),wxarr(nn3,j), wyarr(nn3,j),
     & wzarr(nn3,j)
1235            format(7(d12.6,1x))

194      continue
193   continue


C ****** END OF ASSIGNMENT FOR SAVING DATA OF POTENTIAL ARRAY *******

CLOSE(10)
CLOSE(17)

      STOP
      END

C**************************************************
      SUBROUTINE BUILD
C  Geometry setup routine.

INCLUDE 'common.h'
Integer mtype,kfeed,kaperture,iminf,imaxf,jminf,jmaxf,imina,
     &    imaxa,jmina,jmaxa
Real*8 xapex,yapex,zapex,wbyh,sleng,gamma,fraclen,
     &         al2,be2,gam2,projlen,projfeed,widthf,widtha,heightf,
     &         heighta,tangam2,tanbet2,d1,d2,xmidt,xmidb,xmint,xminb,
     &         xmaxt,xmaxb,ymin,ymax,transit_max,xmaxcapb,
     &         transit_min,freqmax,freqmin,sleng_act
Real*8 xmidcapt,xmidcapb,x,y,z,xmincapt,xmaxcapt,xmincapb
```

```
Integer kmidcap,jmidcap,iminfcap,imaxfcap,jminfcap,
     &    jmaxfcap,kfeedcap,kaperturecap
Real*8 tauc
Real*8 xmaxsf6,xminsf6, zmaxsf6

C  NEW DEFINITION :
Real*8 betaby2,sleng_feed

Common / comcapacloc / iminfcap,imaxfcap,jminfcap,jmaxfcap,
     &    kfeedcap,kaperturecap

Common/ comswitch/ tauc

C     THIS SUBROUTINE IS USED TO DEFINE THE SCATTERING OBJECT WITHIN
C     THE FDTD SOLUTION SPACE.  USER MUST SPECIFY IDONE, IDTWO AND
C     IDTHRE AT DIFFERENT CELL LOCATIONS TO DEFINE THE SCATTERING
C     OBJECT.  SEE THE YEE PAPER (IEEE TRANS. ON AP, MAY 1966) FOR
C     A DESCRIPTION OF THE FDTD ALGORITHM AND THE LOCATION OF FIELD
C     COMPONENTS.
C
C     GEOMETRY DEFINITION
C
C     IDONE, IDTWO, AND IDTHRE ARE USED TO SPECIFY MATERIAL IN CELL I,J,K.
C     IDONE DETERMINES MATERIAL FOR X COMPONENTS OF E
C     IDTWO FOR Y COMPONENTS, IDTHRE FOR Z COMPONENTS
C
C     SET IDONE,IDTWO, AND/OR IDTHRE FOR EACH I,J,K CELL =
C             0       FOR FREE SPACE
C             1       FOR PEC
C             2-9     FOR LOSSY DIELECTRICS
C
C     SUBROUTINE DCUBE BUILDS A CUBE OF DIELECTRIC MATERIAL BY SETTING
C     IDONE, IDTWO, IDTHRE TO THE SAME MATERIAL TYPE.  THE MATERIAL
C     TYPE IS SPECIFIED BY MTYPE.  SPECIFY THE STARTING CELL (LOWER
C     LEFT CORNER (i.e. MINIMUM I,J,K VALUES) AND SPECIFY THE CELL
C     WIDTH IN EACH DIRECTION (USE THE NUMBER OF CELLS IN EACH
C     DIRECTION).  USE NZWIDE=0 FOR A INFINITELY THIN PLATE IN THE
C     XY PLANE.  FOR PEC PLATE USE MTYPE=1.  ISTART, JSTART, KSTART ARE
C     USED TO DEFINE THE STARTING CELL AND NXWIDE, NYWIDE AND NZWIDE EACH
C     SPECIFY THE OBJECT WIDTH IN CELLS IN THE X, Y, AND Z DIRECTIONS.
C     INDIVIDUAL IDONE, TWO OR THRE COMPONENTS CAN BE SET MANUALLY
C     FOR WIRES, ETC.  DCUBE DOES NOT WORK FOR WIRES (I.E. NXWIDE=0 AND
C     NYWIDE=0 FOR EXAMPLE)

C++  User input section  ++

C  Coordinates of apex of TEM Cell
xapex = 15.0d0 * delx
yapex = dely * dfloat(ny/2)
zapex = 12.0d0 * delz

C  Width at aperture (m)
widtha = 2.32d0
C  Height at feed (m)
heightf = 3.0d0 * delx
C  Angle (deg) by which the antenna plate flares-out :
```

```
alpha = 22.76666d0
C Angle converted into radians
alpha = alpha * pi / 1.8d2
C  Angle (deg) by which the top plate is elevated with respect to image-plane
C  (GROUND, which is lower plate in this problem)
        betaby2 = 15.0d0
        betaby2 = betaby2 * pi / 1.8d2
C   Slant length of plate (m) from apex (tip of feed)
sleng = widtha / 2.0d0 / sin(alpha/2.0d0)
C  Length in z-direction from apex to feed point (m):
projfeed = heightf / tan(betaby2)
C Slength from apex to feed point (m):
sleng_feed = projfeed / cos(alpha/2.0d0) / cos(betaby2)
C  Ratio of z-coordinates at feed & aperture
fraclen = sleng_feed / sleng

C  Width at feed :
widthf  = widtha * fraclen

C  End of user-input section

    al2 = alpha * 0.5d0
    be2 = betaby2
    tanbet2 = tan(be2)

C  Length of antenna in z-direction (m) (apex to aperture):
    projlen = sleng * cos(al2) * cos(betaby2)
C  Actual slant length i.e. slant length from feed to aperture :
    sleng_act = sleng - sleng_feed

C  Calculate the angle "gamma" (see Fig.(2) of Shlager):
    gamma = 2.0d0 * atan(tan(al2) / cos(betaby2))
    gam2 = gamma * 0.5d0
    tangam2 = tan(gam2)

C  Width to height (from image plane) ratio :
wbyh = widthf / heightf

c  Axial length from apex to center of aperture (m):
    zleng = projlen - projfeed
    write(14,335) alpha*1.8d2/pi,2.0d0*be2*1.8d2/pi,
     &    wbyh,fraclen,sleng,sleng_act
335    format('(alpha,beta,W/H,fraclen)= ',4(d12.6,1x),/,
     &    ' Slant length from apex (m)= ',e10.4,/,
     &    ' Actual slant length (m)= ',e10.4,/)

write(14,334) gamma,gam2,tangam2,projlen,projfeed,zleng
334    format(' [gamma, gamma/2, Tan(gam/2)]= ',
     &    3(e10.4,1x),/,' (L_a, L_f, Z-length)= ',3(e10.4,1x))

C  Calc. (max. & min. transit times through horn at speed of light):
    transit_min = (projlen - projfeed) / 3.0d8  C In z-direction
    transit_max = sleng_act / 3.0d8    C Along slant length

C  Calc. (min, max) frequencies where this antenna is efficient:
C  The minimum frequency corresponds to a wavelength = twice the actual slant
```

```
C  length. The max. frequency corresponds to the transit-time delay:
freqmin = 3.0d8 / (2.0d0 * sleng_act)
freqmax = 0.604d0 / (transit_max - transit_min)
write(14,333) transit_max,transit_min,freqmin,freqmax
333     format(' (t_max, t_min, F_min, F_max)=',4(e10.4,1x))

C  Calc. width at aperture (specified fully by the parameters fixed above):
C widtha = 2.0d0 * projlen * tan(gam2)
C  Assign heights at feed point and aperture (m):
C  Note: "Height" is in the x-direction.
C heightf = widthf / wbyh
C heighta = widtha / wbyh
heighta = projlen * tan(betaby2)
write(14,555) widthf,widtha,heightf,heighta
555 format(' (Wf, Wa, Hf, Ha)= ',4(e10.4,1x))
write(14,556) xapex,yapex,zapex
556 format(' (x,y,z) of apex= ',3(e10.4,1x))

C+++++  End of initialization block  +++++

mtype = 1

C  Run a loop over all z-locations in the mesh:
do 200 k=1,nz
z = dfloat(k) * delz
d1 = z - zapex

C  Skip the cell if its z-location is smaller than that of the feed:
if (d1.lt.projfeed) then
kfeed = k
go to 200
endif

if (d1.gt.projlen) go to 200
kaperture = k

C**************** Initialization block for SF6 region
C Define zmax for SF6 region
zmaxsf6 = zpaex + 2.0d2 * delz

C Define the x-dimension for filling up SF6
xmaxsf6 = xmidt - 1.0d0*delx
xminsf6 = xmidb + 1.0d0*delx

C********** End of initialization block for SF6 region **********

C  For each "z", calc. the ranges of (x,y) that define the top &
bottom plates:
xmidt = xapex + d1 * tan(betaby2)
xmidb = xapex

C  Make the plate thickness = 1 cells thick in x-direction:
xmaxt = xmidt + 1.0d0*delx
xmint = xmidt - 1.0d0*delx
xmaxb = xmidb + 0.5d0*delx
xminb = xmidb - 0.5d0*delx
```

```
C  The y-range is the same for top & bottom plates:
d2 = d1 * tangam2
ymin = yapex - d2
ymax = yapex + d2

c  Now run a loop over (i,j). Put perfect conductor in cells that lie within
c  (xmin, xmax) and (ymin, ymax):
do 100 i=1,nx
do 100 j=1,ny
x = dfloat(i) * delx
y = dfloat(j) * dely

C Now, first fill the free space of narrow region of the simulator
volume by SF6 :
      if (y.ge.ymin+1.0*dely .and. y.le.ymax-1.0*dely) then
               if (x.ge.xminsf6 .and. x.le.xmaxsf6) then
              if (z .lt. zmaxsf6) then
call dcube(i,j,k,1,1,1,4)
            endif
         endif
      endif

C End of SF6 region

if (y.ge.ymin .and. y.le.ymax) then
if ((x.ge.xmint .and. x.le.xmaxt) .or.
     &     (x.ge.xminb .and. x.le.xmaxb)) then
call dcube(i,j,k,1,1,1,1)
write(12,123) 1.0,x,y,z,i,j,k
123 format(' ',4(e10.4,1x),3(i4,1x))
endif
endif

100  continue
200  continue
close(12)

C  At this point, "kfeed" contains the last z-layer where there is
C  no conductor.  Add "1" to get the first z-layer of the feed:
kfeed = kfeed + 1
C  At this point, "kaperture" contains the z-layer of the aperture.

write(14,355) kfeed,kaperture
355  format(' (kF, kA)= ',2(i4,1x))

C  Now open unit-12 again & read the data. For k=kfeed & k=kaperture,
C  find the max. range of (i) & (j) where conductor has been inserted.
C  These variables are (iminf,imaxf,jminf,jmaxf) & (imina,imaxa,jmina,jmaxa).
C  Also, for the top plate, find the range of "i" which contains conductor,
C  viz., variables (iftop1,iftop2) - typically there would be two cells
C  containing conductor, i.e. (iftop2 - iftop1) = 1. Note that "iftop2"
C  is the same as "imaxf".
C  All these variables are put into COMMON/comhorn/ and are thus available
C  to DATSAV and the Excitation routines (e.g. EXSFLD).
open(12,file='fort.12',status='old')
iminf = nx
```

```
imina = nx
imaxf = 0
imaxa = 0
jminf = ny
jmina = ny
jmaxf = 0
jmaxa = 0
102  continue
read(12,*,end=101) d1,d1,d1,d1,i,j,k
if (k.eq.kfeed) then
    if (jminf.gt.j) jminf = j
    if (jmaxf.lt.j) jmaxf = j
    if (iminf.gt.i) iminf = i
    if (imaxf.lt.i) imaxf = i
endif
if (k.eq.kaperture) then
    if (jmina.gt.j) jmina = j
    if (jmaxa.lt.j) jmaxa = j
    if (imina.gt.i) imina = i
    if (imaxa.lt.i) imaxa = i
endif
go to 102
101  continue

C  Max. & Min. values of "i" which contain conductor on the top plate at the
feed point:
iftop2 = imaxf
iftop1 = iftop2 - 1

write(14,366) iminf,imaxf,jminf,jmaxf,imina,imaxa,jmina,jmaxa,
     &  iftop1,iftop2
366  format(' Feed: (i1,i2,j1,j2)= ',4(i4,1x),/,
     &  ' Aperture: (i1,i2,j1,j2)= ',4(i4,1x),/,' (iftop1, iftop2)= ',
     &  2(i4,1x))

**************************************************************
C SECTION FOR SETTING UP SOURCE -  CAPACITOR BANK :
**************************************************************
C User's section for setting up a parallel plate capacitor :

iminfcap = iminf
imaxfcap = iminfcap + 3
jmidcap = (jmaxf + jminf) / 2
jminfcap = jmidcap - 3
jmaxfcap = jmidcap + 3
kmidcap = kfeed - 5
kfeedcap = kmidcap - 3
kaperturecap = kmidcap + 3

C End of user's section for capacitor setup +++++++++++++

C Fill dielectric material in the region between the plates of the
C capacitor.
do 9  i = iminfcap, imaxfcap - 1
      x = dfloat(i) * delx
do 9  j = jminfcap,jmaxfcap
      y = dfloat(j) * dely
```

```
do 9  k = kfeedcap, kaperturecap
      z = dfloat(k) * delz
    call dcube(i,j,k,1,1,1,2)
    write(9,123) 2.0,x,y,z,i,j,k
9  continue


C Upper plate (m)
xmidcapt = dfloat(imaxfcap) * delx

C Lower plate (m)
xmidcapb = dfloat(iminfcap) * delx

C Allocate the Bottom-plate :
xmincapb = xmidcapb - 0.5d0 * delx
xmaxcapb = xmidcapb + 0.5d0 * delx
      i = iminfcap
      x = dfloat(i) * delx
do 10 j = jminfcap,jmaxfcap
      y = dfloat(j) * dely
do 10 k = kfeedcap, kaperturecap
      z = dfloat(k) * delz
    if ( x .ge. xmincapb .and. x .le. xmaxcapb)then
    call dcube(i,j,k,0,1,1,1)
    write(12,123) 1.0,x,y,z,i,j,k
    endif
10  continue
C
C   Allocate the  Top-plate :
xmincapt = xmidcapt - 0.5d0 * delx
xmaxcapt = xmidcapt + 0.5d0 * delx
   i = imaxfcap
      x = dfloat(i) * delx
do 11 j = jminfcap,jmaxfcap
      y = dfloat(j) * dely
do 11 k = kfeedcap, kaperturecap
      z = dfloat(k) * delz
    if ( x .ge. xmincapt .and. x .le. xmaxcapt)then
    call dcube(i,j,k,0,1,1,1)
    write(12,123) 1.0,x,y,z,i,j,k
    endif
11  continue

****************************************************************
C  SECTION FOR SETTING UP A SWITCH :
****************************************************************
C   Allocate Bottom resistive plate :
i = iminfcap
x = dfloat(i) * delx
k = kaperturecap + 1
z = dfloat(k) * delz
do 12 j = jminfcap,jmaxfcap
y = dfloat(j) * dely
call dcube(i,j,k,0,1,1,3)
write(15,123) 3.0,x,y,z,i,j,k
```

```
12  continue

C   Allocate Top resistive plate :
i = imaxfcap
x = dfloat(i) * delx
k = kaperturecap + 1
z = dfloat(k) * delz
do 13 j = jminfcap,jmaxfcap
call dcube(i,j,k,0,1,1,3)
y = dfloat(j) * dely
write(15,123) 3.0,x,y,z,i,j,k
13  continue

C Test volume :
C Top plate :
      i = imaxa
      x = dfloat(i) * delx
do 16 j = jmina, jmaxa
      y = dfloat(j) * dely
do 16 k = kaperture + 1, kaperture + 270
      z = dfloat(k) * delz
        call dcube(i,j,k,1,1,1,1)
        write(12,123) 1.0,x,y,z,i,j,k
16  continue

C Bottom Plate
      i = imina
      x = dfloat(i) * delx
do 17 j = jmina, jmaxa
      y = dfloat(j) * dely
do 17 k = kaperture + 1, kaperture + 270
      z = dfloat(k) * delz
        call dcube(i,j,k,1,1,1,1)
        write(12,123) 1.0,x,y,z,i,j,k
17  continue
C Resistive sheet termination :

k = kaperture + 271
z = dfloat(k) * delz
do 18 i = imina, imaxa
x = dfloat(i) * delx
do 18 j = jmina, jmaxa
y = dfloat(j) * dely
call dcube(i,j,k,1,1,0,5)
write(12,123) 5.0,x,y,z,i,j,k
18  continue

C   Assignment of characteristic time of charging :
C
tauc = 10.0e-9

C      THE FOLLOWING SECTION OF CODE IS USED TO CHECK IF THE USER HAS
C      SPECIFIED THE PROPER MATERIAL TYPES TO THE PROPER IDXXX ARRAYS.
C
C      FOR MATERIAL TYPE = ?
C         0 FOR FREE SPACE                    USE  IDONE-IDTHRE ARRAYS
```

```
C        1 FOR PEC                          USE  IDONE-IDTHRE ARRAYS
C        2-9 FOR LOSSY DIELECTRICS          USE  IDONE-IDTHRE ARRAYS
C
      DO 1000 K=1,NZ
        DO 900 J=1,NY
          DO 800 I=1,NX
            IF((IDONE(I,J,K).GE.10).OR.(IDTWO(I,J,K).GE.10).OR.
     $ (IDTHRE(I,J,K).GE.10)) THEN
               WRITE (17,*)'ERROR OCCURED. ILLEGAL VALUE FOR'
               WRITE (17,*)'DIELECTRIC TYPE (IDONE-IDTHRE) '
               WRITE (17,*)'AT LOCATION:',I,',',J,',',K
               WRITE (17,*)'EXECUTION HALTED.'
               STOP
            ENDIF
 800      CONTINUE
 900    CONTINUE
 1000 CONTINUE
      RETURN
      END

C*********************************************************
      SUBROUTINE DCUBE (ISTART,JSTART,KSTART,NXWIDE,NYWIDE,NZWIDE,MTYPE)
      INCLUDE 'common.h'
C
C     THIS SUBROUTINE SETS ALL TWELVE IDXXX COMPONENTS FOR ONE CUBE
C     TO THE SAME MATERIAL TYPE SPECIFIED BY MTYPE.  IF NXWIDE, NYWIDE,
C     OR NZWIDE=0, THEN ONLY 4 IDXXX ARRAY COMPONENTS WILL BE SET
C     CORRESPONDING TO AN INFINITELY THIN PLATE.  THE SUBROUTINE IS MOST
C     USEFUL CONSTRUCTING OBJECTS WITH MANY CELLS OF THE SAME MATERIAL
C     (I.E. CUBES, PEC PLATES, ETC.).  THIS SUBROUTINE DOES NOT
C     AUTOMATICALLY DO WIRESC
C
      IMAX=ISTART+NXWIDE-1
      JMAX=JSTART+NYWIDE-1
      KMAX=KSTART+NZWIDE-1
C
      IF (NXWIDE.EQ.0) THEN
DO 20 K=KSTART,KMAX
    DO 10 J=JSTART,JMAX
      IDTWO(ISTART,J,K)=MTYPE
      IDTWO(ISTART,J,K+1)=MTYPE
      IDTHRE(ISTART,J,K)=MTYPE
      IDTHRE(ISTART,J+1,K)=MTYPE
 10         CONTINUE
 20       CONTINUE
      ELSEIF (NYWIDE.EQ.0) THEN
DO 40 K=KSTART,KMAX
    DO 30 I=ISTART,IMAX
      IDONE(I,JSTART,K)=MTYPE
      IDONE(I,JSTART,K+1)=MTYPE
      IDTHRE(I,JSTART,K)=MTYPE
      IDTHRE(I+1,JSTART,K)=MTYPE
 30         CONTINUE
 40       CONTINUE
      ELSEIF (NZWIDE.EQ.0) THEN
DO 60 J=JSTART,JMAX
```

```
      DO 50 I=ISTART,IMAX
        IDONE(I,J,KSTART)=MTYPE
        IDONE(I,J+1,KSTART)=MTYPE
        IDTWO(I,J,KSTART)=MTYPE
        IDTWO(I+1,J,KSTART)=MTYPE
 50         CONTINUE
 60       CONTINUE
        ELSE
DO 90 K=KSTART,KMAX
      DO 80 J=JSTART,JMAX
        DO 70 I=ISTART,IMAX
          IDONE(I,J,K)=MTYPE
          IDONE(I,J,K+1)=MTYPE
          IDONE(I,J+1,K+1)=MTYPE
          IDONE(I,J+1,K)=MTYPE
          IDTWO(I,J,K)=MTYPE
          IDTWO(I+1,J,K)=MTYPE
          IDTWO(I+1,J,K+1)=MTYPE
          IDTWO(I,J,K+1)=MTYPE
          IDTHRE(I,J,K)=MTYPE
          IDTHRE(I+1,J,K)=MTYPE
          IDTHRE(I+1,J+1,K)=MTYPE
          IDTHRE(I,J+1,K)=MTYPE
 70           CONTINUE
 80         CONTINUE
 90       CONTINUE
        ENDIF
        RETURN
        END
C*********************************************************
        SUBROUTINE PMLCUBE
        INCLUDE 'common.h'
C
          DO K=1,NZ
          DO J=1,NY
          DO I=1,NX
C
C           PML CORNERS
C
           IF (I.LE.NXB.AND.J.LE.NYB.AND.K.LE.NZB) THEN
              IDPML(I,J,K)=1
           ELSEIF (I.GE.NXS.AND.J.LE.NYB.AND.K.LE.NZB) THEN
              IDPML(I,J,K)=2
           ELSEIF (I.LE.NXB.AND.J.LE.NYB.AND.K.GE.NZS) THEN
              IDPML(I,J,K)=3
           ELSEIF (I.GE.NXS.AND.J.LE.NYB.AND.K.GE.NZS) THEN
              IDPML(I,J,K)=4
           ELSEIF (I.LE.NXB.AND.J.GE.NYS.AND.K.LE.NZB) THEN
              IDPML(I,J,K)=5
           ELSEIF (I.GE.NXS.AND.J.GE.NYS.AND.K.LE.NZB) THEN
              IDPML(I,J,K)=6
           ELSEIF (I.LE.NXB.AND.J.GE.NYS.AND.K.GE.NZS) THEN
              IDPML(I,J,K)=7
           ELSEIF (I.GE.NXS.AND.J.GE.NYS.AND.K.GE.NZS) THEN
              IDPML(I,J,K)=8
C
```

```
C         PML EDGES
C
       ELSEIF (I.LE.NXB.AND.J.GT.NYB.AND.J.LT.NYS.AND.K.LE.NZB) THEN
          IDPML(I,J,K)=9
       ELSEIF (I.GE.NXS.AND.J.GT.NYB.AND.J.LT.NYS.AND.K.LE.NZB) THEN
          IDPML(I,J,K)=10
       ELSEIF (I.LE.NXB.AND.J.GT.NYB.AND.J.LT.NYS.AND.K.GE.NZS) THEN
          IDPML(I,J,K)=11
       ELSEIF (I.GE.NXS.AND.J.GT.NYB.AND.J.LT.NYS.AND.K.GE.NZS) THEN
          IDPML(I,J,K)=12
C
       ELSEIF (I.GT.NXB.AND.I.LT.NXS.AND.J.GE.NYS.AND.K.LE.NZB) THEN
          IDPML(I,J,K)=13
       ELSEIF (I.GT.NXB.AND.I.LT.NXS.AND.J.GE.NYS.AND.K.GE.NZS) THEN
          IDPML(I,J,K)=14
       ELSEIF (I.GT.NXB.AND.I.LT.NXS.AND.J.LE.NYB.AND.K.LE.NZB) THEN
          IDPML(I,J,K)=15
       ELSEIF (I.GT.NXB.AND.I.LT.NXS.AND.J.LE.NYB.AND.K.GE.NZS) THEN
          IDPML(I,J,K)=16
C
       ELSEIF (I.LE.NXB.AND.J.GE.NYS.AND.K.GT.NZB.AND.K.LT.NZS) THEN
          IDPML(I,J,K)=17
       ELSEIF (I.GE.NXS.AND.J.GE.NYS.AND.K.GT.NZB.AND.K.LT.NZS) THEN
          IDPML(I,J,K)=18
       ELSEIF (I.LE.NXB.AND.J.LE.NYB.AND.K.GT.NZB.AND.K.LT.NZS) THEN
          IDPML(I,J,K)=19
       ELSEIF (I.GE.NXS.AND.J.LE.NYB.AND.K.GT.NZB.AND.K.LT.NZS) THEN
          IDPML(I,J,K)=20
C
C         PML FACES
C
       ELSEIF (I.GT.NXB.AND.I.LT.NXS.AND.J.GT.NYB.AND.J.LT.NYS
     $                                            .AND.K.LE.NZB) THEN
        IDPML(I,J,K)=21
       ELSEIF (I.GT.NXB.AND.I.LT.NXS.AND.J.GT.NYB.AND.J.LT.NYS
     $                                            .AND.K.GE.NZS) THEN
        IDPML(I,J,K)=22
       ELSEIF (I.LE.NXB.AND.J.GT.NYB.AND.J.LT.NYS
     $                              .AND.K.GT.NZB.AND.K.LT.NZS) THEN
        IDPML(I,J,K)=23
       ELSEIF (I.GE.NXS.AND.J.GT.NYB.AND.J.LT.NYS
     $                              .AND.K.GT.NZB.AND.K.LT.NZS) THEN
        IDPML(I,J,K)=24
       ELSEIF (I.GT.NXB.AND.I.LT.NXS.AND.J.LE.NYB
     $                              .AND.K.GT.NZB.AND.K.LT.NZS) THEN
        IDPML(I,J,K)=25
       ELSEIF (I.GT.NXB.AND.I.LT.NXS.AND.J.GE.NYS
     $                              .AND.K.GT.NZB.AND.K.LT.NZS) THEN
        IDPML(I,J,K)=26
       ELSE
         IDPML(I,J,K)=27
       ENDIF
      ENDDO
      ENDDO
      ENDDO

    RETURN
    END
```

```
C******************************************************
         SUBROUTINE SETUP
         INCLUDE 'common.h'

Real*8 tauc,tau0,rho3,rho5,amplitude
Integer iminfcap,imaxfcap,jminfcap,jmaxfcap,kfeedcap,
     &    kaperturecap

Common/ comswitch/ tauc
Common/ comdischarg/ tau0
Common / comcapacloc / iminfcap,imaxfcap,jminfcap,jmaxfcap,
     &    kfeedcap,kaperturecap

C
      C=1.0/SQRT(XMU0*EPS0)
C
      DTXI=C/DELX
      DTYI=C/DELY
      DTZI=C/DELZ
C
      DT=1.0/sqrt(DTXI**2+DTYI**2+DTZI**2)


C ******* CONSTITUTIVE PARAMETERS ************
      DO 10 I=1,9
EPS(I)=EPS0
SIGMA(I)=0.0
EPSP(I)=EPS0
 10   CONTINUE

C DEFINE EPS AND SIGMA FOR EACH MATERIAL HERE

C Material properties of dielectric filled inside the two plates
C of the capacitor.
EPS(2) = 14.5 * EPS0
SIGMA(2) = 0.0
EPSP(3) = EPS(2)

C Material properties of switch :

C Characteristic time for switch closing = taup, which is in this case
C is the characteristic time for capacitor charging

taup = 7.0e-9

if (t .ge. tau0) then
    aa = (4.0d0 / taup)**2
    if (t .ge. (tau0 + taup)) then
      amplitude = 1.0d0
    else
      amplitude = exp(-aa * (t - (tau0+taup))**2)
    endif

    rho3 = 2.0d0 * dfloat(jmaxfcap - jminfcap) * dely * delx
    rho3 = rho3 / delz
            sigma(3) = amplitude / rho3
```

```
else
    sigma(3) = 0.0d0
endif
    EPS(3) = EPS0
    EPSP(4) = EPS(3)

C Material properties of SF6 :
EPS(4) = 1.0623 * EPS0
SIGMA(4) = 1.0d-7
EPSP(5) = EPS(4)

C Resistive sheet termination of 90 ohm.
         rho5 = 90.0d0 * dfloat(jmaxa - jmina) * dely * delz
         rho5  = rho5 / dfloat(imaxa - imina) / delx
   sigma(5) = 1.0d0 / rho5
   EPS(5)=EPS0
   EPSP(6) = EPS(5)

write(19,*)'[tauc,taup,sigma(3),sigma(4),sigma(5)]=',tauc,taup,
     &  sigma(3), sigma(4), sigma(5)

C      FREE SPACE
C
      DTEDX=DT/(EPS0*DELX)
      DTEDY=DT/(EPS0*DELY)
      DTEDZ=DT/(EPS0*DELZ)
      DTMDX=DT/(XMU0*DELX)
      DTMDY=DT/(XMU0*DELY)
      DTMDZ=DT/(XMU0*DELZ)
C
C      LOSSY DIELECTRICS
C
      DO 20 I=2,9
ESB(I)=(2.*EPS(I)-SIGMA(I)*DT)/(2.*EPS(I)+SIGMA(I)  *DT)
ECRLX(I)=2.*DT/((2.*EPS(I)+SIGMA(I)*DT)*DELX)
ECRLY(I)=2.*DT/((2.*EPS(I)+SIGMA(I)*DT)*DELY)
ECRLZ(I)=2.*DT/((2.*EPS(I)+SIGMA(I)*DT)*DELZ)
 20   CONTINUE
      RETURN
      END
C     ****************************************************
      SUBROUTINE PMLSETUP
      INCLUDE 'common.h'

       SIGMAXE(1)=ESIGMAM
       SIGMAXE(2)=ESIGMAM
       SIGMAXE(3)=ESIGMAM
       SIGMAXE(4)=ESIGMAM
       SIGMAXE(5)=ESIGMAM
       SIGMAXE(6)=ESIGMAM
       SIGMAXE(7)=ESIGMAM
       SIGMAXE(8)=ESIGMAM
       SIGMAXE(9)=ESIGMAM
       SIGMAXE(10)=ESIGMAM
       SIGMAXE(11)=ESIGMAM
       SIGMAXE(12)=ESIGMAM
```

```
C
      SIGMAXE(13)=0.
      SIGMAXE(14)=0.
      SIGMAXE(15)=0.
      SIGMAXE(16)=0.
C
      SIGMAXE(17)=ESIGMAM
      SIGMAXE(18)=ESIGMAM
      SIGMAXE(19)=ESIGMAM
      SIGMAXE(20)=ESIGMAM
C
      SIGMAXE(21)=0.
      SIGMAXE(22)=0.
      SIGMAXE(23)=ESIGMAM
      SIGMAXE(24)=ESIGMAM
      SIGMAXE(25)=0.
      SIGMAXE(26)=0.
      SIGMAXE(27)=0.

C
      SIGMAYE(1)=ESIGMAM
      SIGMAYE(2)=ESIGMAM
      SIGMAYE(3)=ESIGMAM
      SIGMAYE(4)=ESIGMAM
      SIGMAYE(5)=ESIGMAM
      SIGMAYE(6)=ESIGMAM
      SIGMAYE(7)=ESIGMAM
      SIGMAYE(8)=ESIGMAM
C
      SIGMAYE(9)=0.
      SIGMAYE(10)=0.
      SIGMAYE(11)=0.
      SIGMAYE(12)=0.
C
      SIGMAYE(13)=ESIGMAM
      SIGMAYE(14)=ESIGMAM
      SIGMAYE(15)=ESIGMAM
      SIGMAYE(16)=ESIGMAM
C
      SIGMAYE(17)=ESIGMAM
      SIGMAYE(18)=ESIGMAM
      SIGMAYE(19)=ESIGMAM
      SIGMAYE(20)=ESIGMAM
C
      SIGMAYE(21)=0.
      SIGMAYE(22)=0.
      SIGMAYE(23)=0.
      SIGMAYE(24)=0.
      SIGMAYE(25)=ESIGMAM
      SIGMAYE(26)=ESIGMAM
      SIGMAYE(27)=0.

C
      SIGMAZE(1)=ESIGMAM
      SIGMAZE(2)=ESIGMAM
      SIGMAZE(3)=ESIGMAM
```

```
      SIGMAZE(4)=ESIGMAM
      SIGMAZE(5)=ESIGMAM
      SIGMAZE(6)=ESIGMAM
      SIGMAZE(7)=ESIGMAM
      SIGMAZE(8)=ESIGMAM
C
      SIGMAZE(9)=ESIGMAM
      SIGMAZE(10)=ESIGMAM
      SIGMAZE(11)=ESIGMAM
      SIGMAZE(12)=ESIGMAM
C
      SIGMAZE(13)=ESIGMAM
      SIGMAZE(14)=ESIGMAM
      SIGMAZE(15)=ESIGMAM
      SIGMAZE(16)=ESIGMAM
C
      SIGMAZE(17)=0.
      SIGMAZE(18)=0.
      SIGMAZE(19)=0.
      SIGMAZE(20)=0.
C
      SIGMAZE(21)=ESIGMAM
      SIGMAZE(22)=ESIGMAM
      SIGMAZE(23)=0.
      SIGMAZE(24)=0.
      SIGMAZE(25)=0.
      SIGMAZE(26)=0.
      SIGMAZE(27)=0.
C
      SIGMAXH(1)=HSIGMAM
      SIGMAXH(2)=HSIGMAM
      SIGMAXH(3)=HSIGMAM
      SIGMAXH(4)=HSIGMAM
      SIGMAXH(5)=HSIGMAM
      SIGMAXH(6)=HSIGMAM
      SIGMAXH(7)=HSIGMAM
      SIGMAXH(8)=HSIGMAM
C
      SIGMAXH(9)=HSIGMAM
      SIGMAXH(10)=HSIGMAM
      SIGMAXH(11)=HSIGMAM
      SIGMAXH(12)=HSIGMAM
C
      SIGMAXH(13)=0.
      SIGMAXH(14)=0.
      SIGMAXH(15)=0.
      SIGMAXH(16)=0.
C
      SIGMAXH(17)=HSIGMAM
      SIGMAXH(18)=HSIGMAM
      SIGMAXH(19)=HSIGMAM
      SIGMAXH(20)=HSIGMAM
C
      SIGMAXH(21)=0.
      SIGMAXH(22)=0.
      SIGMAXH(23)=HSIGMAM
```

```
      SIGMAXH(24)=HSIGMAM
      SIGMAXH(25)=0.
      SIGMAXH(26)=0.
      SIGMAXH(27)=0.
C
      SIGMAYH(1)=HSIGMAM
      SIGMAYH(2)=HSIGMAM
      SIGMAYH(3)=HSIGMAM
      SIGMAYH(4)=HSIGMAM
      SIGMAYH(5)=HSIGMAM
      SIGMAYH(6)=HSIGMAM
      SIGMAYH(7)=HSIGMAM
      SIGMAYH(8)=HSIGMAM
C
      SIGMAYH(9)=0.
      SIGMAYH(10)=0.
      SIGMAYH(11)=0.
      SIGMAYH(12)=0.
C
      SIGMAYH(13)=HSIGMAM
      SIGMAYH(14)=HSIGMAM
      SIGMAYH(15)=HSIGMAM
      SIGMAYH(16)=HSIGMAM
C
      SIGMAYH(17)=HSIGMAM
      SIGMAYH(18)=HSIGMAM
      SIGMAYH(19)=HSIGMAM
      SIGMAYH(20)=HSIGMAM
C
      SIGMAYH(21)=0.
      SIGMAYH(22)=0.
      SIGMAYH(23)=0.
      SIGMAYH(24)=0.
      SIGMAYH(25)=HSIGMAM
      SIGMAYH(26)=HSIGMAM
      SIGMAYH(27)=0.
C
      SIGMAZH(1)=HSIGMAM
      SIGMAZH(2)=HSIGMAM
      SIGMAZH(3)=HSIGMAM
      SIGMAZH(4)=HSIGMAM
      SIGMAZH(5)=HSIGMAM
      SIGMAZH(6)=HSIGMAM
      SIGMAZH(7)=HSIGMAM
      SIGMAZH(8)=HSIGMAM
C
      SIGMAZH(9)=HSIGMAM
      SIGMAZH(10)=HSIGMAM
      SIGMAZH(11)=HSIGMAM
      SIGMAZH(12)=HSIGMAM
C
      SIGMAZH(13)=HSIGMAM
      SIGMAZH(14)=HSIGMAM
      SIGMAZH(15)=HSIGMAM
      SIGMAZH(16)=HSIGMAM
```

```
C
        SIGMAZH(17)=0.
        SIGMAZH(18)=0.
        SIGMAZH(19)=0.
        SIGMAZH(20)=0.
C
        SIGMAZH(21)=HSIGMAM
        SIGMAZH(22)=HSIGMAM
        SIGMAZH(23)=0.
        SIGMAZH(24)=0.
        SIGMAZH(25)=0.
        SIGMAZH(26)=0.
        SIGMAZH(27)=0.

      RETURN
      END
C     ******************************************************

        SUBROUTINE EXFLD
        INCLUDE 'common.h'

Real*8 tauc,aa
Integer iminfcap,imaxfcap,jminfcap,jmaxfcap,kfeedcap,
     &    kaperturecap
Common / comcapacloc / iminfcap,imaxfcap,jminfcap,jmaxfcap,
     &           kfeedcap,kaperturecap
Common/ comswitch/ tauc

C
      DO 30 K=2,NZ1
        KK=K
DO 20 J=2,NY1
         JJ=J
    DO 10 I=1,NX1
         II=I
C
C        FDTD REGION
C
         IF (I.GT.NXB.AND.I.LT.NXS.AND.J.GT.NYBP1.AND.J.LT.NYS.
     $                          AND.K.GT.NZBP1.AND.K.LT.NZS) THEN

C           DETERMINE MATERIAL TYPE
      IF(IDONE(I,J,K).EQ.0) GO TO 100
      IF(IDONE(I,J,K).EQ.1) GO TO 200
      GO TO 300

C           FREE SPACE
 100        EX(I,J,K)=EX(I,J,K)+(HZ(I,J,K)-HZ(I,J-1,K))*DTEDY
     $                          -(HY(I,J,K)-HY(I,J,K-1))*DTEDZ
      GO TO 10

C           PERFECT CONDUCTOR
 200        EX(I,J,K)=0.0
      GO TO 10
```

```
C          LOSSY DIELECTRIC
 300       EX(I,J,K)=EX(I,J,K)*ESB(IDONE(I,J,K))
     $     +(HZ(I,J,K)-HZ(I,J-1,K))*ECRLY(IDONE(I,J,K))
     $     -(HY(I,J,K)-HY(I,J,K-1))*ECRLZ(IDONE(I,J,K))

          ELSE
C
C         PML REGION
C
            CALL PMLEX(II,JJ,KK)

          ENDIF
 10       CONTINUE
 20     CONTINUE
 30   CONTINUE

C Apply Source :--->

if (t .le. tauc)then
     aa = (4.0d0 / tauc)**2

do 111 i = iminfcap, imaxfcap - 1
        do 111 j = jminfcap, jmaxfcap
        do 111 k = kfeedcap, kaperturecap
111         EX(i,j,k) = 18.043e6 * EXP(-aa*(t- tauc)**2)
ENDIF

      RETURN
      END
C     *******************************************************
      SUBROUTINE EYFLD
      INCLUDE 'common.h'
C      REAL SOURCE
C
      DO 30 K=2,NZ1
        KK=K
DO 20 J=1,NY1
          JJ=J
    DO 10 I=2,NX1
          II=I
C
C         FDTD REGION
C
          IF (I.GT.NXBP1.AND.I.LT.NXS.AND.J.GT.NYB.AND.J.LT.NYS.
     $                          AND.K.GT.NZBP1.AND.K.LT.NZS) THEN


C           DETERMINE MATERIAL TYPE
      IF(IDTWO(I,J,K).EQ.0) GO TO 100
      IF(IDTWO(I,J,K).EQ.1) GO TO 200
      GO TO 300

C           FREE SPACE
 100        EY(I,J,K)=EY(I,J,K)+(HX(I,J,K)-HX(I,J,K-1))*DTEDZ
     $                       -(HZ(I,J,K)-HZ(I-1,J,K))*DTEDX
C     $                       +SOURCE(I,J,K)
      GO TO 10
```

```
C           PERFECT CONDUCTOR
 200        EY(I,J,K)=0.0
      GO TO 10

C           LOSSY DIELECTRIC
 300        EY(I,J,K)=EY(I,J,K)*ESB(IDTWO(I,J,K))
     $      +(HX(I,J,K)-HX(I,J,K-1))*ECRLZ(IDTWO(I,J,K))
     $      -(HZ(I,J,K)-HZ(I-1,J,K))*ECRLX(IDTWO(I,J,K))
C    $       +SOURCE(I,J,K)

          ELSE
C
C         PML REGION
C
            CALL PMLEY(II,JJ,KK)
    ENDIF

  10      CONTINUE
  20    CONTINUE
  30  CONTINUE


      RETURN
      END
C     ***************************************************
      SUBROUTINE EZFLD
      INCLUDE 'common.h'
C
      DO 30 K=1,NZ1
        KK=K
DO 20 J=2,NY1
          JJ=J
    DO 10 I=2,NX1
          II=I
C
C         FDTD REGION
C
          IF (I.GT.NXBP1.AND.I.LT.NXS.AND.J.GT.NYBP1.AND.J.LT.NYS.
     $                          AND.K.GT.NZB.AND.K.LT.NZS) THEN


C           DETERMINE MATERIAL TYPE
      IF(IDTHRE(I,J,K).EQ.0) GO TO 100
      IF(IDTHRE(I,J,K).EQ.1) GO TO 200
      GO TO 300
C
C           FREE SPACE
 100        EZ(I,J,K)=EZ(I,J,K)+(HY(I,J,K)-HY(I-1,J,K))*DTEDX
     $                          -(HX(I,J,K)-HX(I,J-1,K))*DTEDY
      GO TO 10

C           PERFECT CONDUCTOR
 200        EZ(I,J,K)=0.0
      GO TO 10
```

```
C           LOSSY DIELECTRIC
 300        EZ(I,J,K)=EZ(I,J,K)*ESB(IDTHRE(I,J,K))
     $      +(HY(I,J,K)-HY(I-1,J,K))*ECRLX(IDTHRE(I,J,K))
     $      -(HX(I,J,K)-HX(I,J-1,K))*ECRLY(IDTHRE(I,J,K))

          ELSE
C
C         PML REGION
C
            CALL PMLEZ(II,JJ,KK)

          ENDIF
 10       CONTINUE
 20     CONTINUE
 30   CONTINUE
C
      RETURN
      END

C     ****************************************************************
      SUBROUTINE HXFLD
      INCLUDE 'common.h'
C
      DO 30 K=1,NZ1
        KK=K
DO 20 J=1,NY1
          JJ=J
    DO 10 I=2,NX1
          II=I
C
C         FDTD REGION
C
          IF (I.GT.NXB.AND.I.LT.NXS.AND.J.GT.NYB.AND.J.LT.NYSM1.
     $                            AND.K.GT.NZB.AND.K.LT.NZSM1) THEN

      HX(I,J,K)=HX(I,J,K)-(EZ(I,J+1,K)-EZ(I,J,K))*DTMDY
     $                        +(EY(I,J,K+1)-EY(I,J,K))*DTMDZ
          ELSE
C
C         PML REGION
C
            CALL PMLHX(II,JJ,KK)

    ENDIF
 10       CONTINUE
 20     CONTINUE
 30   CONTINUE
C
      RETURN
      END
C     *******************************************************
      SUBROUTINE HYFLD
      INCLUDE 'common.h'
C
      DO 30 K=1,NZ1
        KK=K
```

```
DO 20 J=2,NY1
        JJ=J
    DO 10 I=1,NX1
          II=I
C
C         FDTD REGION
C
        IF (I.GT.NXB.AND.I.LT.NXSM1.AND.J.GT.NYB.AND.J.LT.NYS.
     $                          AND.K.GT.NZB.AND.K.LT.NZSM1) THEN

      HY(I,J,K)=HY(I,J,K)-(EX(I,J,K+1)-EX(I,J,K))*DTMDZ
     $                          +(EZ(I+1,J,K)-EZ(I,J,K))*DTMDX
        ELSE
C
C         PML REGION
C
          CALL PMLHY(II,JJ,KK)

        ENDIF
 10       CONTINUE
 20     CONTINUE
 30   CONTINUE
C
      RETURN
      END
C     ******************************************************
      SUBROUTINE HZFLD
      INCLUDE 'common.h'
C
      DO 30 K=2,NZ1
        KK=K
DO 20 J=1,NY1
        JJ=J
    DO 10 I=1,NX1
          II=I
C
C         FDTD REGION
C
        IF (I.GT.NXB.AND.I.LT.NXSM1.AND.J.GT.NYB.AND.J.LT.NYSM1.
     $                          AND.K.GT.NZB.AND.K.LT.NZS) THEN

      HZ(I,J,K)=HZ(I,J,K)-(EY(I+1,J,K)-EY(I,J,K))*DTMDX
     $                          +(EX(I,J+1,K)-EX(I,J,K))*DTMDY
        ELSE
C
C         PML REGION
C
          CALL PMLHZ(II,JJ,KK)

    ENDIF
 10       CONTINUE
 20     CONTINUE
 30   CONTINUE
C
      RETURN
      END
```

```
C       ****************************************************************
        SUBROUTINE ZERO
        INCLUDE 'common.h'
C
        T=0.0
        DO 30 K=1,NZ
DO 20 J=1,NY
      DO 10 I=1,NX
        EX(I,J,K)=0.0
        EY(I,J,K)=0.0
        EZ(I,J,K)=0.0
        HX(I,J,K)=0.0
        HY(I,J,K)=0.0
        HZ(I,J,K)=0.0
        EXY(I,J,K)=0.0
        EYX(I,J,K)=0.0
        EZX(I,J,K)=0.0
        EXZ(I,J,K)=0.0
        EYZ(I,J,K)=0.0
        EZY(I,J,K)=0.0
        HXY(I,J,K)=0.0
        HYX(I,J,K)=0.0
        HZX(I,J,K)=0.0
        HXZ(I,J,K)=0.0
        HYZ(I,J,K)=0.0
        HZY(I,J,K)=0.0
        IDONE(I,J,K)=0.0
        IDTWO(I,J,K)=0.0
        IDTHRE(I,J,K)=0.0
 10         CONTINUE
 20       CONTINUE
 30     CONTINUE


        RETURN
        END
C       *********************************************************
        SUBROUTINE PMLEX(I,J,K)
        INCLUDE 'common.h'

        XJ=1*J
        XK=1*K
        EPSF=EPSP(IDONE(I,J,K)+1)/EPS0
        IF(J.LE.NYB) THEN
          ESIGMAY=SIGMAYE(IDPML(I,J,K))*EPSF*((XNPML-XJ+1.)/XNPML)**2.
        ELSE
          ESIGMAY=SIGMAYE(IDPML(I,J,K))*EPSF*((XJ-YN+XNPML)/XNPML)**2.
        ENDIF

        IF(K.LE.NZB) THEN
          ESIGMAZ=SIGMAZE(IDPML(I,J,K))*EPSF*((XNPML-XK+1.)/XNPML)**2.
        ELSE
          ESIGMAZ=SIGMAZE(IDPML(I,J,K))*EPSF*((XK-ZN+XNPML)/XNPML)**2.
        ENDIF
```

```
      IF (ESIGMAY.LE.0.) THEN
       DTESY=1.
       DTESDY=DT/(DELY*EPSP(IDONE(I,J,K)+1))
      ELSE
       DTESY=EXP(-(ESIGMAY*DT)/EPSP(IDONE(I,J,K)+1))
       DTESDY=(1.-DTESY)/(ESIGMAY*DELY)
      ENDIF

      IF (ESIGMAZ.LE.0.) THEN
       DTESZ=1.
       DTESDZ=DT/(DELZ*EPSP(IDONE(I,J,K)+1))
      ELSE
       DTESZ=EXP(-(ESIGMAZ*DT)/EPSP(IDONE(I,J,K)+1))
       DTESDZ=(1.-DTESZ)/(ESIGMAZ*DELZ)
      ENDIF
C
C         PML INTERFACE
C
          IF (I.GT.NXB.AND.I.LT.NXS.AND.J.EQ.NYBP1
     $                               .AND.K.GT.NZBP1.AND.K.LT.NZS) THEN

            IF (IDONE(I,J,K).EQ.0) THEN

             EX(I,J,K)=EX(I,J,K)+(HZ(I,J,K)
     $               -HZX(I,J-1,K)-HZY(I,J-1,K))*DTEDY
     $               -(HY(I,J,K)-HY(I,J,K-1))*DTEDZ
      ELSEIF (IDONE(I,J,K).EQ.1) THEN
       EX(I,J,K)=0.0
      ELSE
       EX(I,J,K)=EX(I,J,K)*ESB(IDONE(I,J,K))
     $       +(HZ(I,J,K)-HZX(I,J-1,K)-HZY(I,J-1,K))
     $       *ECRLY(IDONE(I,J,K))
     $       -(HY(I,J,K)-HY(I,J,K-1))*ECRLZ(IDONE(I,J,K))

      ENDIF
     ELSEIF (I.GT.NXB.AND.I.LT.NXS.AND.J.GT.NYBP1.AND.J.LT.NYS
     $                                          .AND.K.EQ.NZBP1) THEN

      IF (IDONE(I,J,K).EQ.0) THEN

       EX(I,J,K)=EX(I,J,K)+(HZ(I,J,K)-HZ(I,J-1,K))*DTEDY
     $               -(HY(I,J,K)-HYX(I,J,K-1)-HYZ(I,J,K-1))*DTEDZ
      ELSEIF (IDONE(I,J,K).EQ.1) THEN
       EX(I,J,K)=0.0
      ELSE
       EX(I,J,K)=EX(I,J,K)*ESB(IDONE(I,J,K))
     $       +(HZ(I,J,K)-HZ(I,J-1,K))
     $       *ECRLY(IDONE(I,J,K))
     $       -(HY(I,J,K)-HYX(I,J,K-1)-HYZ(I,J,K-1))*ECRLZ(IDONE(I,J,K))
      ENDIF


     ELSEIF (I.GT.NXB.AND.I.LT.NXS.AND.J.EQ.NYS
     $                               .AND.K.GT.NZB.AND.K.LT.NZS) THEN

      IF(IDONE(I,J,K).EQ.1) THEN
       EXY(I,J,K)=0.0
```

```
          EXZ(I,J,K)=0.0
         ELSE
          EXY(I,J,K)=EXY(I,J,K)*DTESY+(HZX(I,J,K)+HZY(I,J,K)
     $                                -HZ(I,J-1,K))*DTESDY
          EXZ(I,J,K)=EXZ(I,J,K)*DTESZ-(HYX(I,J,K)+HYZ(I,J,K)
     $                                -HYX(I,J,K-1)-HYZ(I,J,K-1))*DTESDZ
         ENDIF

        ELSEIF (I.GT.NXB.AND.I.LT.NXS.AND.J.GT.NYB.AND.J.LT.NYS
     $                                             .AND.K.EQ.NZS) THEN
         IF(IDONE(I,J,K).EQ.1) THEN
          EXY(I,J,K)=0.0
          EXZ(I,J,K)=0.0

         ELSE

          EXZ(I,J,K)=EXZ(I,J,K)*DTESZ-(HYX(I,J,K)+HYZ(I,J,K)
     $                                -HY(I,J,K-1))*DTESDZ
          EXY(I,J,K)=EXY(I,J,K)*DTESY+(HZX(I,J,K)+HZY(I,J,K)
     $                                -HZX(I,J-1,K)-HZY(I,J-1,K))*DTESDY
         ENDIF

        ELSEIF (I.GT.NXB.AND.I.LT.NXS.AND.J.EQ.NYBP1
     $                                   .AND.K.EQ.NZBP1) THEN
          IF (IDONE(I,J,K).EQ.0) THEN

           EX(I,J,K)=EX(I,J,K)+(HZ(I,J,K)
     $         -HZX(I,J-1,K)-HZY(I,J-1,K))*DTEDY
     $        -(HY(I,J,K)-HYX(I,J,K-1)-HYZ(I,J,K-1))*DTEDZ

          ELSEIF (IDONE(I,J,K).EQ.1) THEN
           EX(I,J,K)=0.0

          ELSE
           EX(I,J,K)=EX(I,J,K)*ESB(IDONE(I,J,K))
     $         +(HZ(I,J,K)-HZX(I,J-1,K)-HZY(I,J-1,K))
     $         *ECRLY(IDONE(I,J,K))
     $         -(HY(I,J,K)-HYX(I,J,K-1)-HYZ(I,J,K-1))*ECRLZ(IDONE(I,J,K))

          ENDIF
C
C           PML
C
            ELSE

              IF(IDONE(I,J,K).EQ.1) THEN
                EXY(I,J,K)=0.0
                EXZ(I,J,K)=0.0

              ELSE
                EXY(I,J,K)=EXY(I,J,K)*DTESY+(HZX(I,J,K)+HZY(I,J,K)
     $                -HZX(I,J-1,K)-HZY(I,J-1,K))*DTESDY
                EXZ(I,J,K)=EXZ(I,J,K)*DTESZ-(HYX(I,J,K)+HYZ(I,J,K)
     $                -HYX(I,J,K-1)-HYZ(I,J,K-1))*DTESDZ

              ENDIF
```

```
          ENDIF
      RETURN
      END
C     *******************************************************
      SUBROUTINE PMLEY(I,J,K)
      INCLUDE 'common.h'
C      REAL SOURCE

      XI=1*I
      XK=1*K

      EPSF=EPSP(IDTWO(I,J,K)+1)/EPS0

      IF(I.LE.NXB) THEN
        ESIGMAX=SIGMAXE(IDPML(I,J,K))*EPSF*((XNPML-XI+1.)/XNPML)**2.
      ELSE
        ESIGMAX=SIGMAXE(IDPML(I,J,K))*EPSF*((XI-XN+XNPML)/XNPML)**2.
      ENDIF
      IF(K.LE.NZB) THEN
        ESIGMAZ=SIGMAZE(IDPML(I,J,K))*EPSF*((XNPML-XK+1.)/XNPML)**2.
      ELSE
        ESIGMAZ=SIGMAZE(IDPML(I,J,K))*EPSF*((XK-ZN+XNPML)/XNPML)**2.
      ENDIF

      IF (ESIGMAZ.LE.0.) THEN
       DTESZ=1.
       DTESDZ=DT/(DELZ*EPSP(IDTWO(I,J,K)+1))
      ELSE
       DTESZ=EXP(-(ESIGMAZ*DT)/EPSP(IDTWO(I,J,K)+1))
       DTESDZ=(1.-DTESZ)/(ESIGMAZ*DELZ)
      ENDIF

      IF (ESIGMAX.LE.0.) THEN
       DTESX=1.
       DTESDX=DT/(DELX*EPSP(IDTWO(I,J,K)+1))
      ELSE
       DTESX=EXP(-(ESIGMAX*DT)/EPSP(IDTWO(I,J,K)+1))
       DTESDX=(1.-DTESX)/(ESIGMAX*DELX)
      ENDIF
C
C          PML INTERFACE
C
          IF (I.EQ.NXBP1.AND.J.GT.NYB.AND.J.LT.NYS
     $                                .AND.K.GT.NZBP1.AND.K.LT.NZS) THEN

      IF (IDTWO(I,J,K).EQ.0) THEN
       EY(I,J,K)=EY(I,J,K)+(HX(I,J,K)-HX(I,J,K-1))*DTEDZ
     $       -(HZ(I,J,K)-HZX(I-1,J,K)-HZY(I-1,J,K))*DTEDX
        C+SOURCE(I,J,K)
      ELSEIF (IDTWO(I,J,K).EQ.1) THEN
       EY(I,J,K)=0.0
      ELSE

      EY(I,J,K)=EY(I,J,K)*ESB(IDTWO(I,J,K))
     $       +(HX(I,J,K)-HX(I,J,K-1))*ECRLZ(IDTWO(I,J,K))
```

```
     $       -(HZ(I,J,K)-HZX(I-1,J,K)-HZY(I-1,J,K))*ECRLX(IDTWO(I,J,K))
C     $       +SOURCE(I,J,K)
      ENDIF


     ELSEIF (I.GT.NXBP1.AND.I.LT.NXS.AND.J.GT.NYB.AND.J.LT.NYS
     $                                          .AND.K.EQ.NZBP1) THEN


      IF (IDTWO(I,J,K).EQ.0) THEN
       EY(I,J,K)=EY(I,J,K)+(HX(I,J,K)
     $             -HXY(I,J,K-1)-HXZ(I,J,K-1))*DTEDZ
     $            -(HZ(I,J,K)-HZ(I-1,J,K))*DTEDX   C+SOURCE(I,J,K)
      ELSEIF (IDTWO(I,J,K).EQ.1) THEN
       EY(I,J,K)=0.0
      ELSE

      EY(I,J,K)=EY(I,J,K)*ESB(IDTWO(I,J,K))
     $       +(HX(I,J,K)-HXY(I,J,K-1)-HXZ(I,J,K-1))*ECRLZ(IDTWO(I,J,K))
     $       -(HZ(I,J,K)-HZ(I-1,J,K))*ECRLX(IDTWO(I,J,K))
C     $       +SOURCE(I,J,K)
      ENDIF


     ELSEIF (I.EQ.NXS.AND.J.GT.NYB.AND.J.LT.NYS
     $                                 .AND.K.GT.NZB.AND.K.LT.NZS) THEN
     IF(IDTWO(I,J,K).EQ.1) THEN
      EYX(I,J,K)=0.0
      EYZ(I,J,K)=0.0

     ELSE

      EYX(I,J,K)=EYX(I,J,K)*DTESX-(HZX(I,J,K)+HZY(I,J,K)
     $                              -HZ(I-1,J,K))*DTESDX
      EYZ(I,J,K)=EYZ(I,J,K)*DTESZ+(HXY(I,J,K)+HXZ(I,J,K)
     $                              -HXY(I,J,K-1)-HXZ(I,J,K-1))*DTESDZ
     ENDIF

     ELSEIF (I.GT.NXB.AND.I.LT.NXS.AND.J.GT.NYB.AND.J.LT.NYS
     $                                          .AND.K.EQ.NZS) THEN

     IF(IDTWO(I,J,K).EQ.1) THEN
      EYX(I,J,K)=0.0
      EYZ(I,J,K)=0.0

     ELSE

      EYZ(I,J,K)=EYZ(I,J,K)*DTESZ+(HXY(I,J,K)+HXZ(I,J,K)
     $                              -HX(I,J,K-1))*DTESDZ
      EYX(I,J,K)=EYX(I,J,K)*DTESX-(HZX(I,J,K)+HZY(I,J,K)
     $                              -HZX(I-1,J,K)-HZY(I-1,J,K))*DTESDX
     ENDIF

     ELSEIF (I.EQ.NXBP1.AND.J.GT.NYB.AND.J.LT.NYS
     $                                          .AND.K.EQ.NZBP1) THEN
```

```
      IF (IDTWO(I,J,K).EQ.0) THEN
       EY(I,J,K)=EY(I,J,K)+(HX(I,J,K)
     $      -HXY(I,J,K-1)-HXZ(I,J,K-1))*DTEDZ
     $      -(HZ(I,J,K)-HZX(I-1,J,K)-HZY(I-1,J,K))*DTEDX
C     $      +SOURCE(I,J,K)

      ELSEIF (IDTWO(I,J,K).EQ.1) THEN
       EY(I,J,K)=0.0
      ELSE

      EY(I,J,K)=EY(I,J,K)*ESB(IDTWO(I,J,K))
     $       +(HX(I,J,K)-HXY(I,J,K-1)-HXZ(I,J,K-1))*ECRLZ(IDTWO(I,J,K))
     $       -(HZ(I,J,K)-HZX(I-1,J,K)-HZY(I-1,J,K))*ECRLX(IDTWO(I,J,K))
C     $       +SOURCE(I,J,K)
      ENDIF
C
C          PML
C
          ELSE

            IF(IDTWO(I,J,K).EQ.1) THEN
              EYX(I,J,K)=0.0
              EYZ(I,J,K)=0.0

            ELSE

              EYZ(I,J,K)=EYZ(I,J,K)*DTESZ+(HXY(I,J,K)+HXZ(I,J,K)
     $                -HXY(I,J,K-1)-HXZ(I,J,K-1))*DTESDZ
              EYX(I,J,K)=EYX(I,J,K)*DTESX-(HZX(I,J,K)+HZY(I,J,K)
     $                -HZX(I-1,J,K)-HZY(I-1,J,K))*DTESDX

           ENDIF
          ENDIF

      RETURN
      END

C     ********************************************************
      SUBROUTINE PMLEZ(I,J,K)
      INCLUDE 'common.h'

      XJ=1*J
      XI=1*I

      EPSF=EPSP(IDTHRE(I,J,K)+1)/EPS0
      IF(J.LE.NYB) THEN
        ESIGMAY=SIGMAYE(IDPML(I,J,K))*EPSF*((XNPML-XJ+1.)/XNPML)**2.
      ELSE
        ESIGMAY=SIGMAYE(IDPML(I,J,K))*EPSF*((XJ-YN+XNPML)/XNPML)**2.
      ENDIF

      IF(I.LE.NXB) THEN
        ESIGMAX=SIGMAXE(IDPML(I,J,K))*EPSF*((XNPML-XI+1.)/XNPML)**2.
      ELSE
        ESIGMAX=SIGMAXE(IDPML(I,J,K))*EPSF*((XI-XN+XNPML)/XNPML)**2.
      ENDIF
```

```
        IF (ESIGMAY.LE.0.) THEN
         DTESY=1.
         DTESDY=DT/(DELY*EPSP(IDTHRE(I,J,K)+1))
        ELSE
         DTESY=EXP(-(ESIGMAY*DT)/EPSP(IDTHRE(I,J,K)+1))
         DTESDY=(1.-DTESY)/(ESIGMAY*DELY)
        ENDIF

        IF (ESIGMAX.LE.0.) THEN
         DTESX=1.
         DTESDX=DT/(DELX*EPSP(IDTHRE(I,J,K)+1))
        ELSE
         DTESX=EXP(-(ESIGMAX*DT)/EPSP(IDTHRE(I,J,K)+1))
         DTESDX=(1.-DTESX)/(ESIGMAX*DELX)
        ENDIF
C
C           PML INTERFACE
C
           IF (I.EQ.NXBP1.AND.J.GT.NYBP1.AND.J.LT.NYS
     $                             .AND.K.GT.NZB.AND.K.LT.NZS) THEN

        IF (IDTHRE(I,J,K).EQ.0) THEN
         EZ(I,J,K)=EZ(I,J,K)+(HY(I,J,K)-HYX(I-1,J,K)
     $              -HYZ(I-1,J,K))*DTEDX
     $              -(HX(I,J,K)-HX(I,J-1,K))*DTEDY

        ELSEIF (IDTHRE(I,J,K).EQ.1) THEN
         EZ(I,J,K)=0.0

        ELSE
         EZ(I,J,K)=EZ(I,J,K)*ESB(IDTHRE(I,J,K))
     $     +(HY(I,J,K)-HYX(I-1,J,K)-HYZ(I-1,J,K))*ECRLX(IDTHRE(I,J,K))
     $     -(HX(I,J,K)-HX(I,J-1,K))
     $     *ECRLY(IDTHRE(I,J,K))
        ENDIF


       ELSEIF (I.GT.NXBP1.AND.I.LT.NXS.AND.J.EQ.NYBP1
     $                             .AND.K.GT.NZB.AND.K.LT.NZS) THEN

        IF (IDTHRE(I,J,K).EQ.0) THEN
         EZ(I,J,K)=EZ(I,J,K)+(HY(I,J,K)
     $              -HY(I-1,J,K))*DTEDX
     $              -(HX(I,J,K)-HXY(I,J-1,K)-HXZ(I,J-1,K))*DTEDY

        ELSEIF (IDTHRE(I,J,K).EQ.1) THEN
         EZ(I,J,K)=0.0

        ELSE
         EZ(I,J,K)=EZ(I,J,K)*ESB(IDTHRE(I,J,K))
     $     +(HY(I,J,K)-HY(I-1,J,K))*ECRLX(IDTHRE(I,J,K))
     $     -(HX(I,J,K)-HXY(I,J-1,K)-HXZ(I,J-1,K))
     $     *ECRLY(IDTHRE(I,J,K))
        ENDIF
```

```
      ELSEIF (I.GT.NXB.AND.I.LT.NXS.AND.J.EQ.NYS
       $                              .AND.K.GT.NZB.AND.K.LT.NZS) THEN
       IF(IDTHRE(I,J,K).EQ.1) THEN
        EZX(I,J,K)=0.0
        EZY(I,J,K)=0.0

       ELSE

        EZY(I,J,K)=EZY(I,J,K)*DTESY-(HXY(I,J,K)+HXZ(I,J,K)
       $                           -HX(I,J-1,K))*DTESDY
        EZX(I,J,K)=EZX(I,J,K)*DTESX+(HYX(I,J,K)+HYZ(I,J,K)
       $                -HYX(I-1,J,K)-HYZ(I-1,J,K))*DTESDX
       ENDIF


      ELSEIF (I.EQ.NXS.AND.J.GT.NYB.AND.J.LT.NYS
       $                              .AND.K.GT.NZB.AND.K.LT.NZS) THEN

       IF(IDTHRE(I,J,K).EQ.1) THEN
        EZX(I,J,K)=0.0
        EZY(I,J,K)=0.0

       ELSE

        EZX(I,J,K)=EZX(I,J,K)*DTESX+(HYX(I,J,K)+HYZ(I,J,K)
       $                           -HY(I-1,J,K))*DTESDX
        EZY(I,J,K)=EZY(I,J,K)*DTESY-(HXY(I,J,K)+HXZ(I,J,K)
       $                -HXY(I,J-1,K)-HXZ(I,J-1,K))*DTESDY
       ENDIF

      ELSEIF (I.EQ.NXBP1.AND.J.EQ.NYBP1
       $                              .AND.K.GT.NZB.AND.K.LT.NZS) THEN

       IF(IDTHRE(I,J,K).EQ.0) THEN
        EZ(I,J,K)=EZ(I,J,K)+
       $      (HY(I,J,K)-HYX(I-1,J,K)-HYZ(I-1,J,K))*DTEDX
       $      -(HX(I,J,K)-HXY(I,J-1,K)-HXZ(I,J-1,K))*DTEDY

        ELSEIF(IDTHRE(I,J,K).EQ.1) THEN
         EZ(I,J,K)=0.0
        ELSE
         EZ(I,J,K)=EZ(I,J,K)*ESB(IDTHRE(I,J,K))
       $       +(HY(I,J,K)-HYX(I-1,J,K)-HYZ(I-1,J,K))*ECRLX(IDTHRE(I,J,K))
       $       -(HX(I,J,K)-HXY(I,J-1,K)-HXZ(I,J-1,K))
       $       *ECRLY(IDTHRE(I,J,K))
        ENDIF
C
C          PML
C
       ELSE

              IF(IDTHRE(I,J,K).EQ.1) THEN
                EZX(I,J,K)=0.0
                EZY(I,J,K)=0.0

              ELSE
```

```
            EZX(I,J,K)=EZX(I,J,K)*DTESX+(HYX(I,J,K)+HYZ(I,J,K)
     $                  -HYX(I-1,J,K)-HYZ(I-1,J,K))*DTESDX
            EZY(I,J,K)=EZY(I,J,K)*DTESY-(HXY(I,J,K)+HXZ(I,J,K)
     $                  -HXY(I,J-1,K)-HXZ(I,J-1,K))*DTESDY

          ENDIF
         ENDIF

      RETURN
      END

C     ********************************************************
      SUBROUTINE PMLHX(I,J,K)
      INCLUDE 'common.h'

      XJ=1*J
      XK=1*K

      IF(J.LE.NYB) THEN
        HSIGMAY=SIGMAYH(IDPML(I,J,K))*((XNPML-XJ+0.5)/XNPML)**2.
      ELSE
        HSIGMAY=SIGMAYH(IDPML(I,J,K))*((XJ-YN+XNPML+0.5)/XNPML)**2.
      ENDIF
      IF(K.LE.NZB) THEN
        HSIGMAZ=SIGMAZH(IDPML(I,J,K))*((XNPML-XK+0.5)/XNPML)**2.
      ELSE
        HSIGMAZ=SIGMAZH(IDPML(I,J,K))*((XK-ZN+XNPML+0.5)/XNPML)**2.
      ENDIF

      IF (HSIGMAZ.LE.0.) THEN
       DTMSZ=1.
       DTMSDZ=DT/(DELZ*XMU0)
      ELSE
       DTMSZ=EXP(-(HSIGMAZ*DT)/XMU0)
       DTMSDZ=(1.-DTMSZ)/(HSIGMAZ*DELZ)
      ENDIF

      IF (HSIGMAY.LE.0.) THEN
       DTMSY=1.
       DTMSDY=DT/(DELY*XMU0)
      ELSE
       DTMSY=EXP(-(HSIGMAY*DT)/XMU0)
       DTMSDY=(1.-DTMSY)/(HSIGMAY*DELY)
      ENDIF
C
C          PML INTERFACE
C
         IF (I.GT.NXB.AND.I.LT.NXS.AND.J.EQ.NYSM1
     $                            .AND.K.GT.NZB.AND.K.LT.NZSM1) THEN


      HX(I,J,K)=HX(I,J,K)-(EZX(I,J+1,K)+EZY(I,J+1,K)
     $                -EZ(I,J,K))*DTMDY+(EY(I,J,K+1)-EY(I,J,K))*DTMDZ

    ELSEIF (I.GT.NXB.AND.I.LT.NXS.AND.J.GT.NYB.AND.J.LT.NYSM1
     $                                          .AND.K.EQ.NZSM1) THEN
```

```
      HX(I,J,K)=HX(I,J,K)-(EZ(I,J+1,K)-EZ(I,J,K))*DTMDY
     $                  +(EYX(I,J,K+1)+EYZ(I,J,K+1)-EY(I,J,K))*DTMDZ

     ELSEIF (I.GT.NXB.AND.I.LT.NXS.AND.J.EQ.NYB
     $                             .AND.K.GT.NZB.AND.K.LT.NZS) THEN
     HXY(I,J,K)=HXY(I,J,K)*DTMSY-(EZ(I,J+1,K)
     $                            -EZX(I,J,K)-EZY(I,J,K))*DTMSDY
     HXZ(I,J,K)=HXZ(I,J,K)*DTMSZ+(EYX(I,J,K+1)+EYZ(I,J,K+1)
     $                            -EYX(I,J,K)-EYZ(I,J,K))*DTMSDZ


     ELSEIF (I.GT.NXB.AND.I.LT.NXS.AND.J.GT.NYB.AND.J.LT.NYS
     $                                          .AND.K.EQ.NZB) THEN

           HXZ(I,J,K)=HXZ(I,J,K)*DTMSZ+(EY(I,J,K+1)
     $                            -EYX(I,J,K)-EYZ(I,J,K))*DTMSDZ
     HXY(I,J,K)=HXY(I,J,K)*DTMSY-(EZX(I,J+1,K)+EZY(I,J+1,K)
     $        -EZX(I,J,K)-EZY(I,J,K))*DTMSDY

     ELSEIF (I.GT.NXB.AND.I.LT.NXS.AND.J.EQ.NYSM1
     $                                          .AND.K.EQ.NZSM1) THEN

      HX(I,J,K)=HX(I,J,K)
     $        -(EZX(I,J+1,K)+EZY(I,J+1,K)-EZ(I,J,K))*DTMDY
     $                  +(EYX(I,J,K+1)+EYZ(I,J,K+1)-EY(I,J,K))*DTMDZ
C
C         PML
C
          ELSE

           HXY(I,J,K)=HXY(I,J,K)*DTMSY
     $              -(EZX(I,J+1,K)+EZY(I,J+1,K)
     $              -EZX(I,J,K)-EZY(I,J,K))*DTMSDY
           HXZ(I,J,K)=HXZ(I,J,K)*DTMSZ
     $              +(EYX(I,J,K+1)+EYZ(I,J,K+1)
     $              -EYX(I,J,K)-EYZ(I,J,K))*DTMSDZ

          ENDIF
      RETURN
      END

C     ******************************************************
      SUBROUTINE PMLHY(I,J,K)
      INCLUDE 'common.h'

      XI=1*I
      XK=1*K

      IF(I.LE.NXB) THEN
        HSIGMAX=SIGMAXH(IDPML(I,J,K))*((XNPML-XI+0.5)/XNPML)**2.
      ELSE
        HSIGMAX=SIGMAXH(IDPML(I,J,K))*((XI-XN+XNPML+0.5)/XNPML)**2.
      ENDIF
      IF(K.LE.NZB) THEN
        HSIGMAZ=SIGMAZH(IDPML(I,J,K))*((XNPML-XK+0.5)/XNPML)**2.
      ELSE
```

```
        HSIGMAZ=SIGMAZH(IDPML(I,J,K))*((XK-ZN+XNPML+0.5)/XNPML)**2.
       ENDIF


       IF (HSIGMAZ.LE.0.) THEN
        DTMSZ=1.
        DTMSDZ=DT/(DELZ*XMU0)
       ELSE
        DTMSZ=EXP(-(HSIGMAZ*DT)/XMU0)
        DTMSDZ=(1.-DTMSZ)/(HSIGMAZ*DELZ)
       ENDIF

       IF (HSIGMAX.LE.0.) THEN
        DTMSX=1.
        DTMSDX=DT/(DELX*XMU0)
       ELSE
        DTMSX=EXP(-(HSIGMAX*DT)/XMU0)
        DTMSDX=(1.-DTMSX)/(HSIGMAX*DELX)
       ENDIF
C
C           PML INTERFACE
C
          IF (I.EQ.NXSM1.AND.J.GT.NYB.AND.J.LT.NYS
     $                                  .AND.K.GT.NZB.AND.K.LT.NZSM1) THEN


      HY(I,J,K)=HY(I,J,K)-(EX(I,J,K+1)-EX(I,J,K))*DTMDZ
     $                +(EZX(I+1,J,K)+EZY(I+1,J,K)-EZ(I,J,K))*DTMDX

     ELSEIF (I.GT.NXB.AND.I.LT.NXSM1.AND.J.GT.NYB.AND.J.LT.NYS
     $                                                .AND.K.EQ.NZSM1) THEN

      HY(I,J,K)=HY(I,J,K)-(EXY(I,J,K+1)+EXZ(I,J,K+1)
     $                -EX(I,J,K))*DTMDZ+(EZ(I+1,J,K)-EZ(I,J,K))*DTMDX

         ELSEIF (I.EQ.NXSM1.AND.J.GT.NYB.AND.J.LT.NYS
     $                                                .AND.K.EQ.NZSM1) THEN

      HY(I,J,K)=HY(I,J,K)-(EXY(I,J,K+1)+EXZ(I,J,K+1)
     $               -EX(I,J,K))*DTMDZ
     $               +(EZX(I+1,J,K)+EZY(I+1,J,K)-EZ(I,J,K))*DTMDX


     ELSEIF (I.EQ.NXB.AND.J.GT.NYB.AND.J.LT.NYS
     $                                  .AND.K.GT.NZB.AND.K.LT.NZS) THEN
      HYX(I,J,K)=HYX(I,J,K)*DTMSX+(EZ(I+1,J,K)
     $                               -EZX(I,J,K)-EZY(I,J,K))*DTMSDX
      HYZ(I,J,K)=HYZ(I,J,K)*DTMSZ-(EXY(I,J,K+1)+EXZ(I,J,K+1)
     $                               -EXY(I,J,K)-EXZ(I,J,K))*DTMSDZ


     ELSEIF (I.GT.NXB.AND.I.LT.NXS.AND.J.GT.NYB.AND.J.LT.NYS
     $                                                .AND.K.EQ.NZB) THEN
            HYZ(I,J,K)=HYZ(I,J,K)*DTMSZ-(EX(I,J,K+1)
     $                               -EXY(I,J,K)-EXZ(I,J,K))*DTMSDZ
      HYX(I,J,K)=HYX(I,J,K)*DTMSX+(EZX(I+1,J,K)+EZY(I+1,J,K)
```

```
     $                               -EZX(I,J,K)-EZY(I,J,K))*DTMSDX

C
C          PML
C
    ELSE

           HYX(I,J,K)=HYX(I,J,K)*DTMSX
     $                +(EZX(I+1,J,K)+EZY(I+1,J,K)
     $                -EZX(I,J,K)-EZY(I,J,K))*DTMSDX
           HYZ(I,J,K)=HYZ(I,J,K)*DTMSZ
     $                -(EXY(I,J,K+1)+EXZ(I,J,K+1)
     $                -EXY(I,J,K)-EXZ(I,J,K))*DTMSDZ

          ENDIF

      RETURN
      END

C     ****************************************************
      SUBROUTINE PMLHZ(I,J,K)
      INCLUDE 'common.h'

      XJ=1*J
      XI=1*I

      IF(I.LE.NXB) THEN
        HSIGMAX=SIGMAXH(IDPML(I,J,K))*((XNPML-XI+0.5)/XNPML)**2.
      ELSE
        HSIGMAX=SIGMAXH(IDPML(I,J,K))*((XI-XN+XNPML+0.5)/XNPML)**2.
      ENDIF
      IF(J.LE.NYB) THEN
        HSIGMAY=SIGMAYH(IDPML(I,J,K))*((XNPML-XJ+0.5)/XNPML)**2.
      ELSE
        HSIGMAY=SIGMAYH(IDPML(I,J,K))*((XJ-YN+XNPML+0.5)/XNPML)**2.
      ENDIF

      IF (HSIGMAY.LE.0.) THEN
       DTMSY=1.
       DTMSDY=DT/(DELY*XMU0)
      ELSE
       DTMSY=EXP(-(HSIGMAY*DT)/XMU0)
       DTMSDY=(1.-DTMSY)/(HSIGMAY*DELY)
      ENDIF

      IF (HSIGMAX.LE.0.) THEN
       DTMSX=1.
       DTMSDX=DT/(DELX*XMU0)
      ELSE
       DTMSX=EXP(-(HSIGMAX*DT)/XMU0)
       DTMSDX=(1.-DTMSX)/(HSIGMAX*DELX)
      ENDIF
C
C          PML INTERFACE
C
         IF (I.EQ.NXSM1.AND.J.GT.NYB.AND.J.LT.NYSM1
```

```
     $                                  .AND.K.GT.NZB.AND.K.LT.NZS) THEN

     HZ(I,J,K)=HZ(I,J,K)-(EYX(I+1,J,K)+EYZ(I+1,J,K)
     $                  -EY(I,J,K))*DTMDX+(EX(I,J+1,K)-EX(I,J,K))*DTMDY

    ELSEIF (I.GT.NXB.AND.I.LT.NXSM1.AND.J.EQ.NYSM1
     $                                  .AND.K.GT.NZB.AND.K.LT.NZS) THEN

     HZ(I,J,K)=HZ(I,J,K)-(EY(I+1,J,K)-EY(I,J,K))*DTMDX
     $                  +(EXY(I,J+1,K)+EXZ(I,J+1,K)-EX(I,J,K))*DTMDY

    ELSEIF (I.GT.NXB.AND.I.LT.NXS.AND.J.EQ.NYB
     $                                  .AND.K.GT.NZB.AND.K.LT.NZS) THEN
     HZY(I,J,K)=HZY(I,J,K)*DTMSY+(EX(I,J+1,K)
     $                            -EXY(I,J,K)-EXZ(I,J,K))*DTMSDY
     HZX(I,J,K)=HZX(I,J,K)*DTMSX-(EYX(I+1,J,K)+EYZ(I+1,J,K)
     $                            -EYX(I,J,K)-EYZ(I,J,K))*DTMSDX


    ELSEIF (I.EQ.NXB.AND.J.GT.NYB.AND.J.LT.NYS
     $                                  .AND.K.GT.NZB.AND.K.LT.NZS) THEN

     HZX(I,J,K)=HZX(I,J,K)*DTMSX-(EY(I+1,J,K)
     $                            -EYX(I,J,K)-EYZ(I,J,K))*DTMSDX
     HZY(I,J,K)=HZY(I,J,K)*DTMSY+(EXY(I,J+1,K)+EXZ(I,J+1,K)
     $                            -EXY(I,J,K)-EXZ(I,J,K))*DTMSDY


    ELSEIF (I.EQ.NXSM1.AND.J.EQ.NYSM1
     $                                  .AND.K.GT.NZB.AND.K.LT.NZS) THEN

     HZ(I,J,K)=HZ(I,J,K)-(EYX(I+1,J,K)+EYZ(I+1,J,K)
     $                  -EY(I,J,K))*DTMDX
     $                  +(EXY(I,J+1,K)+EXZ(I,J+1,K)-EX(I,J,K))*DTMDY

C
C        PML
C
    ELSE

           HZX(I,J,K)=HZX(I,J,K)*DTMSX
     $            -(EYX(I+1,J,K)+EYZ(I+1,J,K)
     $            -EYX(I,J,K)-EYZ(I,J,K))*DTMSDX
           HZY(I,J,K)=HZY(I,J,K)*DTMSY
     $            +(EXY(I,J+1,K)+EXZ(I,J+1,K)
     $            -EXY(I,J,K)-EXZ(I,J,K))*DTMSDY
          ENDIF

      RETURN
      END

C*******************************************************************
SUBROUTINE DATSAV

INCLUDE 'common.h'
real*8 dum1C,dum2,dum3
integer i,j,k,ii,jj,nptC,iarr(3),charge(nx)
```

```
integer iminfcap,imaxfcap,jminfcap,jmaxfcap,kfeedcap,
     &    jmidf,jmida,kaperturecap
common / comcapacloc / iminfcap,imaxfcap,jminfcap,jmaxfcap,
     &    kfeedcap,kaperturecap

C ---------------------------------------------------------------------------
C     To completely specify a sampling point, the user must specify
C       4 parameters:  NTYPE, IOBS, JOBS, and KOBS.
C       NTYPE for point L is specified by NTYPE(L).
C         NTYPE controls what quantities are sampled as follows:
C           1 = EX (scattered x-component of electric field)
C           2 = EY (scattered y-component of electric field)
C           3 = EZ (scattered z-component of electric field)
C           4 = HX (scattered x-component of magnetic field)
C           5 = HY (scattered y-component of magnetic field)
C           6 = HZ (scattered z-component of magnetic field)
C           7 = IX (x-component of current through rectangular loop of H)
C           8 = IY (y-component of current through rectangular loop of H)
C           9 = IZ (z-component of current through rectangular loop of H)
C       IOBS(L), JOBS(L) AND KOBS(L) specify the I, J, and K coordinates
C         of the point L.  Thus, for point 1, the user would specify
C         NTYPE(1), IOBS(1), JOBS(1) and KOBS(1).  For point 2, the
C         user would specify NTYPE(2), IOBS(2), JOBS(2), KOBS(2).
C         The exact locations of a component within a cell follows that
C         of the Yee cell.
C       As an example, the following 4 lines save EY at the point
C         (33,26,22):
C           NTYPE(1)=2
C           IOBS(1)=33
C           JOBS(1)=26
C           KOBS(1)=22
C
C     To sample total field, issue a statement to add in the appropriate
C       incident field.  These total field save statements should be added
C       directly after the "3000 CONTINUE" statement near the end of this
C       SUBPROGRAM to override any previous assignment to STORE(NPT).
C       Make certain to back up the incident field by 1 time step (DT)
C       for sampling total E fields and by 1/2 time step (DT/2.0) for
C       sampling total H fields.  For example, to store total EX field
C       at sample point 3, issue the following 3 statements:
C         T=T-DT
C         STORE(3)=EX(IOBS(3),JOBS(3),KOBS(3))+
C    2      EXI(IOBS(3),JOBS(3),KOBS(3))
C         T=T+DT
C ---------------------------------------------------------------------------

C  For the first time through (i.e. N=1), do some initializations:
IF (N.NE.1) GO TO 10

C   User-defined test point cell location(s).  There should be NTEST
C   occurrences of NTYPE and [I,J,K]OBS defined.  Unless total fields
C   are desired, this is probably the only part of this SUBROUTINE
C   the user will need to change.

ntype(1) = 13
ntype(2) = 13
ntype(3) = 13
```

```
ntype(4)  = 13
ntype(5)  = 13
ntype(6)  = 13
ntype(7)  = 13
ntype(8)  = 13
ntype(9)  = 15
ntype(10)  = 16

ntype(11)  = 1
ntype(12)  = 2
ntype(13)  = 3

ntype(14)  = 1
ntype(15)  = 2
ntype(16)  = 3

ntype(17)  = 1
ntype(18)  = 2
ntype(19)  = 3

ntype(20)  = 1
ntype(21)  = 2
ntype(22)  = 3

ntype(23)  = 1
ntype(24)  = 2
ntype(25)  = 3

ntype(26)  = 1
ntype(27)  = 2
ntype(28)  = 3

ntype(29)  = 1
ntype(30)  = 2
ntype(31)  = 3

ntype(32)  = 1
ntype(33)  = 2
ntype(34)  = 3

ntype(35)  = 1
ntype(36)  = 2
ntype(37)  = 3

ntype(38)  = 1
ntype(39)  = 2
ntype(40)  = 3

ntype(41)  = 1
ntype(42)  = 2
ntype(43)  = 3

ntype(44)  = 1
ntype(45)  = 2
ntype(46)  = 3
```

```
ntype(47) = 1
ntype(48) = 2
ntype(49) = 3

ntype(50) = 1
ntype(51) = 2
ntype(52) = 3

ntype(53) = 1
ntype(54) = 2
ntype(55) = 3

ntype(56) = 1
ntype(57) = 2
ntype(58) = 3

C  DEFINE "NTEST" TEST POINT CELL LOCATION(S) HERE:
C  Locations for observing voltage between plates ("iobs" value is not used):
iobs(1) = iminfcap - 2
jobs(1) = (jminfcap + jmaxfcap) / 2
kobs(1) = (kfeedcap + kaperturecap) / 2

iobs(2) = iminfcap - 1
jobs(2) = (jminfcap + jmaxfcap) / 2
kobs(2) = (kfeedcap + kaperturecap) / 2

iobs(3) = iminfcap
jobs(3) = (jminfcap + jmaxfcap) / 2
kobs(3) = (kfeedcap + kaperturecap) / 2

iobs(4) = iminfcap + 1
jobs(4) = (jminfcap + jmaxfcap) / 2
kobs(4) = (kfeedcap + kaperturecap) / 2

iobs(5) = iminfcap + 2
jobs(5) = (jminfcap + jmaxfcap) / 2
kobs(5) = (kfeedcap + kaperturecap) / 2

iobs(6) = iminfcap + 3
jobs(6) = (jminfcap + jmaxfcap) / 2
kobs(6) = (kfeedcap + kaperturecap) / 2

iobs(7) = iminfcap + 4
jobs(7) = (jminfcap + jmaxfcap) / 2
kobs(7) = (kfeedcap + kaperturecap) / 2

iobs(8) = iminfcap + 5
jobs(8) = (jminfcap + jmaxfcap) / 2
kobs(8) = (kfeedcap + kaperturecap) / 2

iobs(9) = iminfcap + 5
jobs(9) = (jminfcap + jmaxfcap) / 2
kobs(9) = (kfeedcap + kaperturecap) / 2

iobs(10) = iminfcap + 6
jobs(10) = (jminfcap + jmaxfcap) / 2
```

```
kobs(10) = (kfeedcap + kaperturecap) / 2

iobs(11) = imina + 2
jobs(11) = (jmina + jmaxa) / 2
kobs(11) = kfeed + 2

iobs(12) = imina + 2
jobs(12) = (jmina + jmaxa) / 2
kobs(12) = kfeed + 2

iobs(13) = imina + 2
jobs(13) = (jmina + jmaxa) / 2
kobs(13) = kfeed + 2

iobs(14) = imina + 2
jobs(14) = (jmina + jmaxa) / 2
kobs(14) = 100

iobs(15) = imina + 2
jobs(15) = (jmina + jmaxa) / 2
kobs(15) = 100

iobs(16) = imina + 2
jobs(16) = (jmina + jmaxa) / 2
kobs(16) = 100

iobs(17) = 25
jobs(17) = (jmina + jmaxa) / 2
kobs(17) = 100

iobs(18) = 25
jobs(18) = (jmina + jmaxa) / 2
kobs(18) = 100

iobs(19) = 25
jobs(19) = (jmina + jmaxa) / 2
kobs(19) = 100

iobs(20) = 35
jobs(20) = (jmina + jmaxa) / 2
kobs(20) = 100

iobs(21) = 35
jobs(21) = (jmina + jmaxa) / 2
kobs(21) = 100

iobs(22) = 35
jobs(22) = (jmina + jmaxa) / 2
kobs(22) = 100

iobs(23) = 17
jobs(23) = (jmina + jmaxa) / 2
kobs(23) =  200

iobs(24) = 17
jobs(24) = (jmina + jmaxa) / 2
kobs(24) =  200
```

```
iobs(25) = 17
jobs(25) = (jmina + jmaxa) / 2
kobs(25) =  200

iobs(26) = 40
jobs(26) = (jmina + jmaxa) / 2
kobs(26) = 200

iobs(27) = 40
jobs(27) = (jmina + jmaxa) / 2
kobs(27) = 200

iobs(28) = 40
jobs(28) = (jmina + jmaxa) / 2
kobs(28) = 200

iobs(29) = 60
jobs(29) = (jmina + jmaxa) / 2
kobs(29) = 200

iobs(30) = 60
jobs(30) = (jmina + jmaxa) / 2
kobs(30) = 200

iobs(31) = 60
jobs(31) = (jmina + jmaxa) / 2
kobs(31) = 200

iobs(32) = 17
jobs(32) = (jmina + jmaxa) / 2
kobs(32) = 300

iobs(33) = 17
jobs(33) = (jmina + jmaxa) / 2
kobs(33) = 300

iobs(34) = 17
jobs(34) = (jmina + jmaxa) / 2
kobs(34) = 300

iobs(35) = 50
jobs(35) = (jmina + jmaxa) / 2
kobs(35) = 300

iobs(36) = 50
jobs(36) = (jmina + jmaxa) / 2
kobs(36) = 300

iobs(37) = 50
jobs(37) = (jmina + jmaxa) / 2
kobs(37) = 300

iobs(38) = 85
jobs(38) = (jmina + jmaxa) / 2
kobs(38) = 300
```

```
iobs(39) = 85
jobs(39) = (jmina + jmaxa) / 2
kobs(39) = 300

iobs(40) = 85
jobs(40) = (jmina + jmaxa) / 2
kobs(40) = 300

iobs(41) = 17
jobs(41) = (jmina + jmaxa) / 2
kobs(41) = 420

iobs(42) = 17
jobs(42) = (jmina + jmaxa) / 2
kobs(42) = 420

iobs(43) = 17
jobs(43) = (jmina + jmaxa) / 2
kobs(43) = 420

iobs(44) = 55
jobs(44) = (jmina + jmaxa) / 2
kobs(44) = 420

iobs(45) = 55
jobs(45) = (jmina + jmaxa) / 2
kobs(45) = 420

iobs(46) = 55
jobs(46) = (jmina + jmaxa) / 2
kobs(46) = 420

iobs(47) = 90
jobs(47) = (jmina + jmaxa) / 2
kobs(47) = 420

iobs(48) = 90
jobs(48) = (jmina + jmaxa) / 2
kobs(48) = 420

iobs(49) = 90
jobs(49) = (jmina + jmaxa) / 2
kobs(49) = 420

iobs(50) = 17
jobs(50) = (jmina + jmaxa) / 2
kobs(50) = 520

iobs(51) = 17
jobs(51) = (jmina + jmaxa) / 2
kobs(51) = 520

iobs(52) = 17
jobs(52) = (jmina + jmaxa) / 2
kobs(52) = 520
```

```
iobs(53) = 55
jobs(53) = (jmina + jmaxa) / 2
kobs(53) = 520

iobs(54) = 55
jobs(54) = (jmina + jmaxa) / 2
kobs(54) = 520

iobs(55) = 55
jobs(55) = (jmina + jmaxa) / 2
kobs(55) = 520

iobs(56) = 90
jobs(56) = (jmina + jmaxa) / 2
kobs(56) = 520

iobs(57) = 90
jobs(57) = (jmina + jmaxa) / 2
kobs(57) = 520

iobs(58) = 90
jobs(58) = (jmina + jmaxa) / 2
kobs(58) = 520

write(14,229) (ntype(j),iobs(j),jobs(j),kobs(j),j=1,ntest)
229  format(' (ntype, iobs, jobs, kobs)= ',/,4(i3,1x))

C   Initialize some variables.
DO 5 II=1,NTEST
5  STORE (II)=0.0d0
NPTS=NTEST

C   Print header info and check [I,J,K]OBS to see that they are in
C      range.  Write header to data file and to DIAGS3D.DAT.
WRITE (17,1400) DELX,DELY,DELZ,DT,NSTOP,NPTS
WRITE (17,1500) NPTS
1400  FORMAT (T2,E12.5,2X,E12.5,2X,E12.5,3X,E14.7,3X,I6,3X,I3)
1500  FORMAT (T2,'QUANTITIES SAMPLED AND SAVED AT',I4, ' LOCATIONS',/)

DO 1700 NPT = 1,NPTS
WRITE (17,1800) NPT,NTYPE(NPT),IOBS(NPT),JOBS(NPT),KOBS(NPT)
IF (IOBS(NPT).GT.NX) GO TO 1600
IF (IOBS(NPT).LT.1) GO TO 1600
IF (JOBS(NPT).GT.NY) GO TO 1600
IF (JOBS(NPT).LT.1) GO TO 1600
IF (KOBS(NPT).GT.NZ) GO TO 1600
IF (KOBS(NPT).LT.1) GO TO 1600
GO TO 1700
1600  WRITE (17,*) 'ERROR IN IOBS, JOBS OR KOBS FOR SAMPLING'
WRITE (17,*) 'POINT ',NPT
WRITE (17,*) 'EXECUTION HALTED.'
CLOSE (UNIT=10)
CLOSE (UNIT=17)
CLOSE (UNIT=30)
STOP
1700  CONTINUE
```

```
1800  FORMAT (T2,'SAMPLE:',I4,',  NTYPE=',I4,', SAMPLED AT CELL I=',
     &  I4,', J=',I4,', K=',I4,/)

C  Check for ERROR in NTYPE specification:
DO 20 NPT=1,NTEST
IF ((NTYPE(NPT).GE.17).OR.(NTYPE(NPTS).LE.0)) THEN
    WRITE (17,*) 'ERROR IN NTYPE FOR SAMPLING POINT ',NPT
    WRITE (17,*) 'EXECUTION HALTED.'
    CLOSE (UNIT=10)
    CLOSE (UNIT=17)
    CLOSE (UNIT=30)
    STOP
ENDIF
20  CONTINUE
C  Finished with initialization.
C*********************************************
C
C  Cycle through all NTEST sample locations:
10 continue

DO 3000 NPT=1,NTEST
C   Put desired quantity for this sample location into STORE(NPT).
I=IOBS(NPT)
J=JOBS(NPT)
K=KOBS(NPT)

   IF (NTYPE(NPT).EQ.1) STORE(NPT)=EX(I,J,K)
   IF (NTYPE(NPT).EQ.2) STORE(NPT)=EY(I,J,K)
   IF (NTYPE(NPT).EQ.3) STORE(NPT)=EZ(I,J,K)
   IF (NTYPE(NPT).EQ.4) STORE(NPT)=HX(I,J,K)
   IF (NTYPE(NPT).EQ.5) STORE(NPT)=HY(I,J,K)
   IF (NTYPE(NPT).EQ.6) STORE(NPT)=HZ(I,J,K)

IF (NTYPE(NPT).EQ.10) THEN
    k = kobs(npt)
    STORE(NPT) = 0.0d0
c  Sum contributions from all cells at feed point on top plate:
    do 123 j = jmina - 4, jmaxa + 4
    do 123 i = iminf -4, iminf + 4
      ii = i
      jj = j
      STORE(NPT)=STORE(NPT)+(-HX(II,JJ,K)+
     &        HX(II,JJ-1,K))* DELX+
     &        (HY(II,JJ,K)-HY(II-1,JJ,K))*DELY
123    continue
ENDIF


IF (NTYPE(NPT).EQ.13) THEN
    store(npt) = 0.0
    i = iobs(npt)
    j = jobs(npt)
    k = kobs(npt)
    dum1 = (EX(i+1,j+1,k+1) - EX(i,j+1,k+1)) * dely*delz
    dum1 = dum1 + (EY(i+1,j+1,k+1) - EY(i+1,j,k+1)) *
     &      delx*delz
```

```
    dum1 = dum1 + (EZ(i+1,j+1,k+1) - EZ(i+1,j+1,k)) *
     &       delx*dely
    store(npt) = store(npt) + dum1*eps0
endif

C  DIAGNOSTIC #15:  Total charge stored in a set of cells (from Gauss law):
C  The cell has I-values starting at "iobs(npt)". (J,K) vary over the range
C  specified in the do-loops below.
IF (NTYPE(NPT).EQ.15) THEN
    store(npt) = 0.0d0
    do 121 i=iminfcap - 2, iminfcap + 1
    do 121 j=jminfcap-1,jmaxfcap+1
    do 121 k=kfeedcap-2,kaperturecap
    dum1 = (EX(i,j,k) - EX(i-1,j,k)) * dely * delz
    dum1 = dum1 + (EY(i,j,k) - EY(i,j-1,k)) * delx * delz
    dum1 = dum1 + (EZ(i,j,k) - EZ(i,j,k-1)) * delx * dely
    store(npt) = store(npt) + dum1
121    continue
    store(npt) = store(npt) * eps0 * EPS(2)
endif

C  DIAGNOSTIC #16:  Voltage between capacitor plates (excluding cells
C  containing the plates themselves):
C  Line integral of "E_x" in the x-direction, at the specified (j,k):
IF (NTYPE(NPT).EQ.16) THEN
    j = jobs(npt)
    k = kobs(npt)
    store(npt) = 0.0d0
    do 129 i = iminfcap,imaxfcap
129    STORE(NPT) = STORE(NPT) + EX(i,j,k)
    store(npt) = store(npt) * delx
ENDIF

3000  CONTINUE

NPTS = NTEST
WRITE(10,12) t,(STORE(II),II=1,NPTS)
12  format(1x,59(e12.6,1x))


      RETURN
      END


C                   COMMON.H
C
C  PARAMETERS AND VARIABLES MOST COMMONLY USED ARE DEFINED IN A HEADER
C  FILE "COMMON.H". THIS HEADER FILE IS THEN SHARED IN EVERY SUBROUTINE
C  BY INCORPORATING THE STATEMENT " INCLUDE 'COMMON.H' ". PLEASE SEE ITS
C IMPLEMENTATION IN THE SUBROUTINE DEFINED ABOVE IN THE COMPUTER
C PROGRAM 'EMPSIM.F'.

      PARAMETER (NPML=8)
      PARAMETER (NX=110,NY=156,NZ=600)
      PARAMETER (NX1=NX-1,NY1=NY-1,NZ1=NZ-1)
      PARAMETER (NTEST=58)
```

```
      PARAMETER (NSTOP=4000)
      PARAMETER (AMP=1.0,PI=3.1415927)
      PARAMETER (EPS0=8.854E-12,XMU0=1.2566306E-6,ETA0=376.733341)
      PARAMETER (NDIM = 10, MXDIM = 4020)

C******************** FOR PML*******************************************
      PARAMETER (NXB=8,NYB=8,NZB=8,NXS=NX-8,NYS=NY-8,NZS=NZ-8)
      PARAMETER (XN=1.*NX,YN=1.*NY,ZN=1.*NZ,XNPML=1.*NPML)
      PARAMETER (NXBM1=NXB-1,NYBM1=NYB-1,NZBM1=NZB-1)
      PARAMETER (NXSM1=NXS-1,NYSM1=NYS-1,NZSM1=NZS-1)
      PARAMETER (NXBP1=NXB+1,NYBP1=NYB+1,NZBP1=NZB+1)
      PARAMETER (NXSP1=NXS+1,NYSP1=NYS+1,NZSP1=NZS+1)

C******************** FOR PML*******************************************
      COMMON/PML/EXY(NX,NY,NZ),EXZ(NX,NY,NZ),EYX(NX,NY,NZ),EYZ(NX,NY,NZ),
      EZX(NX,NY,NZ),EZY(NX,NY,NZ),
     + HXY(NX,NY,NZ),HXZ(NX,NY,NZ),HYX(NX,NY,NZ),HYZ(NX,NY,NZ),HZX(NX,NY,NZ),
      HZY(NX,NY,NZ),IDPML(NX,NY,NZ)

      COMMON/PMLTERM/DTESX,DTESY,DTESZ,DTMSX,DTMSY,DTMSZ, DTESDX,DTESDY,
      DTESDZ,DTMSDX,DTMSDY,DTMSDZ,
     + SIGMAXE(27),SIGMAYE(27),SIGMAZE(27), SIGMAXH(27),SIGMAYH(27),
      SIGMAZH(27), ESIGMAX,ESIGMAY,
     + ESIGMAZ,HSIGMAX,HSIGMAY,HSIGMAZ, EPSF,ESIGMAM,HSIGMAM


C  COMMON VARIABLES USED FOR GEOMETRY SETUP, EM FIELDS, MATERIALS, AND DATA
C  SAVING ARE DEFINED IN THE FOLLOWING:

      COMMON/SAVE/STORE(NTEST),IOBS(NTEST),JOBS(NTEST),KOBS(NTEST),
      NTYPE(NTEST)

      COMMON /COMHORN/KFEED,KAPERTURE,IMINF,IMAXF,JMINF,JMAXF, IMINA,IMAXA,
      JMINA,JMAXA,ZLENG,IFTOP1,IFTOP2

      COMMON/IDS/IDONE(NX,NY,NZ),IDTWO(NX,NY,NZ),IDTHRE(NX,NY,NZ)
      COMMON/EF/EX(NX,NY,NZ),EY(NX,NY,NZ),EZ(NX,NY,NZ)
      COMMON/HF/HX(NX,NY,NZ),HY(NX,NY,NZ),HZ(NX,NY,NZ)
      COMMON/CONSTI/EPS(9),EPSP(9),SIGMA(9),DELX,DELY,DELZ
      COMMON/TERMS/ESB(9),ECRLX(9),ECRLY(9),ECRLZ(9), DTEDX,DTEDY,DTEDZ,
      DTMDX,DTMDY,DTMDZ
      COMMON/EXTRA/N,DT,T,C
```

12.3 Sample Output

For the purpose of illustration, the simulation data (e.g. charge at the lower/upper plates of the capacitor and the x-component of electric-field at the feed of EMP simulator) obtained by running the computer program "EMP Simulator" is depicted below.

```
Sample # 1, NTYPE = 13 (charge), FDTD grid I = 14, J = 78, K = 19
Sample # 2, NTYPE = 13 (charge), FDTD grid I = 17, J = 78, K = 19
Sample # 3, NTYPE = 1 (EX), FDTD grid I = 17, J = 78, K = 26
```

DT	Sample#1	Sample#2	Sample#3
3.5579300E-11	6.1362800E-15	-6.1362800E-15	0.0000000E+00
7.1158600E-11	6.8748600E-15	-6.8748600E-15	0.0000000E+00
1.0673800E-10	7.6992000E-15	-7.6992000E-15	0.0000000E+00
1.4231700E-10	8.6189000E-15	-8.6189000E-15	0.0000000E+00
1.7789700E-10	9.6445600E-15	-9.6445600E-15	1.5319700e-03
2.1347600E-10	1.0787900E-14	-1.0787900E-14	1.0446900e-02
2.4905500E-10	1.2079900E-14	-1.2079900E-14	3.7902300e-02
2.8463400E-10	1.3604300E-14	-1.3604300E-14	9.7157800e-02
3.2021400E-10	1.5488700E-14	-1.5488600E-14	1.9718100e-01
3.5579300E-10	1.7789800E-14	-1.7789400E-14	3.4041500e-01
3.9137200E-10	2.0434800E-14	-2.0433300E-14	5.2161900e-01
4.2695200E-10	2.3249800E-14	-2.3247500E-14	7.3736100e-01
4.6253100E-10	2.6136000E-14	-2.6133900E-14	9.7823400e-01
4.9811000E-10	2.9171000E-14	-2.9172900E-14	1.2168300E+00
5.3369000E-10	3.2565400E-14	-3.2574800E-14	1.4527900E+00
5.6926900E-10	3.6489300E-14	-3.6512900E-14	1.7212200E+00
6.0484800E-10	4.0918800E-14	-4.0977800E-14	2.0258900E+00
6.4042800E-10	4.5724100E-14	-4.5840600E-14	2.3361700E+00
6.7600700E-10	5.0910300E-14	-5.1081600E-14	2.6413900E+00
7.1158600E-10	5.6681600E-14	-5.6883900E-14	2.9559700E+00
7.4716600E-10	6.3236600E-14	-6.3452600E-14	3.2906600E+00
7.8274500E-10	7.0596300E-14	-7.0826100E-14	3.6383900E+00
8.1832400E-10	7.8685600E-14	-7.8929500E-14	3.9845900E+00
8.5390400E-10	8.7513400E-14	-8.7763300E-14	4.3278100E+00
8.8948300E-10	9.7211900E-14	-9.7469100E-14	4.6818600E+00
9.2506200E-10	1.0795700E-13	-1.0824900E-13	5.0656400E+00
9.6064100E-10	1.1987600E-13	-1.2024400E-13	5.4924700E+00
9.9622100E-10	1.3301300E-13	-1.3348800E-13	5.9612800E+00
1.0318000E-09	1.4740100E-13	-1.4798700E-13	6.4710200E+00
1.0673800E-09	1.6318700E-13	-1.6386200E-13	7.0382300E+00
1.1029600E-09	1.8062600E-13	-1.8135900E-13	7.6803600E+00
1.1385400E-09	1.9993500E-13	-2.0070300E-13	8.3974600E+00
1.1741200E-09	2.2123700E-13	-2.2203800E-13	9.1845100E+00
1.2097000E-09	2.4465100E-13	-2.4550600E-13	1.0048600E+01
1.2452800E-09	2.7038200E-13	-2.7132400E-13	1.1007600E+01
1.2808600E-09	2.9872200E-13	-2.9977100E-13	1.2078800E+01
1.3164300E-09	3.2998300E-13	-3.3114200E-13	1.3272100E+01
1.3520100E-09	3.6442300E-13	-3.6569700E-13	1.4594500E+01
1.3875900E-09	4.0226600E-13	-4.0367800E-13	1.6059700E+01
1.4231700E-09	4.4381300E-13	-4.4539100E-13	1.7692800E+01
1.4587500E-09	4.8948100E-13	-4.9124700E-13	1.9522000E+01
1.4943300E-09	5.3971100E-13	-5.4168400E-13	2.1562000E+01
1.5299100E-09	5.9491300E-13	-5.9710200E-13	2.3814100E+01
1.5654900E-09	6.5550300E-13	-6.5790800E-13	2.6289000E+01
1.6010700E-09	7.2195200E-13	-7.2458700E-13	2.9017900E+01
1.6366500E-09	7.9482700E-13	-7.9773300E-13	3.2037800E+01
1.6722300E-09	8.7478100E-13	-8.7800200E-13	3.5376700E+01
1.7078100E-09	9.6248400E-13	-9.6605200E-13	3.9056100E+01
1.7433900E-09	1.0585800E-12	-1.0625300E-12	4.3102800E+01
1.7789700E-09	1.1637900E-12	-1.1681400E-12	4.7551800E+01

1.8145500E-09	1.2789300E-12	-1.2837300E-12	5.2439200E+01
1.8501200E-09	1.4049400E-12	-1.4102300E-12	5.7798700E+01
1.8857000E-09	1.5427900E-12	-1.5486300E-12	6.3666700E+01
1.9212800E-09	1.6935100E-12	-1.6999500E-12	7.0087400E+01
1.9568600E-09	1.8582000E-12	-1.8653100E-12	7.7111500E+01
1.9924400E-09	2.0380400E-12	-2.0459100E-12	8.4789900E+01
2.0280200E-09	2.2344000E-12	-2.2431100E-12	9.3173400E+01
2.0636000E-09	2.4487400E-12	-2.4583600E-12	1.0231700E+02
2.0991800E-09	2.6825700E-12	-2.6931600E-12	1.1228300E+02
2.1347600E-09	2.9375100E-12	-2.9491400E-12	1.2314800E+02
2.1703400E-09	3.2153200E-12	-3.2281000E-12	1.3499600E+02
2.2059200E-09	3.5179600E-12	-3.5320300E-12	1.4791100E+02
2.2414900E-09	3.8475400E-12	-3.8630400E-12	1.6197700E+02
2.2770700E-09	4.2063000E-12	-4.2233700E-12	1.7728800E+02
2.3126500E-09	4.5966600E-12	-4.6154200E-12	1.9394900E+02
2.3482300E-09	5.0211600E-12	-5.0417700E-12	2.1207700E+02
2.3838100E-09	5.4826200E-12	-5.5052700E-12	2.3179700E+02
2.4193900E-09	5.9840800E-12	-6.0089500E-12	2.5323500E+02
2.4549700E-09	6.5287900E-12	-6.5560500E-12	2.7652300E+02
2.4905500E-09	7.1201600E-12	-7.1500400E-12	3.0181100E+02
2.5261300E-09	7.7619100E-12	-7.7946600E-12	3.2926700E+02
2.5617100E-09	8.4580400E-12	-8.4939500E-12	3.5907500E+02
2.5972900E-09	9.2128900E-12	-9.2522500E-12	3.9142300E+02
2.6328600E-09	1.0031100E-11	-1.0074200E-11	4.2650400E+02
2.6684400E-09	1.0917500E-11	-1.0964700E-11	4.6452400E+02
2.7040200E-09	1.1877400E-11	-1.1929000E-11	5.0571500E+02
2.7396000E-09	1.2916500E-11	-1.2972900E-11	5.5032600E+02
2.7751800E-09	1.4040800E-11	-1.4102400E-11	5.9862200E+02
2.8107600E-09	1.5256800E-11	-1.5324200E-11	6.5088000E+02
2.8463400E-09	1.6571400E-11	-1.6645000E-11	7.0739700E+02
2.8819200E-09	1.7992000E-11	-1.8072400E-11	7.6849500E+02
2.9175000E-09	1.9526500E-11	-1.9614200E-11	8.3452100E+02
2.9530800E-09	2.1183300E-11	-2.1279000E-11	9.0584300E+02
2.9886600E-09	2.2971500E-11	-2.3075800E-11	9.8284800E+02
3.0242300E-09	2.4900500E-11	-2.5014200E-11	1.0659500E+03
3.0598100E-09	2.6980700E-11	-2.7104500E-11	1.1555800E+03
3.0953900E-09	2.9222700E-11	-2.9357500E-11	1.2522200E+03
3.1309700E-09	3.1638300E-11	-3.1785100E-11	1.3563700E+03
3.1665500E-09	3.4239800E-11	-3.4399500E-11	1.4685500E+03
3.2021300E-09	3.7040200E-11	-3.7214000E-11	1.5893200E+03
3.2377100E-09	4.0053500E-11	-4.0242400E-11	1.7192700E+03
3.2732900E-09	4.3294400E-11	-4.3499700E-11	1.8590500E+03
3.3088700E-09	4.6778700E-11	-4.7001600E-11	2.0093200E+03
3.3444500E-09	5.0522900E-11	-5.0765000E-11	2.1708000E+03
3.3800300E-09	5.4544800E-11	-5.4807600E-11	2.3442300E+03
3.4156000E-09	5.8863100E-11	-5.9148300E-11	2.5304100E+03
3.4511800E-09	6.3497600E-11	-6.3806800E-11	2.7301900E+03
3.4867600E-09	6.8469200E-11	-6.8804500E-11	2.9444500E+03
3.5223400E-09	7.3800300E-11	-7.4163700E-11	3.1741600E+03
3.5579200E-09	7.9514400E-11	-7.9908000E-11	3.4202900E+03
3.5935000E-09	8.5636200E-11	-8.6062400E-11	3.6838900E+03

3.6290800E-09	9.2192100E-11	-9.2653400E-11	3.9660800E+03
3.6646600E-09	9.9209700E-11	-9.9708800E-11	4.2680100E+03
3.7002400E-09	1.0671800E-10	-1.0725800E-10	4.5909100E+03
3.7358200E-09	1.1474900E-10	-1.1533200E-10	4.9360900E+03
3.7714000E-09	1.2333400E-10	-1.2396400E-10	5.3048800E+03
3.8069800E-09	1.3250800E-10	-1.3318900E-10	5.6987200E+03
3.8425600E-09	1.4230600E-10	-1.4304200E-10	6.1191200E+03
3.8781300E-09	1.5276800E-10	-1.5356100E-10	6.5676300E+03
3.9137100E-09	1.6393200E-10	-1.6478800E-10	7.0459200E+03
3.9492900E-09	1.7584100E-10	-1.7676400E-10	7.5556900E+03
3.9848700E-09	1.8853800E-10	-1.8953400E-10	8.0987700E+03
4.0204500E-09	2.0207100E-10	-2.0314400E-10	8.6770400E+03
4.0560300E-09	2.1648800E-10	-2.1764300E-10	9.2925000E+03
4.0916100E-09	2.3183900E-10	-2.3308300E-10	9.9472000E+03
4.1271900E-09	2.4817900E-10	-2.4951700E-10	1.0643300E+04
4.1627700E-09	2.6556200E-10	-2.6700200E-10	1.1383100E+04
4.1983500E-09	2.8404900E-10	-2.8559600E-10	1.2168900E+04
4.2339300E-09	3.0369900E-10	-3.0536200E-10	1.3003100E+04
4.2695100E-09	3.2457700E-10	-3.2636400E-10	1.3888400E+04
4.3050900E-09	3.4675000E-10	-3.4866900E-10	1.4827300E+04
4.3406700E-09	3.7028800E-10	-3.7234800E-10	1.5822700E+04
4.3762500E-09	3.9526400E-10	-3.9747500E-10	1.6877300E+04
4.4118300E-09	4.2175400E-10	-4.2412500E-10	1.7994200E+04
4.4474100E-09	4.4983600E-10	-4.5237800E-10	1.9176400E+04
4.4829900E-09	4.7959500E-10	-4.8231900E-10	2.0427200E+04
4.5185700E-09	5.1111500E-10	-5.1403300E-10	2.1749700E+04
4.5541500E-09	5.4448600E-10	-5.4761100E-10	2.3147500E+04
4.5897300E-09	5.7980100E-10	-5.8314600E-10	2.4624000E+04
4.6253100E-09	6.1715700E-10	-6.2073600E-10	2.6183000E+04
4.6608800E-09	6.5665300E-10	-6.6048100E-10	2.7828000E+04
4.6964600E-09	6.9839400E-10	-7.0248700E-10	2.9563100E+04
4.7320400E-09	7.4248800E-10	-7.4686100E-10	3.1392200E+04
4.7676200E-09	7.8904600E-10	-7.9371700E-10	3.3319300E+04
4.8032000E-09	8.3818300E-10	-8.4317100E-10	3.5348700E+04
4.8387800E-09	8.9002000E-10	-8.9534400E-10	3.7484800E+04
4.8743600E-09	9.4468000E-10	-9.5036000E-10	3.9731900E+04
4.9099400E-09	1.0022900E-09	-1.0083500E-09	4.2094500E+04
4.9455200E-09	1.0629800E-09	-1.0694400E-09	4.4577200E+04
4.9811000E-09	1.1269000E-09	-1.1337700E-09	4.7185000E+04
5.0166800E-09	1.1941700E-09	-1.2014900E-09	4.9922400E+04
5.0522600E-09	1.2649400E-09	-1.2727400E-09	5.2794500E+04
5.0878400E-09	1.3393600E-09	-1.3476600E-09	5.5806300E+04
5.1234200E-09	1.4175900E-09	-1.4264200E-09	5.8962800E+04
5.1590000E-09	1.4997800E-09	-1.5091700E-09	6.2269200E+04
5.1945800E-09	1.5860900E-09	-1.5960700E-09	6.5730800E+04
5.2301600E-09	1.6766900E-09	-1.6872900E-09	6.9352800E+04
5.2657400E-09	1.7717500E-09	-1.7830100E-09	7.3140700E+04
5.3013200E-09	1.8714300E-09	-1.8833800E-09	7.7099700E+04
5.3369000E-09	1.9759200E-09	-1.9886100E-09	8.1235500E+04
5.3724800E-09	2.0854000E-09	-2.0988600E-09	8.5553300E+04
5.4080500E-09	2.2000500E-09	-2.2143200E-09	9.0058800E+04

5.4436300E-09	2.3200700E-09	-2.3351900E-09	9.4757300E+04
5.4792100E-09	2.4456400E-09	-2.4616600E-09	9.9654600E+04
5.5147900E-09	2.5769600E-09	-2.5939200E-09	1.0475600E+05
5.5503700E-09	2.7142300E-09	-2.7321800E-09	1.1006700E+05
5.5859500E-09	2.8576500E-09	-2.8766500E-09	1.1559300E+05
5.6215300E-09	3.0074300E-09	-3.0275200E-09	1.2133900E+05
5.6571100E-09	3.1637700E-09	-3.1850100E-09	1.2731100E+05
5.6926900E-09	3.3269000E-09	-3.3493400E-09	1.3351400E+05
5.7282700E-09	3.4970100E-09	-3.5207200E-09	1.3995200E+05
5.7638500E-09	3.6743300E-09	-3.6993600E-09	1.4663200E+05
5.7994300E-09	3.8590700E-09	-3.8854900E-09	1.5355800E+05
5.8350100E-09	4.0514500E-09	-4.0793300E-09	1.6073400E+05
5.8705900E-09	4.2517000E-09	-4.2811000E-09	1.6816500E+05
5.9061700E-09	4.4600300E-09	-4.4910200E-09	1.7585600E+05
5.9417500E-09	4.6766800E-09	-4.7093200E-09	1.8381000E+05
5.9773300E-09	4.9018500E-09	-4.9362400E-09	1.9203200E+05
6.0129100E-09	5.1357800E-09	-5.1719800E-09	2.0052400E+05
6.0484900E-09	5.3786800E-09	-5.4167800E-09	2.0929200E+05
6.0840700E-09	5.6307800E-09	-5.6708600E-09	2.1833600E+05
6.1196500E-09	5.8923100E-09	-5.9344500E-09	2.2766200E+05
6.1552300E-09	6.1634700E-09	-6.2077700E-09	2.3727000E+05
6.1908000E-09	6.4445000E-09	-6.4910400E-09	2.4716200E+05
6.2263800E-09	6.7356000E-09	-6.7844700E-09	2.5734200E+05
6.2619600E-09	7.0369800E-09	-7.0882900E-09	2.6780900E+05
6.2975400E-09	7.3488600E-09	-7.4027000E-09	2.7856500E+05
6.3331200E-09	7.6714500E-09	-7.7279100E-09	2.8960900E+05
6.3687000E-09	8.0049300E-09	-8.0641400E-09	3.0094300E+05
6.4042800E-09	8.3495200E-09	-8.4115600E-09	3.1256500E+05
6.4398600E-09	8.7053900E-09	-8.7703900E-09	3.2447400E+05
6.4754400E-09	9.0727500E-09	-9.1408100E-09	3.3666800E+05
6.5110200E-09	9.4517500E-09	-9.5229900E-09	3.4914600E+05
6.5466000E-09	9.8425800E-09	-9.9171200E-09	3.6190400E+05
6.5821800E-09	1.0245400E-08	-1.0323400E-08	3.7493900E+05
6.6177600E-09	1.0660400E-08	-1.0741900E-08	3.8824700E+05
6.6533400E-09	1.1087600E-08	-1.1172800E-08	4.0182300E+05
6.6889200E-09	1.1527300E-08	-1.1616300E-08	4.1566200E+05
6.7245000E-09	1.1979600E-08	-1.2072400E-08	4.2975700E+05
6.7600800E-09	1.2444500E-08	-1.2541400E-08	4.4410200E+05
6.7956600E-09	1.2922200E-08	-1.3023300E-08	4.5868900E+05
6.8312400E-09	1.3412700E-08	-1.3518200E-08	4.7351000E+05
6.8668200E-09	1.3916300E-08	-1.4026200E-08	4.8855500E+05
6.9024000E-09	1.4432800E-08	-1.4547300E-08	5.0381500E+05
6.9379800E-09	1.4962400E-08	-1.5081700E-08	5.1928000E+05
6.9735500E-09	1.5505100E-08	-1.5629300E-08	5.3493700E+05
7.0091300E-09	1.6061000E-08	-1.6190200E-08	5.5077400E+05
7.0447100E-09	1.6630000E-08	-1.6764400E-08	5.6677900E+05
7.0802900E-09	1.7212100E-08	-1.7351900E-08	5.8293800E+05
7.1158700E-09	1.7807400E-08	-1.7952700E-08	5.9923500E+05
7.1514500E-09	1.8415700E-08	-1.8566600E-08	6.1565600E+05
7.1870300E-09	1.9037100E-08	-1.9193800E-08	6.3218300E+05
7.2226100E-09	1.9671400E-08	-1.9834000E-08	6.4880000E+05

7.2581900E-09	2.0318500E-08	-2.0487300E-08	6.6548900E+05
7.2937700E-09	2.0978300E-08	-2.1153400E-08	6.8223100E+05
7.3293500E-09	2.1650800E-08	-2.1832300E-08	6.9900600E+05
7.3649300E-09	2.2335700E-08	-2.2523700E-08	7.1579400E+05
7.4005100E-09	2.3032800E-08	-2.3227600E-08	7.3257400E+05
7.4360900E-09	2.3742000E-08	-2.3943700E-08	7.4932400E+05
7.4716700E-09	2.4463000E-08	-2.4671900E-08	7.6602200E+05
7.5072500E-09	2.5195700E-08	-2.5411700E-08	7.8264300E+05
7.5428300E-09	2.5939700E-08	-2.6163100E-08	7.9916400E+05
7.5784000E-09	2.6694700E-08	-2.6925700E-08	8.1556100E+05
7.6139800E-09	2.7460500E-08	-2.7699200E-08	8.3180700E+05
7.6495600E-09	2.8236700E-08	-2.8483200E-08	8.4787900E+05
7.6851400E-09	2.9023100E-08	-2.9277500E-08	8.6374700E+05
7.7207200E-09	2.9819100E-08	-3.0081700E-08	8.7938600E+05
7.7563000E-09	3.0624400E-08	-3.0895300E-08	8.9476800E+05
7.7918800E-09	3.1438600E-08	-3.1718000E-08	9.0986500E+05
7.8274500E-09	3.2261300E-08	-3.2549200E-08	9.2465000E+05
7.8630300E-09	3.3091900E-08	-3.3388600E-08	9.3909200E+05
7.8986100E-09	3.3930100E-08	-3.4235600E-08	9.5316400E+05
7.9341900E-09	3.4775300E-08	-3.5089800E-08	9.6683600E+05
7.9697700E-09	3.5626900E-08	-3.5950500E-08	9.8007900E+05
8.0053500E-09	3.6484500E-08	-3.6817300E-08	9.9286300E+05
8.0409300E-09	3.7347400E-08	-3.7689600E-08	1.0051600E+06
8.0765000E-09	3.8215100E-08	-3.8566800E-08	1.0169400E+06
8.1120800E-09	3.9086900E-08	-3.9448200E-08	1.0281700E+06
8.1476600E-09	3.9962300E-08	-4.0333300E-08	1.0388300E+06
8.1832400E-09	4.0840500E-08	-4.1221400E-08	1.0488800E+06
8.2188200E-09	4.1721000E-08	-4.2111700E-08	1.0583000E+06
8.2544000E-09	4.2602900E-08	-4.3003700E-08	1.0670500E+06
8.2899800E-09	4.3485800E-08	-4.3896600E-08	1.0751200E+06
8.3255600E-09	4.4368700E-08	-4.4789700E-08	1.0824700E+06
8.3611300E-09	4.5251000E-08	-4.5682200E-08	1.0890700E+06
8.3967100E-09	4.6131900E-08	-4.6573400E-08	1.0949000E+06
8.4322900E-09	4.7010700E-08	-4.7462600E-08	1.0999400E+06
8.4678700E-09	4.7886700E-08	-4.8348900E-08	1.1041500E+06
8.5034500E-09	4.8758900E-08	-4.9231600E-08	1.1075200E+06
8.5390300E-09	4.9626700E-08	-5.0109900E-08	1.1100200E+06
8.5746100E-09	5.0489200E-08	-5.0982900E-08	1.1116300E+06
8.6101800E-09	5.1345700E-08	-5.1849900E-08	1.1123300E+06
8.6457600E-09	5.2195200E-08	-5.2710000E-08	1.1120900E+06
8.6813400E-09	5.3037100E-08	-5.3562400E-08	1.1109000E+06
8.7169200E-09	5.3870400E-08	-5.4406300E-08	1.1087400E+06
8.7525000E-09	5.4694400E-08	-5.5240700E-08	1.1056000E+06
8.7880800E-09	5.5508100E-08	-5.6065000E-08	1.1014500E+06
8.8236600E-09	5.6310900E-08	-5.6878200E-08	1.0962900E+06
8.8592400E-09	5.7101800E-08	-5.7679600E-08	1.0901100E+06
8.8948100E-09	5.7880100E-08	-5.8468200E-08	1.0828800E+06
8.9303900E-09	5.8644900E-08	-5.9243300E-08	1.0746200E+06
8.9659700E-09	5.9395400E-08	-6.0004000E-08	1.0652900E+06
9.0015500E-09	6.0130800E-08	-6.0749600E-08	1.0549100E+06
9.0371300E-09	6.0850300E-08	-6.1479100E-08	1.0434600E+06

9.0727100E-09	6.1553100E-08	-6.2191900E-08	1.0309500E+06
9.1082900E-09	6.2238500E-08	-6.2887100E-08	1.0173700E+06
9.1438600E-09	6.2905600E-08	-6.3563900E-08	1.0027300E+06
9.1794400E-09	6.3553700E-08	-6.4221700E-08	9.8702000E+05
9.2150200E-09	6.4182100E-08	-6.4859500E-08	9.7025600E+05
9.2506000E-09	6.4790100E-08	-6.5476800E-08	9.5244100E+05
9.2861800E-09	6.5377000E-08	-6.6072800E-08	9.3358600E+05
9.3217600E-09	6.5942100E-08	-6.6646800E-08	9.1369900E+05
9.3573400E-09	6.6484600E-08	-6.7198200E-08	8.9279200E+05
9.3929100E-09	6.7004100E-08	-6.7726300E-08	8.7088200E+05
9.4284900E-09	6.7499900E-08	-6.8230400E-08	8.4798500E+05
9.4640700E-09	6.7971300E-08	-6.8710100E-08	8.2411800E+05
9.4996500E-09	6.8417900E-08	-6.9164600E-08	7.9930100E+05
9.5352300E-09	6.8839100E-08	-6.9593600E-08	7.7355500E+05
9.5708100E-09	6.9234400E-08	-6.9996400E-08	7.4690500E+05
9.6063900E-09	6.9603300E-08	-7.0372500E-08	7.1937800E+05
9.6419700E-09	6.9945300E-08	-7.0721600E-08	6.9099800E+05
9.6775400E-09	7.0260100E-08	-7.1043100E-08	6.6179300E+05
9.7131200E-09	7.0547200E-08	-7.1336700E-08	6.3179600E+05
9.7487000E-09	7.0806300E-08	-7.1602100E-08	6.0104000E+05
9.7842800E-09	7.1037100E-08	-7.1838800E-08	5.6955500E+05
9.8198600E-09	7.1239400E-08	-7.2046700E-08	5.3737700E+05
9.8554400E-09	7.1412700E-08	-7.2225400E-08	5.0454200E+05
9.8910200E-09	7.1557000E-08	-7.2374800E-08	4.7108900E+05
9.9265900E-09	7.1672000E-08	-7.2494500E-08	4.3705700E+05
9.9621700E-09	7.1757700E-08	-7.2584600E-08	4.0248500E+05
9.9977500E-09	7.1813800E-08	-7.2644900E-08	3.6741500E+05
1.0033300E-08	7.1840300E-08	-7.2675200E-08	3.3188800E+05
1.0068900E-08	7.1846900E-08	-7.2685200E-08	2.9594700E+05
1.0104500E-08	7.1846000E-08	-7.2687400E-08	2.5963600E+05
1.0140100E-08	7.1837400E-08	-7.2681700E-08	2.2300200E+05
1.0175600E-08	7.1821400E-08	-7.2668200E-08	1.8608600E+05
1.0211200E-08	7.1798100E-08	-7.2647000E-08	1.4893800E+05
1.0246800E-08	7.1767800E-08	-7.2618500E-08	1.1162900E+05
1.0282400E-08	7.1731200E-08	-7.2583200E-08	7.4302300E+04
1.0318000E-08	7.1689300E-08	-7.2542300E-08	3.7256700E+04
1.0353500E-08	7.1644200E-08	-7.2497800E-08	1.0326700E+03
1.0389100E-08	7.1599500E-08	-7.2453200E-08	-3.3564100E+04
1.0424700E-08	7.1559900E-08	-7.2413300E-08	-6.5507200E+04
1.0460300E-08	7.1530500E-08	-7.2383100E-08	-9.3664200E+04
1.0495900E-08	7.1515500E-08	-7.2367000E-08	-1.1692200E+05
1.0531400E-08	7.1517900E-08	-7.2367800E-08	-1.3438400E+05
1.0567000E-08	7.1539800E-08	-7.2388100E-08	-1.4559900E+05
1.0602600E-08	7.1583100E-08	-7.2429500E-08	-1.5045600E+05
1.0638200E-08	7.1648600E-08	-7.2493300E-08	-1.4890800E+05
1.0673700E-08	7.1736000E-08	-7.2579400E-08	-1.4109000E+05
1.0709300E-08	7.1843100E-08	-7.2686300E-08	-1.2761500E+05
1.0744900E-08	7.1966000E-08	-7.2810300E-08	-1.0952000E+05
1.0780500E-08	7.2099800E-08	-7.2946200E-08	-8.8000100E+04
1.0816100E-08	7.2238700E-08	-7.3088100E-08	-6.4264800E+04
1.0851600E-08	7.2376500E-08	-7.3229100E-08	-3.9601500E+04

1.0887200E-08	7.2505500E-08	-7.3361400E-08	-1.5414800E+04
1.0922800E-08	7.2617400E-08	-7.3476300E-08	6.8878700E+03
1.0958400E-08	7.2704000E-08	-7.3565300E-08	2.6085500E+04
1.0994000E-08	7.2758000E-08	-7.3621100E-08	4.1294800E+04
1.1029500E-08	7.2773700E-08	-7.3638200E-08	5.2014700E+04
1.1065100E-08	7.2747300E-08	-7.3612800E-08	5.8045300E+04
1.1100700E-08	7.2677100E-08	-7.3543500E-08	5.9448900E+04
1.1136300E-08	7.2563700E-08	-7.3430900E-08	5.6652800E+04
1.1171800E-08	7.2411000E-08	-7.3278700E-08	5.0456900E+04
1.1207400E-08	7.2225700E-08	-7.3093200E-08	4.1824700E+04
1.1243000E-08	7.2016800E-08	-7.2883400E-08	3.1691300E+04
1.1278600E-08	7.1794700E-08	-7.2659400E-08	2.0936200E+04
1.1314200E-08	7.1570800E-08	-7.2433000E-08	1.0407600E+04
1.1349700E-08	7.1356700E-08	-7.2216100E-08	8.9214000E+02
1.1385300E-08	7.1164100E-08	-7.2020400E-08	-6.9767000E+03
1.1420900E-08	7.1002900E-08	-7.1856300E-08	-1.2813200E+04
1.1456500E-08	7.0880700E-08	-7.1731600E-08	-1.6498800E+04
1.1492100E-08	7.0802100E-08	-7.1651200E-08	-1.8094500E+04
1.1527600E-08	7.0768800E-08	-7.1616700E-08	-1.7746000E+04
1.1563200E-08	7.0779500E-08	-7.1626900E-08	-1.5717000E+04
1.1598800E-08	7.0829900E-08	-7.1677500E-08	-1.2477100E+04
1.1634400E-08	7.0912700E-08	-7.1761000E-08	-8.6664900E+03
1.1669900E-08	7.1018800E-08	-7.1868100E-08	-4.9550900E+03
1.1705500E-08	7.1137800E-08	-7.1988200E-08	-1.9328000E+03
1.1741100E-08	7.1259100E-08	-7.2110900E-08	-7.4807200E+01
1.1776700E-08	7.1373100E-08	-7.2226400E-08	2.6853600E+02
1.1812300E-08	7.1471500E-08	-7.2326400E-08	-1.0911400E+03
1.1847800E-08	7.1547600E-08	-7.2404100E-08	-4.1401200E+03
1.1883400E-08	7.1597300E-08	-7.2455100E-08	-8.6838500E+03
1.1919000E-08	7.1618900E-08	-7.2477900E-08	-1.4410200E+04
1.1954600E-08	7.1613800E-08	-7.2473500E-08	-2.0919600E+04
1.1990200E-08	7.1585300E-08	-7.2445700E-08	-2.7731400E+04
1.2025700E-08	7.1538600E-08	-7.2399700E-08	-3.4334900E+04
1.2061300E-08	7.1480100E-08	-7.2342000E-08	-4.0280000E+04
1.2096900E-08	7.1417000E-08	-7.2279900E-08	-4.5244900E+04
1.2132500E-08	7.1356300E-08	-7.2220700E-08	-4.9054400E+04
1.2168000E-08	7.1304800E-08	-7.2170700E-08	-5.1665000E+04
1.2203600E-08	7.1267700E-08	-7.2134900E-08	-5.3130900E+04
1.2239200E-08	7.1248600E-08	-7.2116900E-08	-5.3578400E+04
1.2274800E-08	7.1249500E-08	-7.2118400E-08	-5.3208500E+04
1.2310400E-08	7.1270700E-08	-7.2139800E-08	-5.2298400E+04
1.2345900E-08	7.1311000E-08	-7.2179900E-08	-5.1176400E+04
1.2381500E-08	7.1367800E-08	-7.2236300E-08	-5.0167500E+04
1.2417100E-08	7.1437400E-08	-7.2305200E-08	-4.9536600E+04
1.2452700E-08	7.1515600E-08	-7.2382500E-08	-4.9467500E+04
1.2488300E-08	7.1597800E-08	-7.2463800E-08	-5.0089000E+04
1.2523800E-08	7.1679900E-08	-7.2544600E-08	-5.1501100E+04
1.2559400E-08	7.1757700E-08	-7.2621000E-08	-5.3758000E+04
1.2595000E-08	7.1827700E-08	-7.2689600E-08	-5.6816700E+04
1.2630600E-08	7.1887300E-08	-7.2748100E-08	-6.0505600E+04
1.2666200E-08	7.1934900E-08	-7.2795100E-08	-6.4548600E+04

1.2701700E-08	7.1970100E-08	-7.2830300E-08	-6.8627800E+04
1.2737300E-08	7.1993500E-08	-7.2854400E-08	-7.2431900E+04
1.2772900E-08	7.2006500E-08	-7.2868700E-08	-7.5671800E+04
1.2808500E-08	7.2011100E-08	-7.2875100E-08	-7.8084800E+04
1.2844000E-08	7.2010200E-08	-7.2876500E-08	-7.9457200E+04
1.2879600E-08	7.2007200E-08	-7.2876100E-08	-7.9666600E+04
1.2915200E-08	7.2005400E-08	-7.2877000E-08	-7.8704400E+04
1.2950800E-08	7.2007800E-08	-7.2882200E-08	-7.6653900E+04
1.2986400E-08	7.2017200E-08	-7.2894100E-08	-7.3655900E+04
1.3021900E-08	7.2035500E-08	-7.2914800E-08	-6.9895800E+04
1.3057500E-08	7.2063900E-08	-7.2945300E-08	-6.5605900E+04
1.3093100E-08	7.2102400E-08	-7.2985500E-08	-6.1046100E+04
1.3128700E-08	7.2149600E-08	-7.3033900E-08	-5.6456100E+04
1.3164300E-08	7.2202900E-08	-7.3087800E-08	-5.2025500E+04
1.3199800E-08	7.2258200E-08	-7.3143400E-08	-4.7901700E+04
1.3235400E-08	7.2310900E-08	-7.3196000E-08	-4.4199800E+04
1.3271000E-08	7.2355400E-08	-7.3240200E-08	-4.0994300E+04
1.3306600E-08	7.2386200E-08	-7.3270300E-08	-3.8316000E+04
1.3342100E-08	7.2397600E-08	-7.3280900E-08	-3.6160000E+04
1.3377700E-08	7.2384800E-08	-7.3267000E-08	-3.4495400E+04
1.3413300E-08	7.2344300E-08	-7.3225500E-08	-3.3271000E+04
1.3448900E-08	7.2274700E-08	-7.3154700E-08	-3.2410300E+04
1.3484500E-08	7.2175900E-08	-7.3055000E-08	-3.1819000E+04
1.3520000E-08	7.2050500E-08	-7.2928700E-08	-3.1407300E+04
1.3555600E-08	7.1903000E-08	-7.2780700E-08	-3.1098100E+04
1.3591200E-08	7.1739900E-08	-7.2617500E-08	-3.0813600E+04
1.3626800E-08	7.1569500E-08	-7.2447600E-08	-3.0476300E+04
1.3662400E-08	7.1400900E-08	-7.2279800E-08	-3.0031600E+04
1.3697900E-08	7.1243400E-08	-7.2123400E-08	-2.9466200E+04
1.3733500E-08	7.1105600E-08	-7.1987100E-08	-2.8808000E+04
1.3769100E-08	7.0995300E-08	-7.1878200E-08	-2.8124700E+04
1.3804700E-08	7.0918500E-08	-7.1802800E-08	-2.7525900E+04
1.3840200E-08	7.0878600E-08	-7.1764100E-08	-2.7159800E+04
1.3875800E-08	7.0876600E-08	-7.1763100E-08	-2.7199500E+04
1.3911400E-08	7.0911100E-08	-7.1798100E-08	-2.7825900E+04
1.3947000E-08	7.0978000E-08	-7.1865200E-08	-2.9216800E+04
1.3982600E-08	7.1071700E-08	-7.1958600E-08	-3.1542900E+04
1.4018100E-08	7.1184800E-08	-7.2071100E-08	-3.4971400E+04
1.4053700E-08	7.1309100E-08	-7.2194500E-08	-3.9650400E+04
1.4089300E-08	7.1436400E-08	-7.2320700E-08	-4.5671900E+04
1.4124900E-08	7.1559100E-08	-7.2442100E-08	-5.3067300E+04
1.4160500E-08	7.1670500E-08	-7.2552200E-08	-6.1838100E+04
1.4196000E-08	7.1765800E-08	-7.2646400E-08	-7.1970600E+04
1.4231600E-08	7.1841400E-08	-7.2721300E-08	-8.3416600E+04
1.4267200E-08	7.1896100E-08	-7.2775500E-08	-9.6057500E+04
1.4302800E-08	7.1930000E-08	-7.2809400E-08	-1.0968300E+05
1.4338300E-08	7.1945300E-08	-7.2824900E-08	-1.2401700E+05
1.4373900E-08	7.1944800E-08	-7.2825000E-08	-1.3876000E+05
1.4409500E-08	7.1932500E-08	-7.2813500E-08	-1.5358700E+05
1.4445100E-08	7.1912900E-08	-7.2794500E-08	-1.6814200E+05
1.4480700E-08	7.1890100E-08	-7.2772400E-08	-1.8203200E+05

1.4516200E-08	7.1868500E-08	-7.2751300E-08	-1.9485600E+05
1.4551800E-08	7.1851600E-08	-7.2734700E-08	-2.0623400E+05
1.4587400E-08	7.1842300E-08	-7.2725300E-08	-2.1579800E+05
1.4623000E-08	7.1842400E-08	-7.2725100E-08	-2.2318600E+05
1.4658600E-08	7.1853000E-08	-7.2735000E-08	-2.2806100E+05
1.4694100E-08	7.1874000E-08	-7.2754900E-08	-2.3014300E+05
1.4729700E-08	7.1904400E-08	-7.2784100E-08	-2.2923900E+05
1.4765300E-08	7.1942400E-08	-7.2820700E-08	-2.2526600E+05
1.4800900E-08	7.1985400E-08	-7.2862200E-08	-2.1824200E+05
1.4836400E-08	7.2030200E-08	-7.2905500E-08	-2.0826700E+05
1.4872000E-08	7.2073300E-08	-7.2947300E-08	-1.9554600E+05
1.4907600E-08	7.2111100E-08	-7.2983900E-08	-1.8038200E+05
1.4943200E-08	7.2140200E-08	-7.3012000E-08	-1.6316000E+05
1.4978800E-08	7.2157200E-08	-7.3028300E-08	-1.4433900E+05
1.5014300E-08	7.2159400E-08	-7.3030200E-08	-1.2444800E+05
1.5049900E-08	7.2145000E-08	-7.3015700E-08	-1.0407300E+05
1.5085500E-08	7.2112800E-08	-7.2983900E-08	-8.3856600E+04
1.5121100E-08	7.2062800E-08	-7.2934700E-08	-6.4501000E+04
1.5156700E-08	7.1996000E-08	-7.2868900E-08	-4.6732500E+04
1.5192200E-08	7.1914100E-08	-7.2788400E-08	-3.1243400E+04
1.5227800E-08	7.1819800E-08	-7.2695800E-08	-1.8647200E+04
1.5263400E-08	7.1716600E-08	-7.2594400E-08	-9.4567000E+03
1.5299000E-08	7.1608400E-08	-7.2487900E-08	-4.0615200E+03
1.5334600E-08	7.1498800E-08	-7.2380200E-08	-2.7003400E+03
1.5370200E-08	7.1391900E-08	-7.2275000E-08	-5.4378600E+03
1.5405700E-08	7.1291100E-08	-7.2175700E-08	-1.2149300E+04
1.5441300E-08	7.1199700E-08	-7.2085500E-08	-2.2516100E+04
1.5476900E-08	7.1120200E-08	-7.2006800E-08	-3.6044100E+04
1.5512500E-08	7.1054500E-08	-7.1941600E-08	-5.2102100E+04
1.5548100E-08	7.1003900E-08	-7.1891100E-08	-6.9952200E+04
1.5583600E-08	7.0969200E-08	-7.1855900E-08	-8.8765900E+04
1.5619200E-08	7.0950700E-08	-7.1836600E-08	-1.0765700E+05
1.5654800E-08	7.0948400E-08	-7.1832900E-08	-1.2573000E+05
1.5690400E-08	7.0961700E-08	-7.1844500E-08	-1.4213900E+05
1.5726000E-08	7.0989900E-08	-7.1870500E-08	-1.5613000E+05
1.5761500E-08	7.1031900E-08	-7.1910300E-08	-1.6707800E+05
1.5797100E-08	7.1086700E-08	-7.1962600E-08	-1.7452600E+05
1.5832700E-08	7.1152800E-08	-7.2026300E-08	-1.7821200E+05
1.5868300E-08	7.1228600E-08	-7.2099800E-08	-1.7807400E+05
1.5903900E-08	7.1312000E-08	-7.2181100E-08	-1.7424400E+05
1.5939400E-08	7.1400600E-08	-7.2268100E-08	-1.6703000E+05
1.5975000E-08	7.1491900E-08	-7.2358300E-08	-1.5691700E+05
1.6010600E-08	7.1582900E-08	-7.2448800E-08	-1.4455700E+05
1.6046200E-08	7.1670500E-08	-7.2536600E-08	-1.3073000E+05
1.6081800E-08	7.1751300E-08	-7.2618400E-08	-1.1627500E+05
1.6117300E-08	7.1822200E-08	-7.2691200E-08	-1.0203200E+05
1.6152900E-08	7.1880700E-08	-7.2752200E-08	-8.8815400E+04
1.6188500E-08	7.1924500E-08	-7.2799100E-08	-7.7374500E+04
1.6224100E-08	7.1952200E-08	-7.2830500E-08	-6.8356000E+04
1.6259700E-08	7.1963300E-08	-7.2845700E-08	-6.2263200E+04
1.6295200E-08	7.1958500E-08	-7.2845000E-08	-5.9400400E+04

1.6330800E-08	7.1939500E-08	-7.2830000E-08	-5.9850500E+04
1.6366400E-08	7.1909100E-08	-7.2803300E-08	-6.3492700E+04
1.6402000E-08	7.1870800E-08	-7.2768100E-08	-7.0031500E+04
1.6437600E-08	7.1828600E-08	-7.2728400E-08	-7.9021400E+04
1.6473100E-08	7.1787200E-08	-7.2688600E-08	-8.9888400E+04
1.6508700E-08	7.1751200E-08	-7.2653200E-08	-1.0196000E+05
1.6544300E-08	7.1724800E-08	-7.2626500E-08	-1.1450000E+05
1.6579900E-08	7.1711500E-08	-7.2612100E-08	-1.2673800E+05
1.6615500E-08	7.1714000E-08	-7.2612600E-08	-1.3792500E+05
1.6651000E-08	7.1733800E-08	-7.2629800E-08	-1.4737800E+05
1.6686600E-08	7.1771000E-08	-7.2664000E-08	-1.5452600E+05
1.6722200E-08	7.1824800E-08	-7.2714400E-08	-1.5894700E+05
1.6757800E-08	7.1892500E-08	-7.2778900E-08	-1.6039400E+05
1.6793400E-08	7.1970700E-08	-7.2854100E-08	-1.5879500E+05
1.6828900E-08	7.2055100E-08	-7.2935800E-08	-1.5426100E+05
1.6864500E-08	7.2140400E-08	-7.3019200E-08	-1.4709900E+05
1.6900100E-08	7.2221500E-08	-7.3099100E-08	-1.3778300E+05
1.6935700E-08	7.2293200E-08	-7.3170200E-08	-1.2689300E+05
1.6971300E-08	7.2350400E-08	-7.3227900E-08	-1.1506800E+05
1.7006900E-08	7.2389100E-08	-7.3267700E-08	-1.0297500E+05
1.7042400E-08	7.2406200E-08	-7.3286700E-08	-9.1280700E+04
1.7078000E-08	7.2399900E-08	-7.3282700E-08	-8.0628400E+04
1.7113600E-08	7.2369600E-08	-7.3255100E-08	-7.1600700E+04
1.7149200E-08	7.2316200E-08	-7.3204600E-08	-6.4672200E+04
1.7184800E-08	7.2241800E-08	-7.3133200E-08	-6.0172100E+04
1.7220300E-08	7.2150000E-08	-7.3044300E-08	-5.8266400E+04
1.7255900E-08	7.2045300E-08	-7.2942100E-08	-5.8953700E+04
1.7291500E-08	7.1933100E-08	-7.2831900E-08	-6.2067700E+04
1.7327100E-08	7.1818700E-08	-7.2719100E-08	-6.7291100E+04
1.7362700E-08	7.1707800E-08	-7.2609100E-08	-7.4185000E+04
1.7398200E-08	7.1605600E-08	-7.2507100E-08	-8.2224500E+04
1.7433800E-08	7.1516300E-08	-7.2417500E-08	-9.0837300E+04
1.7469400E-08	7.1443400E-08	-7.2343700E-08	-9.9453400E+04
1.7505000E-08	7.1388800E-08	-7.2287800E-08	-1.0754800E+05
1.7540600E-08	7.1353000E-08	-7.2250400E-08	-1.1466000E+05
1.7576100E-08	7.1335300E-08	-7.2230900E-08	-1.2041500E+05
1.7611700E-08	7.1333700E-08	-7.2227500E-08	-1.2456100E+05
1.7647300E-08	7.1345300E-08	-7.2237300E-08	-1.2699400E+05
1.7682900E-08	7.1366300E-08	-7.2256700E-08	-1.2776500E+05
1.7718500E-08	7.1392600E-08	-7.2281700E-08	-1.2707900E+05
1.7754000E-08	7.1420300E-08	-7.2308300E-08	-1.2527500E+05
1.7789600E-08	7.1445800E-08	-7.2333100E-08	-1.2281400E+05
1.7825200E-08	7.1466300E-08	-7.2353400E-08	-1.2023700E+05
1.7860800E-08	7.1480100E-08	-7.2367200E-08	-1.1810700E+05
1.7896400E-08	7.1486200E-08	-7.2373800E-08	-1.1696000E+05
1.7931900E-08	7.1485300E-08	-7.2373500E-08	-1.1726000E+05
1.7967500E-08	7.1478800E-08	-7.2367600E-08	-1.1936900E+05
1.8003100E-08	7.1468800E-08	-7.2358400E-08	-1.2353000E+05
1.8038700E-08	7.1458000E-08	-7.2348200E-08	-1.2984300E+05
1.8074300E-08	7.1449400E-08	-7.2339900E-08	-1.3823400E+05
1.8109800E-08	7.1445600E-08	-7.2336100E-08	-1.4847700E+05

1.8145400E-08	7.1449000E-08	-7.2339000E-08	-1.6021500E+05
1.8181000E-08	7.1461100E-08	-7.2350300E-08	-1.7296900E+05
1.8216600E-08	7.1482800E-08	-7.2370800E-08	-1.8615900E+05
1.8252200E-08	7.1513900E-08	-7.2400200E-08	-1.9914600E+05
1.8287700E-08	7.1553700E-08	-7.2437900E-08	-2.1125700E+05
1.8323300E-08	7.1600300E-08	-7.2482300E-08	-2.2184400E+05
1.8358900E-08	7.1651900E-08	-7.2531400E-08	-2.3033700E+05
1.8394500E-08	7.1706000E-08	-7.2583100E-08	-2.3627200E+05
1.8430100E-08	7.1760300E-08	-7.2635100E-08	-2.3931300E+05
1.8465600E-08	7.1812400E-08	-7.2685200E-08	-2.3927600E+05
1.8501200E-08	7.1860600E-08	-7.2732000E-08	-2.3615700E+05
1.8536800E-08	7.1903600E-08	-7.2774100E-08	-2.3014500E+05
1.8572400E-08	7.1940700E-08	-7.2810900E-08	-2.2157400E+05
1.8608000E-08	7.1971400E-08	-7.2842100E-08	-2.1089000E+05
1.8643600E-08	7.1996000E-08	-7.2867900E-08	-1.9861900E+05
1.8679100E-08	7.2014800E-08	-7.2888500E-08	-1.8535000E+05
1.8714700E-08	7.2028500E-08	-7.2904500E-08	-1.7169000E+05
1.8750300E-08	7.2037200E-08	-7.2916000E-08	-1.5823100E+05
1.8785900E-08	7.2041300E-08	-7.2923200E-08	-1.4551000E+05
1.8821500E-08	7.2040700E-08	-7.2925700E-08	-1.3396200E+05
1.8857000E-08	7.2034900E-08	-7.2922900E-08	-1.2388200E+05
1.8892600E-08	7.2023300E-08	-7.2914000E-08	-1.1543000E+05
1.8928200E-08	7.2004900E-08	-7.2898100E-08	-1.0863400E+05
1.8963800E-08	7.1978900E-08	-7.2873900E-08	-1.0339600E+05
1.8999400E-08	7.1944400E-08	-7.2840700E-08	-9.9502500E+04
1.9034900E-08	7.1900900E-08	-7.2797900E-08	-9.6653500E+04
1.9070500E-08	7.1848400E-08	-7.2745400E-08	-9.4504400E+04
1.9106100E-08	7.1787400E-08	-7.2683800E-08	-9.2702500E+04
1.9141700E-08	7.1718900E-08	-7.2614100E-08	-9.0929700E+04
1.9177300E-08	7.1644700E-08	-7.2538300E-08	-8.8933800E+04
1.9212800E-08	7.1567100E-08	-7.2458800E-08	-8.6553700E+04
1.9248400E-08	7.1488700E-08	-7.2378300E-08	-8.3739500E+04
1.9284000E-08	7.1412500E-08	-7.2300000E-08	-8.0555200E+04
1.9319600E-08	7.1341400E-08	-7.2226800E-08	-7.7173300E+04
1.9355200E-08	7.1278000E-08	-7.2161600E-08	-7.3867900E+04
1.9390700E-08	7.1224800E-08	-7.2106700E-08	-7.0992700E+04
1.9426300E-08	7.1183500E-08	-7.2064000E-08	-6.8951900E+04
1.9461900E-08	7.1155200E-08	-7.2034800E-08	-6.8167500E+04
1.9497500E-08	7.1140300E-08	-7.2019400E-08	-6.9026100E+04
1.9533100E-08	7.1138600E-08	-7.2017500E-08	-7.1834900E+04
1.9568600E-08	7.1149300E-08	-7.2028300E-08	-7.6792600E+04
1.9604200E-08	7.1170900E-08	-7.2050300E-08	-8.3958600E+04
1.9639800E-08	7.1201900E-08	-7.2082000E-08	-9.3230200E+04
1.9675400E-08	7.1240400E-08	-7.2121500E-08	-1.0433600E+05
1.9711000E-08	7.1284800E-08	-7.2166800E-08	-1.1684200E+05
1.9746500E-08	7.1333100E-08	-7.2216200E-08	-1.3017600E+05
1.9782100E-08	7.1384100E-08	-7.2268100E-08	-1.4367800E+05
1.9817700E-08	7.1436500E-08	-7.2321500E-08	-1.5664000E+05
1.9853300E-08	7.1489500E-08	-7.2375400E-08	-1.6835100E+05
1.9888900E-08	7.1542600E-08	-7.2429300E-08	-1.7813900E+05
1.9924400E-08	7.1595500E-08	-7.2483000E-08	-1.8542600E+05

1.9960000E-08	7.1648200E-08	-7.2536400E-08	-1.8976200E+05
1.9995600E-08	7.1700700E-08	-7.2589600E-08	-1.9087000E+05
2.0031200E-08	7.1753200E-08	-7.2642600E-08	-1.8866400E+05
2.0066800E-08	7.1805700E-08	-7.2695600E-08	-1.8325500E+05
2.0102300E-08	7.1857900E-08	-7.2748300E-08	-1.7494000E+05
2.0137900E-08	7.1909400E-08	-7.2800500E-08	-1.6419800E+05
2.0173500E-08	7.1959800E-08	-7.2851500E-08	-1.5166200E+05
2.0209100E-08	7.2008100E-08	-7.2900500E-08	-1.3807300E+05
2.0244700E-08	7.2053200E-08	-7.2946400E-08	-1.2422400E+05
2.0280300E-08	7.2093800E-08	-7.2988000E-08	-1.1090600E+05
2.0315800E-08	7.2128600E-08	-7.3023700E-08	-9.8862200E+04
2.0351400E-08	7.2156200E-08	-7.3052300E-08	-8.8749300E+04
2.0387000E-08	7.2175200E-08	-7.3072300E-08	-8.1096800E+04
2.0422600E-08	7.2184500E-08	-7.3082400E-08	-7.6276400E+04
2.0458200E-08	7.2183100E-08	-7.3081900E-08	-7.4487400E+04
2.0493700E-08	7.2170700E-08	-7.3070100E-08	-7.5746600E+04
2.0529300E-08	7.2147200E-08	-7.3047000E-08	-7.9896400E+04
2.0564900E-08	7.2113300E-08	-7.3013300E-08	-8.6618500E+04
2.0600500E-08	7.2070200E-08	-7.2970200E-08	-9.5452500E+04
2.0636100E-08	7.2019700E-08	-7.2919400E-08	-1.0582400E+05
2.0671600E-08	7.1964100E-08	-7.2863300E-08	-1.1707900E+05
2.0707200E-08	7.1906300E-08	-7.2804700E-08	-1.2853100E+05
2.0742800E-08	7.1849500E-08	-7.2746900E-08	-1.3951500E+05
2.0778400E-08	7.1797000E-08	-7.2693200E-08	-1.4941700E+05
2.0814000E-08	7.1752000E-08	-7.2646900E-08	-1.5770600E+05
2.0849500E-08	7.1717400E-08	-7.2610900E-08	-1.6397200E+05
2.0885100E-08	7.1695800E-08	-7.2587900E-08	-1.6793200E+05
2.0920700E-08	7.1688700E-08	-7.2579400E-08	-1.6943800E+05
2.0956300E-08	7.1696900E-08	-7.2586400E-08	-1.6847900E+05
2.0991900E-08	7.1720200E-08	-7.2608600E-08	-1.6518400E+05
2.1027400E-08	7.1757200E-08	-7.2644600E-08	-1.5980100E+05
2.1063000E-08	7.1805300E-08	-7.2692000E-08	-1.5269000E+05
2.1098600E-08	7.1861100E-08	-7.2747300E-08	-1.4429000E+05
2.1134200E-08	7.1920200E-08	-7.2806200E-08	-1.3510200E+05
2.1169800E-08	7.1977500E-08	-7.2863600E-08	-1.2565600E+05
2.1205300E-08	7.2027800E-08	-7.2914400E-08	-1.1647200E+05
2.1240900E-08	7.2065900E-08	-7.2953200E-08	-1.0804500E+05
2.1276500E-08	7.2087100E-08	-7.2975400E-08	-1.0080400E+05
2.1312100E-08	7.2087300E-08	-7.2977000E-08	-9.5095400E+04
2.1347700E-08	7.2063800E-08	-7.2955100E-08	-9.1167300E+04
2.1383200E-08	7.2015200E-08	-7.2908300E-08	-8.9160800E+04
2.1418800E-08	7.1941700E-08	-7.2836700E-08	-8.9102500E+04
2.1454400E-08	7.1845300E-08	-7.2742200E-08	-9.0919200E+04
2.1490000E-08	7.1729700E-08	-7.2628200E-08	-9.4449100E+04
2.1525600E-08	7.1599800E-08	-7.2499700E-08	-9.9444500E+04
2.1561100E-08	7.1462200E-08	-7.2362900E-08	-1.0558800E+05
2.1596700E-08	7.1323900E-08	-7.2224900E-08	-1.1251200E+05
2.1632300E-08	7.1192600E-08	-7.2093300E-08	-1.1983000E+05
2.1667900E-08	7.1075800E-08	-7.1975400E-08	-1.2715300E+05
2.1703500E-08	7.0980500E-08	-7.1878000E-08	-1.3410000E+05
2.1739000E-08	7.0912300E-08	-7.1807000E-08	-1.4031300E+05

2.1774600E-08	7.0875400E-08	-7.1766500E-08	-1.4547400E+05
2.1810200E-08	7.0872300E-08	-7.1759100E-08	-1.4930100E+05
2.1845800E-08	7.0903200E-08	-7.1785100E-08	-1.5154500E+05
2.1881400E-08	7.0966100E-08	-7.1842700E-08	-1.5200700E+05
2.1917000E-08	7.1057100E-08	-7.1928200E-08	-1.5052000E+05
2.1952500E-08	7.1170400E-08	-7.2036000E-08	-1.4693000E+05
2.1988100E-08	7.1298700E-08	-7.2158900E-08	-1.4111200E+05
2.2023700E-08	7.1433800E-08	-7.2289100E-08	-1.3295800E+05
2.2059300E-08	7.1567200E-08	-7.2417900E-08	-1.2238000E+05
2.2094900E-08	7.1690500E-08	-7.2537200E-08	-1.0930400E+05
2.2130400E-08	7.1796200E-08	-7.2639500E-08	-9.3670500E+04
2.2166000E-08	7.1878200E-08	-7.2718700E-08	-7.5435400E+04
2.2201600E-08	7.1932300E-08	-7.2770700E-08	-5.4573300E+04
2.2237200E-08	7.1956500E-08	-7.2793200E-08	-3.1067300E+04
2.2272800E-08	7.1950900E-08	-7.2786500E-08	-4.9057200E+03
2.2308300E-08	7.1918100E-08	-7.2752800E-08	2.3927000E+04
2.2343900E-08	7.1862600E-08	-7.2696400E-08	5.5453900E+04
2.2379500E-08	7.1790500E-08	-7.2623400E-08	8.9701600E+04
2.2415100E-08	7.1708900E-08	-7.2540500E-08	1.2672400E+05
2.2450700E-08	7.1625200E-08	-7.2455100E-08	1.6662300E+05
2.2486200E-08	7.1546600E-08	-7.2374000E-08	2.0956200E+05
2.2521800E-08	7.1479300E-08	-7.2303100E-08	2.5577500E+05
2.2557400E-08	7.1427600E-08	-7.2246800E-08	3.0558600E+05
2.2593000E-08	7.1394200E-08	-7.2207200E-08	3.5940400E+05
2.2628600E-08	7.1379100E-08	-7.2184400E-08	4.1774000E+05
2.2664100E-08	7.1380100E-08	-7.2176100E-08	4.8119500E+05
2.2699700E-08	7.1393000E-08	-7.2177700E-08	5.5043600E+05
2.2735300E-08	7.1411600E-08	-7.2183200E-08	6.2619400E+05
2.2770900E-08	7.1428700E-08	-7.2185300E-08	7.0923700E+05
2.2806500E-08	7.1436800E-08	-7.2176500E-08	8.0034900E+05
2.2842000E-08	7.1428300E-08	-7.2149300E-08	9.0032700E+05
2.2877600E-08	7.1397000E-08	-7.2097600E-08	1.0099500E+06
2.2913200E-08	7.1338200E-08	-7.2016700E-08	1.1299600E+06
2.2948800E-08	7.1249000E-08	-7.1904100E-08	1.2610300E+06
2.2984400E-08	7.1129300E-08	-7.1759500E-08	1.4037600E+06
2.3019900E-08	7.0980800E-08	-7.1585100E-08	1.5586900E+06
2.3055500E-08	7.0807700E-08	-7.1384900E-08	1.7262400E+06
2.3091100E-08	7.0615600E-08	-7.1164600E-08	1.9067700E+06
2.3126700E-08	7.0411300E-08	-7.0931200E-08	2.1005500E+06
2.3162300E-08	7.0202000E-08	-7.0691700E-08	2.3077900E+06
2.3197800E-08	6.9994600E-08	-7.0452900E-08	2.5286300E+06
2.3233400E-08	6.9795000E-08	-7.0220800E-08	2.7631700E+06
2.3269000E-08	6.9607900E-08	-6.9999600E-08	3.0114700E+06
2.3304600E-08	6.9435900E-08	-6.9791900E-08	3.2735700E+06
2.3340200E-08	6.9279600E-08	-6.9598000E-08	3.5494700E+06
2.3375700E-08	6.9137400E-08	-6.9416200E-08	3.8392100E+06
2.3411300E-08	6.9006200E-08	-6.9243200E-08	4.1427500E+06
2.3446900E-08	6.8880900E-08	-6.9074000E-08	4.4600900E+06
2.3482500E-08	6.8755800E-08	-6.8902600E-08	4.7911700E+06
2.3518100E-08	6.8624600E-08	-6.8722900E-08	5.1358800E+06
2.3553600E-08	6.8480800E-08	-6.8528600E-08	5.4940500E+06

2.3589200E-08	6.8318800E-08	-6.8314200E-08	5.8654200E+06
2.3624800E-08	6.8134100E-08	-6.8075400E-08	6.2496100E+06
2.3660400E-08	6.7923300E-08	-6.7809100E-08	6.6460700E+06
2.3696000E-08	6.7684700E-08	-6.7514200E-08	7.0540500E+06
2.3731600E-08	6.7418500E-08	-6.7191000E-08	7.4726700E+06
2.3767100E-08	6.7126100E-08	-6.6841400E-08	7.9008000E+06
2.3802700E-08	6.6810700E-08	-6.6468900E-08	8.3371500E+06
2.3838300E-08	6.6476200E-08	-6.6077800E-08	8.7802300E+06
2.3873900E-08	6.6127600E-08	-6.5673100E-08	9.2283800E+06
2.3909500E-08	6.5769700E-08	-6.5260200E-08	9.6798100E+06
2.3945000E-08	6.5407700E-08	-6.4844200E-08	1.0132600E+07
2.3980600E-08	6.5045900E-08	-6.4429500E-08	1.0584800E+07
2.4016200E-08	6.4687900E-08	-6.4019800E-08	1.1034300E+07
2.4051800E-08	6.4336200E-08	-6.3617500E-08	1.1479200E+07
2.4087400E-08	6.3991900E-08	-6.3223900E-08	1.1917500E+07
2.4122900E-08	6.3655100E-08	-6.2838600E-08	1.2347400E+07
2.4158500E-08	6.3324200E-08	-6.2460300E-08	1.2767100E+07
2.4194100E-08	6.2997100E-08	-6.2086700E-08	1.3175200E+07
2.4229700E-08	6.2670600E-08	-6.1714400E-08	1.3570200E+07
2.4265300E-08	6.2341100E-08	-6.1340100E-08	1.3951100E+07
2.4300800E-08	6.2004800E-08	-6.0959900E-08	1.4316900E+07
2.4336400E-08	6.1658400E-08	-6.0570600E-08	1.4666900E+07
2.4372000E-08	6.1299300E-08	-6.0169700E-08	1.5000800E+07
2.4407600E-08	6.0925500E-08	-5.9755600E-08	1.5318400E+07
2.4443200E-08	6.0536500E-08	-5.9327700E-08	1.5619700E+07
2.4478700E-08	6.0132700E-08	-5.8887000E-08	1.5905000E+07
2.4514300E-08	5.9716100E-08	-5.8435500E-08	1.6174800E+07
2.4549900E-08	5.9289500E-08	-5.7976200E-08	1.6429500E+07
2.4585500E-08	5.8856800E-08	-5.7513200E-08	1.6670000E+07
2.4621100E-08	5.8422300E-08	-5.7050900E-08	1.6896800E+07
2.4656600E-08	5.7990800E-08	-5.6594100E-08	1.7110900E+07
2.4692200E-08	5.7566700E-08	-5.6147100E-08	1.7312900E+07
2.4727800E-08	5.7154100E-08	-5.5713900E-08	1.7503500E+07
2.4763400E-08	5.6756300E-08	-5.5297400E-08	1.7683300E+07
2.4799000E-08	5.6375200E-08	-5.4899500E-08	1.7852900E+07
2.4834500E-08	5.6011700E-08	-5.4520800E-08	1.8012800E+07
2.4870100E-08	5.5665500E-08	-5.4160500E-08	1.8163200E+07
2.4905700E-08	5.5335000E-08	-5.3816700E-08	1.8304600E+07
2.4941300E-08	5.5017600E-08	-5.3486700E-08	1.8437000E+07
2.4976900E-08	5.4709900E-08	-5.3166800E-08	1.8560800E+07
2.5012400E-08	5.4408100E-08	-5.2853200E-08	1.8676100E+07
2.5048000E-08	5.4108600E-08	-5.2542000E-08	1.8783200E+07
2.5083600E-08	5.3808000E-08	-5.2230000E-08	1.8882500E+07
2.5119200E-08	5.3503700E-08	-5.1914600E-08	1.8974400E+07
2.5154800E-08	5.3194000E-08	-5.1594200E-08	1.9059400E+07
2.5190300E-08	5.2878500E-08	-5.1268700E-08	1.9138200E+07
2.5225900E-08	5.2558100E-08	-5.0938900E-08	1.9211500E+07
2.5261500E-08	5.2234700E-08	-5.0607200E-08	1.9279900E+07
2.5297100E-08	5.1911300E-08	-5.0276700E-08	1.9344400E+07
2.5332700E-08	5.1591500E-08	-4.9951300E-08	1.9405400E+07
2.5368300E-08	5.1279500E-08	-4.9635000E-08	1.9463800E+07

2.5403800E-08	5.0979100E-08	-4.9331900E-08	1.9520000E+07
2.5439400E-08	5.0693600E-08	-4.9045200E-08	1.9574200E+07
2.5475000E-08	5.0425400E-08	-4.8777500E-08	1.9626700E+07
2.5510600E-08	5.0175800E-08	-4.8529800E-08	1.9677400E+07
2.5546200E-08	4.9944500E-08	-4.8302000E-08	1.9725800E+07
2.5581700E-08	4.9729700E-08	-4.8092000E-08	1.9771600E+07
2.5617300E-08	4.9528300E-08	-4.7896700E-08	1.9814000E+07
2.5652900E-08	4.9336100E-08	-4.7711700E-08	1.9852300E+07
2.5688500E-08	4.9147900E-08	-4.7531700E-08	1.9885500E+07
2.5724100E-08	4.8958300E-08	-4.7351300E-08	1.9912800E+07
2.5759600E-08	4.8762100E-08	-4.7164800E-08	1.9933500E+07
2.5795200E-08	4.8554400E-08	-4.6967600E-08	1.9946900E+07
2.5830800E-08	4.8331700E-08	-4.6755800E-08	1.9952500E+07
2.5866400E-08	4.8091800E-08	-4.6527200E-08	1.9950000E+07
2.5902000E-08	4.7834000E-08	-4.6280900E-08	1.9939600E+07
2.5937500E-08	4.7559200E-08	-4.6017700E-08	1.9921500E+07
2.5973100E-08	4.7270000E-08	-4.5740100E-08	1.9896200E+07
2.6008700E-08	4.6969800E-08	-4.5451500E-08	1.9864400E+07
2.6044300E-08	4.6663100E-08	-4.5156300E-08	1.9827200E+07
2.6079900E-08	4.6354600E-08	-4.4859000E-08	1.9785400E+07
2.6115400E-08	4.6048600E-08	-4.4564000E-08	1.9740300E+07
2.6151000E-08	4.5748800E-08	-4.4274800E-08	1.9692800E+07
2.6186600E-08	4.5457500E-08	-4.3993800E-08	1.9644000E+07
2.6222200E-08	4.5175600E-08	-4.3722100E-08	1.9594600E+07
2.6257800E-08	4.4902500E-08	-4.3458900E-08	1.9545100E+07
2.6293300E-08	4.4635800E-08	-4.3202000E-08	1.9495900E+07
2.6328900E-08	4.4371800E-08	-4.2947700E-08	1.9446800E+07
2.6364500E-08	4.4105800E-08	-4.2691500E-08	1.9397500E+07
2.6400100E-08	4.3832700E-08	-4.2428400E-08	1.9347500E+07
2.6435700E-08	4.3547200E-08	-4.2153100E-08	1.9295700E+07
2.6471200E-08	4.3244500E-08	-4.1861100E-08	1.9241100E+07
2.6506800E-08	4.2921000E-08	-4.1548800E-08	1.9182500E+07
2.6542400E-08	4.2574400E-08	-4.1214100E-08	1.9118500E+07
2.6578000E-08	4.2204200E-08	-4.0856600E-08	1.9047900E+07
2.6613600E-08	4.1811500E-08	-4.0477400E-08	1.8969400E+07
2.6649100E-08	4.1399400E-08	-4.0079600E-08	1.8881900E+07
2.6684700E-08	4.0972300E-08	-3.9667700E-08	1.8784700E+07
2.6720300E-08	4.0535800E-08	-3.9247300E-08	1.8677200E+07
2.6755900E-08	4.0096500E-08	-3.8824800E-08	1.8559100E+07
2.6791500E-08	3.9661300E-08	-3.8407000E-08	1.8430600E+07
2.6827000E-08	3.9237000E-08	-3.8000500E-08	1.8291900E+07
2.6862600E-08	3.8830000E-08	-3.7611500E-08	1.8143600E+07
2.6898200E-08	3.8445800E-08	-3.7245100E-08	1.7986700E+07
2.6933800E-08	3.8088900E-08	-3.6905600E-08	1.7822100E+07
2.6969400E-08	3.7762500E-08	-3.6595900E-08	1.7650900E+07
2.7005000E-08	3.7468600E-08	-3.6317500E-08	1.7474400E+07
2.7040500E-08	3.7207700E-08	-3.6070700E-08	1.7293600E+07
2.7076100E-08	3.6978800E-08	-3.5854400E-08	1.7109600E+07
2.7111700E-08	3.6780100E-08	-3.5666500E-08	1.6923400E+07
2.7147300E-08	3.6608300E-08	-3.5503800E-08	1.6735900E+07
2.7182900E-08	3.6459300E-08	-3.5362000E-08	1.6547600E+07

2.7218400E-08	3.6328200E-08	-3.5236500E-08	1.6359100E+07
2.7254000E-08	3.6209400E-08	-3.5121900E-08	1.6170700E+07
2.7289600E-08	3.6097200E-08	-3.5012600E-08	1.5982600E+07
2.7325200E-08	3.5985400E-08	-3.4903000E-08	1.5794900E+07
2.7360800E-08	3.5868700E-08	-3.4787900E-08	1.5607500E+07
2.7396300E-08	3.5742200E-08	-3.4662800E-08	1.5420500E+07
2.7431900E-08	3.5601900E-08	-3.4524300E-08	1.5233700E+07
2.7467500E-08	3.5445100E-08	-3.4370300E-08	1.5047000E+07
2.7503100E-08	3.5270700E-08	-3.4199800E-08	1.4860600E+07
2.7538700E-08	3.5079000E-08	-3.4013700E-08	1.4674500E+07
2.7574200E-08	3.4872300E-08	-3.3814300E-08	1.4489200E+07
2.7609800E-08	3.4654600E-08	-3.3605700E-08	1.4304800E+07
2.7645400E-08	3.4431300E-08	-3.3393300E-08	1.4122000E+07
2.7681000E-08	3.4208700E-08	-3.3183200E-08	1.3941600E+07
2.7716600E-08	3.3993300E-08	-3.2982100E-08	1.3764300E+07
2.7752100E-08	3.3791900E-08	-3.2796300E-08	1.3591200E+07
2.7787700E-08	3.3610100E-08	-3.2631100E-08	1.3423200E+07
2.7823300E-08	3.3452200E-08	-3.2490500E-08	1.3261400E+07
2.7858900E-08	3.3320600E-08	-3.2376500E-08	1.3106800E+07
2.7894500E-08	3.3215400E-08	-3.2288700E-08	1.2960100E+07
2.7930000E-08	3.3134000E-08	-3.2224300E-08	1.2822400E+07
2.7965600E-08	3.3072000E-08	-3.2178500E-08	1.2694200E+07
2.8001200E-08	3.3022800E-08	-3.2144200E-08	1.2576200E+07
2.8036800E-08	3.2978600E-08	-3.2113400E-08	1.2468600E+07
2.8072400E-08	3.2930800E-08	-3.2077300E-08	1.2371600E+07
2.8107900E-08	3.2871000E-08	-3.2027400E-08	1.2285100E+07
2.8143500E-08	3.2791700E-08	-3.1956200E-08	1.2208800E+07
2.8179100E-08	3.2687100E-08	-3.1857800E-08	1.2142300E+07
2.8214700E-08	3.2553500E-08	-3.1728900E-08	1.2084800E+07
2.8250300E-08	3.2389700E-08	-3.1568400E-08	1.2035600E+07
2.8285800E-08	3.2197000E-08	-3.1378000E-08	1.1993900E+07
2.8321400E-08	3.1979200E-08	-3.1161700E-08	1.1958500E+07
2.8357000E-08	3.1741700E-08	-3.0925700E-08	1.1928400E+07
2.8392600E-08	3.1491800E-08	-3.0677300E-08	1.1902700E+07
2.8428200E-08	3.1236800E-08	-3.0424400E-08	1.1880400E+07
2.8463700E-08	3.0984500E-08	-3.0174800E-08	1.1860600E+07
2.8499300E-08	3.0741500E-08	-2.9935600E-08	1.1842500E+07
2.8534900E-08	3.0513400E-08	-2.9712400E-08	1.1825400E+07
2.8570500E-08	3.0303900E-08	-2.9508800E-08	1.1808800E+07
2.8606100E-08	3.0114900E-08	-2.9326900E-08	1.1792300E+07
2.8641700E-08	2.9946400E-08	-2.9166300E-08	1.1775800E+07
2.8677200E-08	2.9796700E-08	-2.9025300E-08	1.1759100E+07
2.8712800E-08	2.9663000E-08	-2.8900700E-08	1.1742200E+07
2.8748400E-08	2.9541500E-08	-2.8788400E-08	1.1725500E+07
2.8784000E-08	2.9427900E-08	-2.8683900E-08	1.1709100E+07
2.8819600E-08	2.9318100E-08	-2.8583000E-08	1.1693300E+07
2.8855100E-08	2.9208500E-08	-2.8481600E-08	1.1678600E+07
2.8890700E-08	2.9096100E-08	-2.8376700E-08	1.1665100E+07
2.8926300E-08	2.8978900E-08	-2.8266300E-08	1.1653300E+07
2.8961900E-08	2.8856100E-08	-2.8149300E-08	1.1643400E+07
2.8997500E-08	2.8727300E-08	-2.8025600E-08	1.1635600E+07

2.9033000E-08	2.8593400E-08	-2.7895800E-08	1.1629800E+07
2.9068600E-08	2.8455600E-08	-2.7761500E-08	1.1626100E+07
2.9104200E-08	2.8315700E-08	-2.7624300E-08	1.1624300E+07
2.9139800E-08	2.8175500E-08	-2.7486400E-08	1.1624100E+07
2.9175400E-08	2.8037000E-08	-2.7349900E-08	1.1625200E+07
2.9210900E-08	2.7901800E-08	-2.7216500E-08	1.1627200E+07
2.9246500E-08	2.7771500E-08	-2.7088000E-08	1.1629600E+07
2.9282100E-08	2.7647300E-08	-2.6965600E-08	1.1631800E+07
2.9317700E-08	2.7529900E-08	-2.6850300E-08	1.1633200E+07
2.9353300E-08	2.7419600E-08	-2.6742600E-08	1.1633300E+07
2.9388800E-08	2.7316500E-08	-2.6642500E-08	1.1631600E+07
2.9424400E-08	2.7220000E-08	-2.6549700E-08	1.1627400E+07
2.9460000E-08	2.7129400E-08	-2.6463300E-08	1.1620500E+07
2.9495600E-08	2.7043400E-08	-2.6382300E-08	1.1610500E+07
2.9531200E-08	2.6960600E-08	-2.6305100E-08	1.1597100E+07
2.9566700E-08	2.6879500E-08	-2.6230100E-08	1.1580200E+07
2.9602300E-08	2.6798200E-08	-2.6155400E-08	1.1559900E+07
2.9637900E-08	2.6715100E-08	-2.6079300E-08	1.1536100E+07
2.9673500E-08	2.6628600E-08	-2.6000000E-08	1.1508900E+07
2.9709100E-08	2.6537600E-08	-2.5916300E-08	1.1478600E+07
2.9744600E-08	2.6441000E-08	-2.5827000E-08	1.1445400E+07
2.9780200E-08	2.6338300E-08	-2.5731400E-08	1.1409400E+07
2.9815800E-08	2.6229400E-08	-2.5629400E-08	1.1370900E+07
2.9851400E-08	2.6114400E-08	-2.5520800E-08	1.1330000E+07
2.9887000E-08	2.5993700E-08	-2.5406200E-08	1.1287000E+07
2.9922500E-08	2.5867800E-08	-2.5285800E-08	1.1241900E+07
2.9958100E-08	2.5736900E-08	-2.5160100E-08	1.1194800E+07
2.9993700E-08	2.5601400E-08	-2.5029300E-08	1.1145700E+07
3.0029300E-08	2.5461000E-08	-2.4893400E-08	1.1094600E+07
3.0064800E-08	2.5315400E-08	-2.4751900E-08	1.1041600E+07
3.0100400E-08	2.5163600E-08	-2.4604100E-08	1.0986500E+07
3.0136000E-08	2.5004700E-08	-2.4449200E-08	1.0929300E+07
3.0171600E-08	2.4837700E-08	-2.4286100E-08	1.0870000E+07
3.0207200E-08	2.4661700E-08	-2.4114000E-08	1.0808500E+07
3.0242700E-08	2.4476200E-08	-2.3932700E-08	1.0744700E+07
3.0278300E-08	2.4281500E-08	-2.3742200E-08	1.0678600E+07
3.0313900E-08	2.4078400E-08	-2.3543700E-08	1.0610100E+07
3.0349500E-08	2.3868900E-08	-2.3338900E-08	1.0539200E+07
3.0385000E-08	2.3655700E-08	-2.3130600E-08	1.0465900E+07
3.0420600E-08	2.3442100E-08	-2.2922100E-08	1.0390300E+07
3.0456200E-08	2.3232000E-08	-2.2717300E-08	1.0312400E+07
3.0491800E-08	2.3029200E-08	-2.2519900E-08	1.0232300E+07
3.0527300E-08	2.2837500E-08	-2.2333600E-08	1.0149900E+07
3.0562900E-08	2.2660000E-08	-2.2161500E-08	1.0065600E+07
3.0598500E-08	2.2498900E-08	-2.2005700E-08	9.9793000E+06
3.0634100E-08	2.2355100E-08	-2.1867200E-08	9.8912300E+06
3.0669600E-08	2.2228500E-08	-2.1745600E-08	9.8015200E+06
3.0705200E-08	2.2117600E-08	-2.1639400E-08	9.7103500E+06
3.0740800E-08	2.2020000E-08	-2.1546100E-08	9.6179200E+06
3.0776400E-08	2.1932400E-08	-2.1462200E-08	9.5244600E+06
3.0812000E-08	2.1851100E-08	-2.1384000E-08	9.4302200E+06

3.0847500E-08	2.1772100E-08	-2.1307600E-08	9.3354700E+06
3.0883100E-08	2.1692100E-08	-2.1229400E-08	9.2404700E+06
3.0918700E-08	2.1607900E-08	-2.1146300E-08	9.1454900E+06
3.0954300E-08	2.1517700E-08	-2.1056400E-08	9.0508000E+06
3.0989800E-08	2.1420200E-08	-2.0958600E-08	8.9567100E+06
3.1025400E-08	2.1315300E-08	-2.0852800E-08	8.8635100E+06
3.1061000E-08	2.1203900E-08	-2.0740200E-08	8.7715100E+06
3.1096600E-08	2.1087700E-08	-2.0622600E-08	8.6809800E+06
3.1132100E-08	2.0969000E-08	-2.0502500E-08	8.5921800E+06
3.1167700E-08	2.0850500E-08	-2.0382800E-08	8.5053800E+06
3.1203300E-08	2.0735100E-08	-2.0266700E-08	8.4208500E+06
3.1238900E-08	2.0625600E-08	-2.0157100E-08	8.3388300E+06
3.1274500E-08	2.0524500E-08	-2.0056900E-08	8.2595800E+06
3.1310000E-08	2.0434100E-08	-1.9968400E-08	8.1833500E+06
3.1345600E-08	2.0355900E-08	-1.9893300E-08	8.1103400E+06
3.1381200E-08	2.0291100E-08	-1.9832500E-08	8.0407800E+06
3.1416800E-08	2.0240000E-08	-1.9786500E-08	7.9748300E+06
3.1452300E-08	2.0202100E-08	-1.9754600E-08	7.9126500E+06
3.1487900E-08	2.0176300E-08	-1.9735600E-08	7.8543700E+06
3.1523500E-08	2.0160900E-08	-1.9727400E-08	7.8000500E+06
3.1559100E-08	2.0153300E-08	-1.9727300E-08	7.7497500E+06
3.1594600E-08	2.0150400E-08	-1.9731800E-08	7.7034300E+06
3.1630200E-08	2.0148800E-08	-1.9737400E-08	7.6610500E+06
3.1665800E-08	2.0144600E-08	-1.9740100E-08	7.6224900E+06
3.1701400E-08	2.0134300E-08	-1.9736000E-08	7.5876000E+06
3.1737000E-08	2.0114400E-08	-1.9721800E-08	7.5562000E+06
3.1772500E-08	2.0082000E-08	-1.9694500E-08	7.5280700E+06
3.1808100E-08	2.0035200E-08	-1.9652100E-08	7.5029700E+06
3.1843700E-08	1.9972900E-08	-1.9593700E-08	7.4806200E+06
3.1879300E-08	1.9895200E-08	-1.9519400E-08	7.4607500E+06
3.1914800E-08	1.9803200E-08	-1.9430600E-08	7.4430700E+06
3.1950400E-08	1.9699200E-08	-1.9329300E-08	7.4273200E+06
3.1986000E-08	1.9586300E-08	-1.9218800E-08	7.4132500E+06
3.2021600E-08	1.9468000E-08	-1.9102700E-08	7.4006300E+06
3.2057100E-08	1.9348200E-08	-1.8984500E-08	7.3892600E+06
3.2092700E-08	1.9230400E-08	-1.8868000E-08	7.3789700E+06
3.2128300E-08	1.9117700E-08	-1.8756000E-08	7.3696300E+06
3.2163900E-08	1.9012100E-08	-1.8650600E-08	7.3611400E+06
3.2199500E-08	1.8915000E-08	-1.8553100E-08	7.3534600E+06
3.2235000E-08	1.8826300E-08	-1.8463400E-08	7.3465800E+06
3.2270600E-08	1.8744800E-08	-1.8380600E-08	7.3405400E+06
3.2306200E-08	1.8668400E-08	-1.8302600E-08	7.3353900E+06
3.2341800E-08	1.8594400E-08	-1.8226900E-08	7.3312100E+06
3.2377300E-08	1.8519600E-08	-1.8150700E-08	7.3280600E+06
3.2412900E-08	1.8440800E-08	-1.8071000E-08	7.3260000E+06
3.2448500E-08	1.8355200E-08	-1.7985200E-08	7.3250300E+06
3.2484100E-08	1.8260600E-08	-1.7891500E-08	7.3251400E+06
3.2519600E-08	1.8155800E-08	-1.7788500E-08	7.3262500E+06
3.2555200E-08	1.8040400E-08	-1.7676300E-08	7.3282100E+06
3.2590800E-08	1.7915500E-08	-1.7555700E-08	7.3308200E+06
3.2626400E-08	1.7782900E-08	-1.7428200E-08	7.3337800E+06

3.2662000E-08	1.7645400E-08	-1.7296400E-08	7.3367600E+06
3.2697500E-08	1.7506300E-08	-1.7163300E-08	7.3393300E+06
3.2733100E-08	1.7369300E-08	-1.7032300E-08	7.3410600E+06
3.2768700E-08	1.7238100E-08	-1.6906500E-08	7.3415100E+06
3.2804300E-08	1.7116100E-08	-1.6789100E-08	7.3402400E+06
3.2839800E-08	1.7006200E-08	-1.6682600E-08	7.3368800E+06
3.2875400E-08	1.6910500E-08	-1.6589100E-08	7.3311100E+06
3.2911000E-08	1.6830200E-08	-1.6509600E-08	7.3227400E+06
3.2946600E-08	1.6765700E-08	-1.6444600E-08	7.3116700E+06
3.2982100E-08	1.6716400E-08	-1.6393800E-08	7.2979700E+06
3.3017700E-08	1.6680900E-08	-1.6355900E-08	7.2818000E+06
3.3053300E-08	1.6657200E-08	-1.6329300E-08	7.2634900E+06
3.3088900E-08	1.6642700E-08	-1.6311900E-08	7.2434700E+06
3.3124500E-08	1.6634700E-08	-1.6301200E-08	7.2222200E+06
3.3160000E-08	1.6630300E-08	-1.6294500E-08	7.2003000E+06
3.3195600E-08	1.6626600E-08	-1.6289300E-08	7.1782600E+06
3.3231200E-08	1.6621000E-08	-1.6283100E-08	7.1566100E+06
3.3266800E-08	1.6611300E-08	-1.6273600E-08	7.1358000E+06
3.3302300E-08	1.6595300E-08	-1.6258800E-08	7.1161500E+06
3.3337900E-08	1.6571700E-08	-1.6237000E-08	7.0978100E+06
3.3373500E-08	1.6539400E-08	-1.6207100E-08	7.0807600E+06
3.3409100E-08	1.6497700E-08	-1.6168000E-08	7.0647700E+06
3.3444600E-08	1.6446200E-08	-1.6119500E-08	7.0494800E+06
3.3480200E-08	1.6385200E-08	-1.6061400E-08	7.0343200E+06
3.3515800E-08	1.6314900E-08	-1.5994100E-08	7.0186100E+06
3.3551400E-08	1.6236200E-08	-1.5918200E-08	7.0015800E+06
3.3587000E-08	1.6149900E-08	-1.5834700E-08	6.9823800E+06
3.3622500E-08	1.6057300E-08	-1.5745000E-08	6.9601900E+06
3.3658100E-08	1.5959700E-08	-1.5650400E-08	6.9342000E+06
3.3693700E-08	1.5858600E-08	-1.5552700E-08	6.9037500E+06
3.3729300E-08	1.5755300E-08	-1.5453200E-08	6.8683200E+06
3.3764800E-08	1.5651400E-08	-1.5353600E-08	6.8276000E+06
3.3800400E-08	1.5548000E-08	-1.5255200E-08	6.7814800E+06
3.3836000E-08	1.5446100E-08	-1.5158900E-08	6.7301300E+06
3.3871600E-08	1.5346500E-08	-1.5065500E-08	6.6739200E+06
3.3907100E-08	1.5249500E-08	-1.4975100E-08	6.6134500E+06
3.3942700E-08	1.5155300E-08	-1.4887500E-08	6.5495000E+06
3.3978300E-08	1.5063500E-08	-1.4802200E-08	6.4830100E+06
3.4013900E-08	1.4973500E-08	-1.4718400E-08	6.4150000E+06
3.4049500E-08	1.4884700E-08	-1.4634900E-08	6.3465300E+06
3.4085000E-08	1.4796200E-08	-1.4550800E-08	6.2786200E+06
3.4120600E-08	1.4707300E-08	-1.4464900E-08	6.2122100E+06
3.4156200E-08	1.4617200E-08	-1.4376400E-08	6.1481300E+06
3.4191800E-08	1.4525500E-08	-1.4284900E-08	6.0869800E+06
3.4227300E-08	1.4432100E-08	-1.4190100E-08	6.0291800E+06
3.4262900E-08	1.4337100E-08	-1.4092500E-08	5.9749100E+06
3.4298500E-08	1.4240900E-08	-1.3992600E-08	5.9241300E+06
3.4334100E-08	1.4144300E-08	-1.3891500E-08	5.8765900E+06
3.4369600E-08	1.4048100E-08	-1.3790600E-08	5.8318700E+06
3.4405200E-08	1.3953400E-08	-1.3691200E-08	5.7894300E+06
3.4440800E-08	1.3861100E-08	-1.3594800E-08	5.7486300E+06

3.4476400E-08	1.3772300E-08	-1.3502700E-08	5.7088100E+06
3.4512000E-08	1.3687600E-08	-1.3416000E-08	5.6693700E+06
3.4547500E-08	1.3607700E-08	-1.3335500E-08	5.6297600E+06
3.4583100E-08	1.3532800E-08	-1.3261600E-08	5.5895800E+06
3.4618700E-08	1.3463100E-08	-1.3194500E-08	5.5485500E+06
3.4654300E-08	1.3398500E-08	-1.3133700E-08	5.5066000E+06
3.4689800E-08	1.3338600E-08	-1.3078800E-08	5.4638400E+06
3.4725400E-08	1.3283200E-08	-1.3029100E-08	5.4205200E+06
3.4761000E-08	1.3231800E-08	-1.2983600E-08	5.3770900E+06
3.4796600E-08	1.3184100E-08	-1.2941600E-08	5.3340900E+06
3.4832100E-08	1.3139700E-08	-1.2902300E-08	5.2921400E+06
3.4867700E-08	1.3098600E-08	-1.2865200E-08	5.2518800E+06
3.4903300E-08	1.3060700E-08	-1.2830200E-08	5.2139100E+06
3.4938900E-08	1.3026300E-08	-1.2797100E-08	5.1787400E+06
3.4974500E-08	1.2995600E-08	-1.2766300E-08	5.1468100E+06
3.5010000E-08	1.2968800E-08	-1.2738200E-08	5.1183700E+06
3.5045600E-08	1.2946200E-08	-1.2713300E-08	5.0935200E+06
3.5081200E-08	1.2927800E-08	-1.2692200E-08	5.0722000E+06
3.5116800E-08	1.2913700E-08	-1.2675000E-08	5.0541800E+06
3.5152300E-08	1.2903200E-08	-1.2661900E-08	5.0390500E+06
3.5187900E-08	1.2895700E-08	-1.2652500E-08	5.0263200E+06
3.5223500E-08	1.2890100E-08	-1.2646000E-08	5.0154000E+06
3.5259100E-08	1.2885000E-08	-1.2641300E-08	5.0056900E+06
3.5294600E-08	1.2878600E-08	-1.2636800E-08	4.9965600E+06
3.5330200E-08	1.2869400E-08	-1.2630700E-08	4.9874700E+06
3.5365800E-08	1.2855700E-08	-1.2621300E-08	4.9779700E+06
3.5401400E-08	1.2836000E-08	-1.2606800E-08	4.9677000E+06
3.5437000E-08	1.2809200E-08	-1.2585700E-08	4.9564800E+06
3.5472500E-08	1.2774900E-08	-1.2557200E-08	4.9442600E+06
3.5508100E-08	1.2733200E-08	-1.2520900E-08	4.9311200E+06
3.5543700E-08	1.2684700E-08	-1.2477100E-08	4.9172800E+06
3.5579300E-08	1.2630700E-08	-1.2426800E-08	4.9030800E+06
3.5614800E-08	1.2573000E-08	-1.2371700E-08	4.8888800E+06
3.5650400E-08	1.2513700E-08	-1.2313800E-08	4.8751300E+06
3.5686000E-08	1.2455000E-08	-1.2255200E-08	4.8622200E+06
3.5721600E-08	1.2398900E-08	-1.2198100E-08	4.8505300E+06
3.5757100E-08	1.2346900E-08	-1.2144300E-08	4.8403600E+06
3.5792700E-08	1.2300100E-08	-1.2095100E-08	4.8318900E+06
3.5828300E-08	1.2258700E-08	-1.2050900E-08	4.8252200E+06
3.5863900E-08	1.2222100E-08	-1.2011600E-08	4.8203000E+06
3.5899500E-08	1.2188900E-08	-1.1975900E-08	4.8169600E+06
3.5935000E-08	1.2157000E-08	-1.1941900E-08	4.8149300E+06
3.5970600E-08	1.2123600E-08	-1.1907100E-08	4.8138900E+06
3.6006200E-08	1.2085800E-08	-1.1868600E-08	4.8134400E+06
3.6041800E-08	1.2040500E-08	-1.1823400E-08	4.8131800E+06
3.6077300E-08	1.1985000E-08	-1.1768800E-08	4.8127100E+06
3.6112900E-08	1.1917000E-08	-1.1702500E-08	4.8116700E+06
3.6148500E-08	1.1835100E-08	-1.1623100E-08	4.8097700E+06
3.6184100E-08	1.1738900E-08	-1.1530200E-08	4.8068000E+06
3.6219600E-08	1.1629100E-08	-1.1424400E-08	4.8026300E+06
3.6255200E-08	1.1507400E-08	-1.1307300E-08	4.7972400E+06

3.6290800E-08	1.1376700E-08	-1.1181700E-08	4.7907100E+06
3.6326400E-08	1.1240700E-08	-1.1051200E-08	4.7831800E+06
3.6361900E-08	1.1103800E-08	-1.0919700E-08	4.7748700E+06
3.6397500E-08	1.0970600E-08	-1.0792000E-08	4.7660100E+06
3.6433100E-08	1.0846000E-08	-1.0672400E-08	4.7568400E+06
3.6468700E-08	1.0734500E-08	-1.0565200E-08	4.7476000E+06
3.6504300E-08	1.0640100E-08	-1.0474100E-08	4.7384900E+06
3.6539800E-08	1.0566000E-08	-1.0401900E-08	4.7296500E+06
3.6575400E-08	1.0514300E-08	-1.0350400E-08	4.7211600E+06
3.6611000E-08	1.0485800E-08	-1.0320400E-08	4.7130200E+06
3.6646600E-08	1.0480000E-08	-1.0311300E-08	4.7051600E+06
3.6682100E-08	1.0495100E-08	-1.0321300E-08	4.6974500E+06
3.6717700E-08	1.0528100E-08	-1.0347800E-08	4.6896900E+06
3.6753300E-08	1.0575100E-08	-1.0387000E-08	4.6816400E+06
3.6788900E-08	1.0631100E-08	-1.0434600E-08	4.6730500E+06
3.6824400E-08	1.0690800E-08	-1.0485800E-08	4.6636400E+06
3.6860000E-08	1.0748500E-08	-1.0535700E-08	4.6531800E+06
3.6895600E-08	1.0799100E-08	-1.0579600E-08	4.6414200E+06
3.6931200E-08	1.0837700E-08	-1.0613400E-08	4.6282300E+06
3.6966800E-08	1.0860300E-08	-1.0633600E-08	4.6134700E+06
3.7002300E-08	1.0864200E-08	-1.0637800E-08	4.5971000E+06
3.7037900E-08	1.0847900E-08	-1.0624600E-08	4.5791400E+06
3.7073500E-08	1.0811100E-08	-1.0593900E-08	4.5596300E+06
3.7109100E-08	1.0755200E-08	-1.0546600E-08	4.5386700E+06
3.7144600E-08	1.0682700E-08	-1.0484900E-08	4.5163800E+06
3.7180200E-08	1.0597200E-08	-1.0411800E-08	4.4929100E+06
3.7215800E-08	1.0502900E-08	-1.0330700E-08	4.4683800E+06
3.7251400E-08	1.0404500E-08	-1.0245800E-08	4.4429200E+06
3.7286900E-08	1.0306900E-08	-1.0160900E-08	4.4166100E+06
3.7322500E-08	1.0214400E-08	-1.0079800E-08	4.3895300E+06
3.7358100E-08	1.0130600E-08	-1.0005500E-08	4.3616800E+06
3.7393700E-08	1.0058300E-08	-9.9403500E-09	4.3330400E+06
3.7429300E-08	9.9989600E-09	-9.8855600E-09	4.3035400E+06
3.7464800E-08	9.9529600E-09	-9.8413600E-09	4.2731300E+06
3.7500400E-08	9.9192500E-09	-9.8068900E-09	4.2417100E+06
3.7536000E-08	9.8957800E-09	-9.7803200E-09	4.2092000E+06
3.7571600E-08	9.8795700E-09	-9.7590500E-09	4.1755700E+06
3.7607100E-08	9.8669700E-09	-9.7399500E-09	4.1407900E+06
3.7642700E-08	9.8541100E-09	-9.7196700E-09	4.1048900E+06
3.7678300E-08	9.8371900E-09	-9.6949500E-09	4.0679500E+06
3.7713900E-08	9.8128100E-09	-9.6629100E-09	4.0301000E+06
3.7749400E-08	9.7782800E-09	-9.6213100E-09	3.9915400E+06
3.7785000E-08	9.7318500E-09	-9.5687900E-09	3.9525000E+06
3.7820600E-08	9.6728300E-09	-9.5049600E-09	3.9132800E+06
3.7856200E-08	9.6016300E-09	-9.4304600E-09	3.8742000E+06
3.7891800E-08	9.5197700E-09	-9.3469000E-09	3.8356000E+06
3.7927300E-08	9.4296800E-09	-9.2567200E-09	3.7978300E+06
3.7962900E-08	9.3344500E-09	-9.1629800E-09	3.7612100E+06
3.7998500E-08	9.2376700E-09	-9.0691400E-09	3.7260600E+06
3.8034100E-08	9.1430800E-09	-8.9787300E-09	3.6926400E+06
3.8069600E-08	9.0542300E-09	-8.8950800E-09	3.6611800E+06

3.8105200E-08	8.9742100E-09	-8.8209900E-09	3.6318300E+06
3.8140800E-08	8.9054600E-09	-8.7585800E-09	3.6047200E+06
3.8176400E-08	8.8495000E-09	-8.7090300E-09	3.5799100E+06
3.8211900E-08	8.8068600E-09	-8.6725500E-09	3.5573900E+06
3.8247500E-08	8.7770600E-09	-8.6483100E-09	3.5371500E+06
3.8283100E-08	8.7586600E-09	-8.6345600E-09	3.5191100E+06
3.8318700E-08	8.7493600E-09	-8.6287400E-09	3.5031600E+06
3.8354300E-08	8.7462100E-09	-8.6277400E-09	3.4891900E+06
3.8389800E-08	8.7459600E-09	-8.6281600E-09	3.4770400E+06
3.8425400E-08	8.7452700E-09	-8.6266400E-09	3.4665600E+06
3.8461000E-08	8.7409700E-09	-8.6201200E-09	3.4576000E+06
3.8496600E-08	8.7304400E-09	-8.6061700E-09	3.4500000E+06
3.8532100E-08	8.7118600E-09	-8.5831900E-09	3.4436200E+06
3.8567700E-08	8.6843100E-09	-8.5506200E-09	3.4383100E+06
3.8603300E-08	8.6479000E-09	-8.5089500E-09	3.4338900E+06
3.8638900E-08	8.6038000E-09	-8.4596900E-09	3.4302100E+06
3.8674400E-08	8.5540900E-09	-8.4052900E-09	3.4270900E+06
3.8710000E-08	8.5015800E-09	-8.3488700E-09	3.4243300E+06
3.8745600E-08	8.4495600E-09	-8.2939200E-09	3.4217400E+06
3.8781200E-08	8.4014400E-09	-8.2439600E-09	3.4191200E+06
3.8816800E-08	8.3603300E-09	-8.2021900E-09	3.4162600E+06
3.8852300E-08	8.3288900E-09	-8.1711600E-09	3.4129600E+06
3.8887900E-08	8.3088700E-09	-8.1524900E-09	3.4090300E+06
3.8923500E-08	8.3009400E-09	-8.1466900E-09	3.4043000E+06
3.8959100E-08	8.3046700E-09	-8.1530900E-09	3.3986100E+06
3.8994600E-08	8.3184800E-09	-8.1699100E-09	3.3918700E+06
3.9030200E-08	8.3398000E-09	-8.1944200E-09	3.3840300E+06
3.9065800E-08	8.3654000E-09	-8.2232200E-09	3.3750900E+06
3.9101400E-08	8.3916400E-09	-8.2525700E-09	3.3651100E+06
3.9136900E-08	8.4148000E-09	-8.2787900E-09	3.3542200E+06
3.9172500E-08	8.4316600E-09	-8.2986200E-09	3.3426100E+06
3.9208100E-08	8.4396900E-09	-8.3095200E-09	3.3305100E+06
3.9243700E-08	8.4372700E-09	-8.3099300E-09	3.3181900E+06
3.9279300E-08	8.4239300E-09	-8.2993800E-09	3.3059400E+06
3.9314800E-08	8.4002900E-09	-8.2784200E-09	3.2940500E+06
3.9350400E-08	8.3679700E-09	-8.2485600E-09	3.2827900E+06
3.9386000E-08	8.3292500E-09	-8.2119600E-09	3.2724000E+06
3.9421600E-08	8.2868500E-09	-8.1711300E-09	3.2630500E+06
3.9457100E-08	8.2434400E-09	-8.1285600E-09	3.2548600E+06
3.9492700E-08	8.2013400E-09	-8.0864000E-09	3.2478500E+06
3.9528300E-08	8.1621300E-09	-8.0461400E-09	3.2419800E+06
3.9563900E-08	8.1265100E-09	-8.0084100E-09	3.2371000E+06
3.9599400E-08	8.0941700E-09	-7.9729600E-09	3.2330200E+06
3.9635000E-08	8.0637600E-09	-7.9386300E-09	3.2294500E+06
3.9670600E-08	8.0331500E-09	-7.9036300E-09	3.2260900E+06
3.9706200E-08	7.9997500E-09	-7.8657400E-09	3.2226100E+06
3.9741800E-08	7.9607900E-09	-7.8226400E-09	3.2186900E+06
3.9777300E-08	7.9137700E-09	-7.7723400E-09	3.2140500E+06
3.9812900E-08	7.8568600E-09	-7.7134400E-09	3.2084600E+06
3.9848500E-08	7.7892000E-09	-7.6454400E-09	3.2017600E+06
3.9884100E-08	7.7110500E-09	-7.5688200E-09	3.1938800E+06

3.9919600E-08	7.6239500E-09	-7.4851400E-09	3.1848300E+06
3.9955200E-08	7.5305600E-09	-7.3969000E-09	3.1747200E+06
3.9990800E-08	7.4344000E-09	-7.3073200E-09	3.1637500E+06
4.0026400E-08	7.3396100E-09	-7.2200400E-09	3.1521900E+06
4.0061900E-08	7.2505600E-09	-7.1387600E-09	3.1403500E+06
4.0097500E-08	7.1712200E-09	-7.0668200E-09	3.1286000E+06
4.0133100E-08	7.1049500E-09	-7.0069300E-09	3.1172800E+06
4.0168700E-08	7.0541500E-09	-6.9608600E-09	3.1067300E+06
4.0204300E-08	7.0199500E-09	-6.9292800E-09	3.0972400E+06
4.0239800E-08	7.0021200E-09	-6.9117400E-09	3.0890200E+06
4.0275400E-08	6.9992200E-09	-6.9067000E-09	3.0822100E+06
4.0311000E-08	7.0086900E-09	-6.9117700E-09	3.0768500E+06
4.0346600E-08	7.0271100E-09	-6.9239200E-09	3.0728600E+06
4.0382100E-08	7.0506400E-09	-6.9398600E-09	3.0700700E+06
4.0417700E-08	7.0753200E-09	-6.9563000E-09	3.0682300E+06
4.0453300E-08	7.0974800E-09	-6.9703100E-09	3.0669800E+06
4.0488900E-08	7.1140700E-09	-6.9795200E-09	3.0659200E+06
4.0524400E-08	7.1229000E-09	-6.9823600E-09	3.0646000E+06
4.0560000E-08	7.1227000E-09	-6.9781000E-09	3.0625700E+06
4.0595600E-08	7.1132500E-09	-6.9668400E-09	3.0593900E+06
4.0631200E-08	7.0953000E-09	-6.9494300E-09	3.0546300E+06
4.0666800E-08	7.0703300E-09	-6.9272900E-09	3.0479600E+06
4.0702300E-08	7.0403900E-09	-6.9022500E-09	3.0391300E+06
4.0737900E-08	7.0078600E-09	-6.8762700E-09	3.0279500E+06
4.0773500E-08	6.9751300E-09	-6.8512400E-09	3.0143700E+06
4.0809100E-08	6.9443600E-09	-6.8288000E-09	2.9984000E+06
4.0844600E-08	6.9173500E-09	-6.8101700E-09	2.9801800E+06
4.0880200E-08	6.8953100E-09	-6.7960600E-09	2.9599200E+06
4.0915800E-08	6.8788300E-09	-6.7866300E-09	2.9378900E+06
4.0951400E-08	6.8678600E-09	-6.7815000E-09	2.9144100E+06
4.0986900E-08	6.8617500E-09	-6.7798200E-09	2.8898500E+06
4.1022500E-08	6.8592900E-09	-6.7803200E-09	2.8645700E+06
4.1058100E-08	6.8589200E-09	-6.7814900E-09	2.8389200E+06
4.1093700E-08	6.8588600E-09	-6.7816500E-09	2.8132200E+06
4.1129300E-08	6.8571500E-09	-6.7790900E-09	2.7877300E+06
4.1164800E-08	6.8518800E-09	-6.7721500E-09	2.7626500E+06
4.1200400E-08	6.8413300E-09	-6.7593500E-09	2.7381400E+06
4.1236000E-08	6.8239800E-09	-6.7394500E-09	2.7142800E+06
4.1271600E-08	6.7986400E-09	-6.7114900E-09	2.6910700E+06
4.1307100E-08	6.7645200E-09	-6.6748500E-09	2.6685000E+06
4.1342700E-08	6.7212400E-09	-6.6292900E-09	2.6465000E+06
4.1378300E-08	6.6688700E-09	-6.5749500E-09	2.6249600E+06
4.1413900E-08	6.6079400E-09	-6.5124000E-09	2.6037700E+06
4.1449400E-08	6.5394400E-09	-6.4425900E-09	2.5828200E+06
4.1485000E-08	6.4648100E-09	-6.3669300E-09	2.5620400E+06
4.1520600E-08	6.3858900E-09	-6.2871700E-09	2.5413600E+06
4.1556200E-08	6.3049100E-09	-6.2054200E-09	2.5207800E+06
4.1591800E-08	6.2242900E-09	-6.1240700E-09	2.5003300E+06
4.1627300E-08	6.1466500E-09	-6.0456600E-09	2.4801100E+06
4.1662900E-08	6.0746100E-09	-5.9727700E-09	2.4602500E+06
4.1698500E-08	6.0106200E-09	-5.9078400E-09	2.4409400E+06

4.1734100E-08	5.9568200E-09	-5.8530300E-09	2.4223900E+06
4.1769600E-08	5.9148100E-09	-5.8100000E-09	2.4048500E+06
4.1805200E-08	5.8855300E-09	-5.7797800E-09	2.3885700E+06
4.1840800E-08	5.8691500E-09	-5.7626300E-09	2.3737800E+06
4.1876400E-08	5.8649400E-09	-5.7579500E-09	2.3607200E+06
4.1911900E-08	5.8713800E-09	-5.7643000E-09	2.3495700E+06
4.1947500E-08	5.8861100E-09	-5.7793900E-09	2.3404900E+06
4.1983100E-08	5.9061000E-09	-5.8002700E-09	2.3335600E+06
4.2018700E-08	5.9279300E-09	-5.8235300E-09	2.3288100E+06
4.2054200E-08	5.9480400E-09	-5.8455400E-09	2.3261900E+06
4.2089800E-08	5.9629800E-09	-5.8627700E-09	2.3256000E+06
4.2125400E-08	5.9697200E-09	-5.8720800E-09	2.3268400E+06
4.2161000E-08	5.9660000E-09	-5.8709900E-09	2.3296900E+06
4.2196600E-08	5.9504200E-09	-5.8579600E-09	2.3338400E+06
4.2232100E-08	5.9227400E-09	-5.8324800E-09	2.3389600E+06
4.2267700E-08	5.8838100E-09	-5.7952000E-09	2.3447000E+06
4.2303300E-08	5.8355100E-09	-5.7478500E-09	2.3506900E+06
4.2338900E-08	5.7806300E-09	-5.6931000E-09	2.3565500E+06
4.2374400E-08	5.7226400E-09	-5.6343400E-09	2.3619300E+06
4.2410000E-08	5.6653700E-09	-5.5754000E-09	2.3665300E+06
4.2445600E-08	5.6126300E-09	-5.5201800E-09	2.3700700E+06
4.2481200E-08	5.5679400E-09	-5.4723400E-09	2.3723400E+06
4.2516700E-08	5.5341600E-09	-5.4349400E-09	2.3732000E+06
4.2552300E-08	5.5133000E-09	-5.4102600E-09	2.3725800E+06
4.2587900E-08	5.5063700E-09	-5.3995500E-09	2.3704800E+06
4.2623500E-08	5.5132800E-09	-5.4030100E-09	2.3669400E+06
4.2659100E-08	5.5328500E-09	-5.4197400E-09	2.3620800E+06
4.2694600E-08	5.5630500E-09	-5.4479300E-09	2.3560500E+06
4.2730200E-08	5.6011000E-09	-5.4849500E-09	2.3490500E+06
4.2765800E-08	5.6437200E-09	-5.5276800E-09	2.3412700E+06
4.2801400E-08	5.6875000E-09	-5.5727300E-09	2.3329400E+06
4.2836900E-08	5.7290400E-09	-5.6166800E-09	2.3242500E+06
4.2872500E-08	5.7652300E-09	-5.6563300E-09	2.3153700E+06
4.2908100E-08	5.7934600E-09	-5.6889400E-09	2.3064600E+06
4.2943700E-08	5.8117300E-09	-5.7123100E-09	2.2976100E+06
4.2979200E-08	5.8187600E-09	-5.7249000E-09	2.2888500E+06
4.3014800E-08	5.8139000E-09	-5.7258200E-09	2.2801900E+06
4.3050400E-08	5.7971700E-09	-5.7148400E-09	2.2715800E+06
4.3086000E-08	5.7691600E-09	-5.6922800E-09	2.2629300E+06
4.3121600E-08	5.7309600E-09	-5.6589600E-09	2.2541200E+06
4.3157100E-08	5.6840200E-09	-5.6161100E-09	2.2450400E+06
4.3192700E-08	5.6300100E-09	-5.5652200E-09	2.2355600E+06
4.3228300E-08	5.5707900E-09	-5.5079800E-09	2.2255700E+06
4.3263900E-08	5.5082400E-09	-5.4461900E-09	2.2150300E+06
4.3299400E-08	5.4442600E-09	-5.3817100E-09	2.2039000E+06
4.3335000E-08	5.3806700E-09	-5.3163500E-09	2.1922200E+06
4.3370600E-08	5.3190600E-09	-5.2518200E-09	2.1801100E+06
4.3406200E-08	5.2608100E-09	-5.1896400E-09	2.1677100E+06
4.3441700E-08	5.2070500E-09	-5.1311500E-09	2.1552700E+06
4.3477300E-08	5.1585400E-09	-5.0773600E-09	2.1430400E+06
4.3512900E-08	5.1156700E-09	-5.0289800E-09	2.1313400E+06

4.3548500E-08	5.0784400E-09	-4.9863500E-09	2.1204600E+06
4.3584100E-08	5.0464900E-09	-4.9494400E-09	2.1106900E+06
4.3619600E-08	5.0191100E-09	-4.9178900E-09	2.1022800E+06
4.3655200E-08	4.9954100E-09	-4.8910700E-09	2.0954100E+06
4.3690800E-08	4.9743300E-09	-4.8681900E-09	2.0901900E+06
4.3726400E-08	4.9547900E-09	-4.8483500E-09	2.0866400E+06
4.3761900E-08	4.9358900E-09	-4.8307100E-09	2.0846700E+06
4.3797500E-08	4.9169500E-09	-4.8145700E-09	2.0841200E+06
4.3833100E-08	4.8976400E-09	-4.7994800E-09	2.0847300E+06
4.3868700E-08	4.8780300E-09	-4.7852600E-09	2.0861700E+06
4.3904200E-08	4.8585400E-09	-4.7720400E-09	2.0880500E+06
4.3939800E-08	4.8399400E-09	-4.7601700E-09	2.0899700E+06
4.3975400E-08	4.8231400E-09	-4.7501400E-09	2.0915200E+06
4.4011000E-08	4.8091100E-09	-4.7424600E-09	2.0923400E+06
4.4046600E-08	4.7986900E-09	-4.7375300E-09	2.0920900E+06
4.4082100E-08	4.7923900E-09	-4.7354800E-09	2.0905500E+06
4.4117700E-08	4.7903100E-09	-4.7361000E-09	2.0875700E+06
4.4153300E-08	4.7920200E-09	-4.7387800E-09	2.0830800E+06
4.4188900E-08	4.7966000E-09	-4.7425100E-09	2.0771400E+06
4.4224400E-08	4.8026700E-09	-4.7460000E-09	2.0698800E+06
4.4260000E-08	4.8085600E-09	-4.7477900E-09	2.0614900E+06
4.4295600E-08	4.8125500E-09	-4.7464500E-09	2.0522400E+06
4.4331200E-08	4.8130100E-09	-4.7408000E-09	2.0423900E+06
4.4366700E-08	4.8086900E-09	-4.7300700E-09	2.0322200E+06
4.4402300E-08	4.7989200E-09	-4.7140900E-09	2.0219800E+06
4.4437900E-08	4.7836600E-09	-4.6933400E-09	2.0118700E+06
4.4473500E-08	4.7636100E-09	-4.6689500E-09	2.0020100E+06
4.4509100E-08	4.7401600E-09	-4.6426200E-09	1.9924700E+06
4.4544600E-08	4.7151200E-09	-4.6164400E-09	1.9831800E+06
4.4580200E-08	4.6906700E-09	-4.5926400E-09	1.9740500E+06
4.4615800E-08	4.6690100E-09	-4.5733100E-09	1.9648600E+06
4.4651400E-08	4.6519500E-09	-4.5601000E-09	1.9553900E+06
4.4686900E-08	4.6408100E-09	-4.5540000E-09	1.9453500E+06
4.4722500E-08	4.6362000E-09	-4.5551400E-09	1.9344700E+06
4.4758100E-08	4.6377500E-09	-4.5627000E-09	1.9224800E+06
4.4793700E-08	4.6442100E-09	-4.5749800E-09	1.9091900E+06
4.4829200E-08	4.6535600E-09	-4.5894900E-09	1.8944500E+06
4.4864800E-08	4.6632000E-09	-4.6032300E-09	1.8782000E+06
4.4900400E-08	4.6701000E-09	-4.6129600E-09	1.8604600E+06
4.4936000E-08	4.6713200E-09	-4.6155700E-09	1.8413900E+06
4.4971600E-08	4.6642200E-09	-4.6084200E-09	1.8212100E+06
4.5007100E-08	4.6468000E-09	-4.5896400E-09	1.8002600E+06
4.5042700E-08	4.6179500E-09	-4.5583400E-09	1.7789300E+06
4.5078300E-08	4.5775900E-09	-4.5147500E-09	1.7576900E+06
4.5113900E-08	4.5266700E-09	-4.4601900E-09	1.7370000E+06
4.5149400E-08	4.4671700E-09	-4.3970000E-09	1.7173500E+06
4.5185000E-08	4.4019200E-09	-4.3283200E-09	1.6991600E+06
4.5220600E-08	4.3342700E-09	-4.2577700E-09	1.6828300E+06
4.5256200E-08	4.2678300E-09	-4.1891600E-09	1.6686600E+06
4.5291700E-08	4.2061400E-09	-4.1261000E-09	1.6568400E+06
4.5327300E-08	4.1523300E-09	-4.0717100E-09	1.6474900E+06

4.5362900E-08	4.1088400E-09	-4.0283400E-09	1.6405800E+06
4.5398500E-08	4.0772300E-09	-3.9973800E-09	1.6360100E+06
4.5434100E-08	4.0580600E-09	-3.9791700E-09	1.6335500E+06
4.5469600E-08	4.0508800E-09	-3.9730100E-09	1.6329000E+06
4.5505200E-08	4.0542100E-09	-3.9772400E-09	1.6337200E+06
4.5540800E-08	4.0657800E-09	-3.9894100E-09	1.6356300E+06
4.5576400E-08	4.0827300E-09	-4.0065500E-09	1.6382500E+06
4.5611900E-08	4.1018300E-09	-4.0253800E-09	1.6412200E+06
4.5647500E-08	4.1198200E-09	-4.0426300E-09	1.6442400E+06
4.5683100E-08	4.1335500E-09	-4.0553000E-09	1.6470600E+06
4.5718700E-08	4.1404300E-09	-4.0609300E-09	1.6495200E+06
4.5754200E-08	4.1385300E-09	-4.0577500E-09	1.6515300E+06
4.5789800E-08	4.1267500E-09	-4.0448800E-09	1.6531100E+06
4.5825400E-08	4.1049400E-09	-4.0223500E-09	1.6543000E+06
4.5861000E-08	4.0739400E-09	-3.9911200E-09	1.6552200E+06
4.5896600E-08	4.0354800E-09	-3.9530300E-09	1.6560400E+06
4.5932100E-08	3.9921300E-09	-3.9106700E-09	1.6569300E+06
4.5967700E-08	3.9470600E-09	-3.8671300E-09	1.6580300E+06
4.6003300E-08	3.9037400E-09	-3.8258000E-09	1.6594900E+06
4.6038900E-08	3.8657800E-09	-3.7901000E-09	1.6613800E+06
4.6074400E-08	3.8365200E-09	-3.7631600E-09	1.6637100E+06
4.6110000E-08	3.8187300E-09	-3.7475200E-09	1.6664400E+06
4.6145600E-08	3.8143400E-09	-3.7449200E-09	1.6694300E+06
4.6181200E-08	3.8242300E-09	-3.7560400E-09	1.6725100E+06
4.6216700E-08	3.8480900E-09	-3.7804400E-09	1.6754400E+06
4.6252300E-08	3.8843000E-09	-3.8165000E-09	1.6779700E+06
4.6287900E-08	3.9300800E-09	-3.8614900E-09	1.6798200E+06
4.6323500E-08	3.9816900E-09	-3.9117600E-09	1.6807500E+06
4.6359100E-08	4.0346200E-09	-3.9630800E-09	1.6805300E+06
4.6394600E-08	4.0840900E-09	-4.0109100E-09	1.6790100E+06
4.6430200E-08	4.1254000E-09	-4.0508900E-09	1.6760700E+06
4.6465800E-08	4.1544200E-09	-4.0792000E-09	1.6716800E+06
4.6501400E-08	4.1680000E-09	-4.0929400E-09	1.6658900E+06
4.6536900E-08	4.1642500E-09	-4.0904300E-09	1.6588100E+06
4.6572500E-08	4.1427400E-09	-4.0713800E-09	1.6506100E+06
4.6608100E-08	4.1046700E-09	-4.0369400E-09	1.6415100E+06
4.6643700E-08	4.0526300E-09	-3.9895900E-09	1.6317500E+06
4.6679200E-08	3.9904400E-09	-3.9328900E-09	1.6215800E+06
4.6714800E-08	3.9227400E-09	-3.8711200E-09	1.6112300E+06
4.6750400E-08	3.8545500E-09	-3.8088800E-09	1.6009300E+06
4.6786000E-08	3.7907500E-09	-3.7506100E-09	1.5908400E+06
4.6821600E-08	3.7356200E-09	-3.7001300E-09	1.5810900E+06
4.6857100E-08	3.6924000E-09	-3.6603400E-09	1.5717600E+06
4.6892700E-08	3.6630400E-09	-3.6328900E-09	1.5628700E+06
4.6928300E-08	3.6480200E-09	-3.6181300E-09	1.5544100E+06
4.6963900E-08	3.6463600E-09	-3.6151100E-09	1.5463300E+06
4.6999400E-08	3.6558400E-09	-3.6217400E-09	1.5385500E+06
4.7035000E-08	3.6732200E-09	-3.6351000E-09	1.5309800E+06
4.7070600E-08	3.6946700E-09	-3.6517700E-09	1.5235500E+06
4.7106200E-08	3.7161700E-09	-3.6682100E-09	1.5162100E+06
4.7141700E-08	3.7339500E-09	-3.6811400E-09	1.5089200E+06

4.7177300E-08	3.7448000E-09	-3.6878200E-09	1.5017200E+06
4.7212900E-08	3.7463800E-09	-3.6863400E-09	1.4946500E+06
4.7248500E-08	3.7373500E-09	-3.6756300E-09	1.4878500E+06
4.7284100E-08	3.7174500E-09	-3.6556000E-09	1.4814700E+06
4.7319600E-08	3.6873400E-09	-3.6269400E-09	1.4756800E+06
4.7355200E-08	3.6485100E-09	-3.5910400E-09	1.4706900E+06
4.7390800E-08	3.6030300E-09	-3.5497500E-09	1.4666800E+06
4.7426400E-08	3.5533700E-09	-3.5052100E-09	1.4638200E+06
4.7461900E-08	3.5021000E-09	-3.4596200E-09	1.4622300E+06
4.7497500E-08	3.4516800E-09	-3.4150700E-09	1.4619700E+06
4.7533100E-08	3.4043700E-09	-3.3734400E-09	1.4630400E+06
4.7568700E-08	3.3621000E-09	-3.3363400E-09	1.4653500E+06
4.7604200E-08	3.3264300E-09	-3.3050700E-09	1.4687200E+06
4.7639800E-08	3.2985300E-09	-3.2806200E-09	1.4728900E+06
4.7675400E-08	3.2791800E-09	-3.2636500E-09	1.4775500E+06
4.7711000E-08	3.2688100E-09	-3.2545800E-09	1.4823200E+06
4.7746500E-08	3.2674900E-09	-3.2534900E-09	1.4868200E+06
4.7782100E-08	3.2749700E-09	-3.2601900E-09	1.4906500E+06
4.7817700E-08	3.2905800E-09	-3.2741500E-09	1.4934400E+06
4.7853300E-08	3.3133000E-09	-3.2944700E-09	1.4949000E+06
4.7888900E-08	3.3417400E-09	-3.3199100E-09	1.4947900E+06
4.7924400E-08	3.3740800E-09	-3.3488100E-09	1.4929700E+06
4.7960000E-08	3.4081900E-09	-3.3792500E-09	1.4894100E+06
4.7995600E-08	3.4417500E-09	-3.4090400E-09	1.4841700E+06
4.8031200E-08	3.4722800E-09	-3.4359500E-09	1.4774200E+06
4.8066700E-08	3.4974300E-09	-3.4578000E-09	1.4693900E+06
4.8102300E-08	3.5150900E-09	-3.4726800E-09	1.4604000E+06
4.8137900E-08	3.5236200E-09	-3.4791500E-09	1.4507800E+06
4.8173500E-08	3.5220100E-09	-3.4763700E-09	1.4408600E+06
4.8209000E-08	3.5100600E-09	-3.4642200E-09	1.4309700E+06
4.8244600E-08	3.4883200E-09	-3.4433300E-09	1.4213600E+06
4.8280200E-08	3.4581300E-09	-3.4150600E-09	1.4122200E+06
4.8315800E-08	3.4215600E-09	-3.3813600E-09	1.4036300E+06
4.8351400E-08	3.3811900E-09	-3.3446400E-09	1.3956000E+06
4.8386900E-08	3.3398300E-09	-3.3074600E-09	1.3880300E+06
4.8422500E-08	3.3003200E-09	-3.2723600E-09	1.3807500E+06
4.8458100E-08	3.2652600E-09	-3.2415900E-09	1.3735100E+06
4.8493700E-08	3.2367300E-09	-3.2169100E-09	1.3660400E+06
4.8529200E-08	3.2161200E-09	-3.1994100E-09	1.3580800E+06
4.8564800E-08	3.2040800E-09	-3.1894700E-09	1.3493700E+06
4.8600400E-08	3.2004000E-09	-3.1867100E-09	1.3397100E+06
4.8636000E-08	3.2040900E-09	-3.1901200E-09	1.3289900E+06
4.8671500E-08	3.2135800E-09	-3.1981600E-09	1.3171900E+06
4.8707100E-08	3.2267800E-09	-3.2089300E-09	1.3044000E+06
4.8742700E-08	3.2414200E-09	-3.2204100E-09	1.2908200E+06
4.8778300E-08	3.2552500E-09	-3.2306200E-09	1.2767700E+06
4.8813900E-08	3.2661200E-09	-3.2378300E-09	1.2626200E+06
4.8849400E-08	3.2723600E-09	-3.2406900E-09	1.2488300E+06
4.8885000E-08	3.2728100E-09	-3.2383100E-09	1.2358700E+06
4.8920600E-08	3.2668300E-09	-3.2303100E-09	1.2241900E+06
4.8956200E-08	3.2543800E-09	-3.2167700E-09	1.2142200E+06

4.8991700E-08	3.2359600E-09	-3.1981900E-09	1.2063000E+06
4.9027300E-08	3.2124400E-09	-3.1754100E-09	1.2006700E+06
4.9062900E-08	3.1850500E-09	-3.1494600E-09	1.1974700E+06
4.9098500E-08	3.1551600E-09	-3.1214400E-09	1.1966900E+06
4.9134000E-08	3.1241200E-09	-3.0924600E-09	1.1981900E+06
4.9169600E-08	3.0932400E-09	-3.0634700E-09	1.2017000E+06
4.9205200E-08	3.0635600E-09	-3.0352700E-09	1.2068800E+06
4.9240800E-08	3.0358700E-09	-3.0084300E-09	1.2132800E+06
4.9276400E-08	3.0106800E-09	-2.9832900E-09	1.2204100E+06
4.9311900E-08	2.9882200E-09	-2.9599900E-09	1.2277800E+06
4.9347500E-08	2.9684100E-09	-2.9385100E-09	1.2349100E+06
4.9383100E-08	2.9510300E-09	-2.9187300E-09	1.2413900E+06
4.9418700E-08	2.9357200E-09	-2.9005100E-09	1.2469000E+06
4.9454200E-08	2.9221200E-09	-2.8837000E-09	1.2512300E+06
4.9489800E-08	2.9099200E-09	-2.8682900E-09	1.2542500E+06
4.9525400E-08	2.8989300E-09	-2.8544100E-09	1.2559900E+06
4.9561000E-08	2.8891600E-09	-2.8423400E-09	1.2565600E+06
4.9596500E-08	2.8808800E-09	-2.8325600E-09	1.2561700E+06
4.9632100E-08	2.8745400E-09	-2.8257200E-09	1.2551200E+06
4.9667700E-08	2.8708500E-09	-2.8225800E-09	1.2537000E+06
4.9703300E-08	2.8706300E-09	-2.8239700E-09	1.2522600E+06
4.9738900E-08	2.8747200E-09	-2.8306700E-09	1.2510600E+06
4.9774400E-08	2.8840000E-09	-2.8433600E-09	1.2503600E+06
4.9810000E-08	2.8991500E-09	-2.8624200E-09	1.2502900E+06
4.9845600E-08	2.9205200E-09	-2.8879600E-09	1.2509200E+06
4.9881200E-08	2.9480800E-09	-2.9196000E-09	1.2521800E+06
4.9916700E-08	2.9813400E-09	-2.9565000E-09	1.2539200E+06
4.9952300E-08	3.0192300E-09	-2.9972900E-09	1.2558700E+06
4.9987900E-08	3.0602000E-09	-3.0401400E-09	1.2577300E+06
5.0023500E-08	3.1021300E-09	-3.0827800E-09	1.2591100E+06
5.0059000E-08	3.1425500E-09	-3.1226600E-09	1.2596400E+06
5.0094600E-08	3.1787800E-09	-3.1571000E-09	1.2589600E+06
5.0130200E-08	3.2080700E-09	-3.1835100E-09	1.2567400E+06
5.0165800E-08	3.2278800E-09	-3.1995800E-09	1.2527600E+06
5.0201400E-08	3.2361000E-09	-3.2035500E-09	1.2468800E+06
5.0236900E-08	3.2313100E-09	-3.1943300E-09	1.2390800E+06
5.0272500E-08	3.2128200E-09	-3.1717100E-09	1.2294700E+06
5.0308100E-08	3.1809800E-09	-3.1363800E-09	1.2182600E+06
5.0343700E-08	3.1370700E-09	-3.0899900E-09	1.2057900E+06
5.0379200E-08	3.0833200E-09	-3.0350200E-09	1.1924500E+06
5.0414800E-08	3.0227700E-09	-2.9746500E-09	1.1787000E+06
5.0450400E-08	2.9590000E-09	-2.9124800E-09	1.1650100E+06
5.0486000E-08	2.8959300E-09	-2.8522700E-09	1.1518400E+06
5.0521500E-08	2.8374300E-09	-2.7976200E-09	1.1396200E+06
5.0557100E-08	2.7869800E-09	-2.7516500E-09	1.1286700E+06
5.0592700E-08	2.7474400E-09	-2.7167400E-09	1.1192700E+06
5.0628300E-08	2.7207100E-09	-2.6943000E-09	1.1115500E+06
5.0663900E-08	2.7076300E-09	-2.6846600E-09	1.1055300E+06
5.0699400E-08	2.7078500E-09	-2.6870600E-09	1.1011600E+06
5.0735000E-08	2.7199000E-09	-2.6997100E-09	1.0982500E+06
5.0770600E-08	2.7413300E-09	-2.7199700E-09	1.0965500E+06

5.0806200E-08	2.7689600E-09	-2.7446300E-09	1.0957600E+06
5.0841700E-08	2.7991100E-09	-2.7701800E-09	1.0955600E+06
5.0877300E-08	2.8279700E-09	-2.7931500E-09	1.0956100E+06
5.0912900E-08	2.8519900E-09	-2.8104000E-09	1.0956300E+06
5.0948500E-08	2.8681200E-09	-2.8194400E-09	1.0953900E+06
5.0984000E-08	2.8740700E-09	-2.8185700E-09	1.0947400E+06
5.1019600E-08	2.8685000E-09	-2.8070400E-09	1.0935800E+06
5.1055200E-08	2.8511600E-09	-2.7850800E-09	1.0919400E+06
5.1090800E-08	2.8228400E-09	-2.7538500E-09	1.0899000E+06
5.1126400E-08	2.7852700E-09	-2.7153200E-09	1.0876400E+06
5.1161900E-08	2.7409900E-09	-2.6720900E-09	1.0853500E+06
5.1197500E-08	2.6931800E-09	-2.6271600E-09	1.0832900E+06
5.1233100E-08	2.6452500E-09	-2.5836500E-09	1.0817000E+06
5.1268700E-08	2.6006900E-09	-2.5446200E-09	1.0808000E+06
5.1304200E-08	2.5628200E-09	-2.5127900E-09	1.0807800E+06
5.1339800E-08	2.5343800E-09	-2.4903300E-09	1.0817500E+06
5.1375400E-08	2.5174300E-09	-2.4787400E-09	1.0837400E+06
5.1411000E-08	2.5131400E-09	-2.4786800E-09	1.0867100E+06
5.1446500E-08	2.5217200E-09	-2.4899400E-09	1.0905100E+06
5.1482100E-08	2.5423500E-09	-2.5114400E-09	1.0949400E+06
5.1517700E-08	2.5731900E-09	-2.5413000E-09	1.0997200E+06
5.1553300E-08	2.6115500E-09	-2.5769100E-09	1.1045300E+06
5.1588900E-08	2.6540600E-09	-2.6151300E-09	1.1090400E+06
5.1624400E-08	2.6969500E-09	-2.6525300E-09	1.1129000E+06
5.1660000E-08	2.7362900E-09	-2.6856500E-09	1.1158100E+06
5.1695600E-08	2.7683300E-09	-2.7112500E-09	1.1175200E+06
5.1731200E-08	2.7898800E-09	-2.7266100E-09	1.1178100E+06
5.1766700E-08	2.7985700E-09	-2.7298100E-09	1.1165600E+06
5.1802300E-08	2.7930000E-09	-2.7198700E-09	1.1137500E+06
5.1837900E-08	2.7730600E-09	-2.6969400E-09	1.1094200E+06
5.1873500E-08	2.7399000E-09	-2.6622900E-09	1.1036800E+06
5.1909000E-08	2.6957200E-09	-2.6182500E-09	1.0967100E+06
5.1944600E-08	2.6438000E-09	-2.5680400E-09	1.0887400E+06
5.1980200E-08	2.5881400E-09	-2.5154500E-09	1.0800400E+06
5.2015800E-08	2.5330100E-09	-2.4646000E-09	1.0708500E+06
5.2051400E-08	2.4826600E-09	-2.4194600E-09	1.0614300E+06
5.2086900E-08	2.4408800E-09	-2.3835000E-09	1.0520000E+06
5.2122500E-08	2.4106300E-09	-2.3593700E-09	1.0427500E+06
5.2158100E-08	2.3938300E-09	-2.3486500E-09	1.0338000E+06
5.2193700E-08	2.3910400E-09	-2.3516500E-09	1.0252600E+06
5.2229200E-08	2.4015900E-09	-2.3674000E-09	1.0171400E+06
5.2264800E-08	2.4236500E-09	-2.3937700E-09	1.0094500E+06
5.2300400E-08	2.4543300E-09	-2.4276500E-09	1.0021400E+06
5.2336000E-08	2.4899800E-09	-2.4652600E-09	9.9515400E+05
5.2371500E-08	2.5267200E-09	-2.5025500E-09	9.8841700E+05
5.2407100E-08	2.5606700E-09	-2.5355600E-09	9.8186600E+05
5.2442700E-08	2.5883200E-09	-2.5607800E-09	9.7545900E+05
5.2478300E-08	2.6068300E-09	-2.5754600E-09	9.6918400E+05
5.2513900E-08	2.6143000E-09	-2.5778700E-09	9.6306500E+05
5.2549400E-08	2.6098600E-09	-2.5673600E-09	9.5715900E+05
5.2585000E-08	2.5936500E-09	-2.5444000E-09	9.5155800E+05

5.2620600E-08	2.5667400E-09	-2.5105100E-09	9.4638400E+05
5.2656200E-08	2.5310600E-09	-2.4680200E-09	9.4177800E+05
5.2691700E-08	2.4890600E-09	-2.4198400E-09	9.3788400E+05
5.2727300E-08	2.4435500E-09	-2.3692500E-09	9.3483500E+05
5.2762900E-08	2.3974500E-09	-2.3195000E-09	9.3274600E+05
5.2798500E-08	2.3534900E-09	-2.2736400E-09	9.3170300E+05
5.2834000E-08	2.3141000E-09	-2.2342600E-09	9.3174600E+05
5.2869600E-08	2.2811700E-09	-2.2033000E-09	9.3286800E+05
5.2905200E-08	2.2560700E-09	-2.1819700E-09	9.3500700E+05
5.2940800E-08	2.2395000E-09	-2.1707400E-09	9.3805000E+05
5.2976400E-08	2.2315400E-09	-2.1693000E-09	9.4184300E+05
5.3011900E-08	2.2316600E-09	-2.1766400E-09	9.4619600E+05
5.3047500E-08	2.2388000E-09	-2.1912200E-09	9.5089700E+05
5.3083100E-08	2.2515200E-09	-2.2110700E-09	9.5572900E+05
5.3118700E-08	2.2680700E-09	-2.2339500E-09	9.6047700E+05
5.3154200E-08	2.2865000E-09	-2.2575700E-09	9.6495300E+05
5.3189800E-08	2.3048800E-09	-2.2797100E-09	9.6899400E+05
5.3225400E-08	2.3214300E-09	-2.2984300E-09	9.7247900E+05
5.3261000E-08	2.3345900E-09	-2.3121900E-09	9.7532700E+05
5.3296500E-08	2.3432100E-09	-2.3200000E-09	9.7750200E+05
5.3332100E-08	2.3467000E-09	-2.3214400E-09	9.7900600E+05
5.3367700E-08	2.3449800E-09	-2.3167200E-09	9.7987000E+05
5.3403300E-08	2.3385400E-09	-2.3066900E-09	9.8015100E+05
5.3438800E-08	2.3284500E-09	-2.2927100E-09	9.7991700E+05
5.3474400E-08	2.3161900E-09	-2.2765500E-09	9.7923600E+05
5.3510000E-08	2.3035200E-09	-2.2602400E-09	9.7816100E+05
5.3545600E-08	2.2923900E-09	-2.2458300E-09	9.7672300E+05
5.3581200E-08	2.2845300E-09	-2.2351800E-09	9.7493100E+05
5.3616700E-08	2.2814300E-09	-2.2297800E-09	9.7276800E+05
5.3652300E-08	2.2840400E-09	-2.2305600E-09	9.7018900E+05
5.3687900E-08	2.2926500E-09	-2.2377600E-09	9.6713200E+05
5.3723500E-08	2.3067900E-09	-2.2508800E-09	9.6352300E+05
5.3759000E-08	2.3252800E-09	-2.2687000E-09	9.5928900E+05
5.3794600E-08	2.3463100E-09	-2.2893700E-09	9.5436800E+05
5.3830200E-08	2.3675900E-09	-2.3106200E-09	9.4872300E+05
5.3865800E-08	2.3865900E-09	-2.3299800E-09	9.4234100E+05
5.3901300E-08	2.4008000E-09	-2.3450200E-09	9.3524500E+05
5.3936900E-08	2.4081100E-09	-2.3536400E-09	9.2750200E+05
5.3972500E-08	2.4069400E-09	-2.3542900E-09	9.1921600E+05
5.4008100E-08	2.3964600E-09	-2.3461600E-09	9.1052400E+05
5.4043700E-08	2.3767500E-09	-2.3292400E-09	9.0159400E+05
5.4079200E-08	2.3487200E-09	-2.3043400E-09	8.9260700E+05
5.4114800E-08	2.3141000E-09	-2.2730000E-09	8.8375100E+05
5.4150400E-08	2.2751500E-09	-2.2372800E-09	8.7520800E+05
5.4186000E-08	2.2345200E-09	-2.1995600E-09	8.6714000E+05
5.4221500E-08	2.1948500E-09	-2.1622600E-09	8.5967800E+05
5.4257100E-08	2.1585000E-09	-2.1275500E-09	8.5291800E+05
5.4292700E-08	2.1274200E-09	-2.0971700E-09	8.4691800E+05
5.4328300E-08	2.1028000E-09	-2.0722300E-09	8.4169600E+05
5.4363800E-08	2.0850600E-09	-2.0531500E-09	8.3723300E+05
5.4399400E-08	2.0739100E-09	-2.0396500E-09	8.3347700E+05

5.4435000E-08	2.0683000E-09	-2.0308600E-09	8.3035600E+05
5.4470600E-08	2.0667500E-09	-2.0254800E-09	8.2778200E+05
5.4506200E-08	2.0675000E-09	-2.0219700E-09	8.2565900E+05
5.4541700E-08	2.0687800E-09	-2.0188200E-09	8.2390200E+05
5.4577300E-08	2.0689900E-09	-2.0146800E-09	8.2243600E+05
5.4612900E-08	2.0669500E-09	-2.0086100E-09	8.2120700E+05
5.4648500E-08	2.0619700E-09	-2.0001100E-09	8.2017900E+05
5.4684000E-08	2.0539600E-09	-1.9892500E-09	8.1934400E+05
5.4719600E-08	2.0433300E-09	-1.9765500E-09	8.1870900E+05
5.4755200E-08	2.0309200E-09	-1.9629400E-09	8.1830100E+05
5.4790800E-08	2.0179300E-09	-1.9496400E-09	8.1815700E+05
5.4826300E-08	2.0056600E-09	-1.9379700E-09	8.1831600E+05
5.4861900E-08	1.9954000E-09	-1.9292200E-09	8.1881200E+05
5.4897500E-08	1.9883300E-09	-1.9244700E-09	8.1967100E+05
5.4933100E-08	1.9853000E-09	-1.9245000E-09	8.2090200E+05
5.4968700E-08	1.9868500E-09	-1.9297200E-09	8.2249500E+05
5.5004200E-08	1.9930700E-09	-1.9401100E-09	8.2441700E+05
5.5039800E-08	2.0037400E-09	-1.9551800E-09	8.2661200E+05
5.5075400E-08	2.0182600E-09	-1.9740800E-09	8.2900100E+05
5.5111000E-08	2.0356700E-09	-1.9955600E-09	8.3148400E+05
5.5146500E-08	2.0547700E-09	-2.0181300E-09	8.3394600E+05
5.5182100E-08	2.0741800E-09	-2.0400800E-09	8.3626000E+05
5.5217700E-08	2.0923700E-09	-2.0595900E-09	8.3829200E+05
5.5253300E-08	2.1077800E-09	-2.0748700E-09	8.3991400E+05
5.5288800E-08	2.1188600E-09	-2.0842400E-09	8.4100800E+05
5.5324400E-08	2.1241800E-09	-2.0862700E-09	8.4146600E+05
5.5360000E-08	2.1225600E-09	-2.0799200E-09	8.4120100E+05
5.5395600E-08	2.1131800E-09	-2.0646000E-09	8.4015000E+05
5.5431200E-08	2.0956400E-09	-2.0403500E-09	8.3827900E+05
5.5466700E-08	2.0701100E-09	-2.0078400E-09	8.3557800E+05
5.5502300E-08	2.0373800E-09	-1.9684400E-09	8.3206600E+05
5.5537900E-08	1.9988600E-09	-1.9241800E-09	8.2778800E+05
5.5573500E-08	1.9565700E-09	-1.8776200E-09	8.2281400E+05
5.5609000E-08	1.9130300E-09	-1.8317500E-09	8.1723000E+05
5.5644600E-08	1.8710500E-09	-1.7897000E-09	8.1113800E+05
5.5680200E-08	1.8336000E-09	-1.7545700E-09	8.0464900E+05
5.5715800E-08	1.8034400E-09	-1.7290800E-09	7.9787800E+05
5.5751300E-08	1.7829300E-09	-1.7153000E-09	7.9093800E+05
5.5786900E-08	1.7737700E-09	-1.7144800E-09	7.8394200E+05
5.5822500E-08	1.7767300E-09	-1.7268200E-09	7.7699100E+05
5.5858100E-08	1.7916100E-09	-1.7514000E-09	7.7018500E+05
5.5893700E-08	1.8171600E-09	-1.7862700E-09	7.6361200E+05
5.5929200E-08	1.8511500E-09	-1.8284900E-09	7.5735300E+05
5.5964800E-08	1.8905500E-09	-1.8744500E-09	7.5147600E+05
5.6000400E-08	1.9317700E-09	-1.9201000E-09	7.4604500E+05
5.6036000E-08	1.9710500E-09	-1.9613800E-09	7.4111400E+05
5.6071500E-08	2.0046900E-09	-1.9945300E-09	7.3672300E+05
5.6107100E-08	2.0295100E-09	-2.0164700E-09	7.3290600E+05
5.6142700E-08	2.0431300E-09	-2.0250400E-09	7.2968200E+05
5.6178300E-08	2.0440700E-09	-2.0192500E-09	7.2705700E+05
5.6213800E-08	2.0320100E-09	-1.9992800E-09	7.2502700E+05

5.6249400E-08	2.0077700E-09	-1.9665300E-09	7.2357100E+05
5.6285000E-08	1.9731900E-09	-1.9234300E-09	7.2265300E+05
5.6320600E-08	1.9309300E-09	-1.8732000E-09	7.2222200E+05
5.6356200E-08	1.8842100E-09	-1.8195700E-09	7.2221500E+05
5.6391700E-08	1.8365100E-09	-1.7664100E-09	7.2256200E+05
5.6427300E-08	1.7912900E-09	-1.7174300E-09	7.2318300E+05
5.6462900E-08	1.7515300E-09	-1.6758300E-09	7.2399500E+05
5.6498500E-08	1.7196900E-09	-1.6440500E-09	7.2491900E+05
5.6534000E-08	1.6973900E-09	-1.6236200E-09	7.2588500E+05
5.6569600E-08	1.6853400E-09	-1.6150700E-09	7.2683800E+05
5.6605200E-08	1.6833500E-09	-1.6179600E-09	7.2773700E+05
5.6640800E-08	1.6904400E-09	-1.6309400E-09	7.2856100E+05
5.6676300E-08	1.7049600E-09	-1.6519600E-09	7.2930400E+05
5.6711900E-08	1.7247300E-09	-1.6784900E-09	7.2998500E+05
5.6747500E-08	1.7473900E-09	-1.7077100E-09	7.3063900E+05
5.6783100E-08	1.7705200E-09	-1.7368500E-09	7.3131200E+05
5.6818700E-08	1.7918700E-09	-1.7633500E-09	7.3205300E+05
5.6854200E-08	1.8095600E-09	-1.7850800E-09	7.3291300E+05
5.6889800E-08	1.8222000E-09	-1.8004700E-09	7.3393000E+05
5.6925400E-08	1.8289500E-09	-1.8085900E-09	7.3512900E+05
5.6961000E-08	1.8295200E-09	-1.8091900E-09	7.3651000E+05
5.6996500E-08	1.8242000E-09	-1.8026800E-09	7.3804400E+05
5.7032100E-08	1.8137600E-09	-1.7899800E-09	7.3967400E+05
5.7067700E-08	1.7993200E-09	-1.7725100E-09	7.4131200E+05
5.7103300E-08	1.7823100E-09	-1.7519700E-09	7.4284100E+05
5.7138800E-08	1.7643200E-09	-1.7302200E-09	7.4412700E+05
5.7174400E-08	1.7469400E-09	-1.7091500E-09	7.4502200E+05
5.7210000E-08	1.7316700E-09	-1.6905100E-09	7.4537000E+05
5.7245600E-08	1.7198800E-09	-1.6758000E-09	7.4502600E+05
5.7281200E-08	1.7126100E-09	-1.6661600E-09	7.4386100E+05
5.7316700E-08	1.7105300E-09	-1.6623100E-09	7.4176800E+05
5.7352300E-08	1.7139600E-09	-1.6645000E-09	7.3867600E+05
5.7387900E-08	1.7228200E-09	-1.6725000E-09	7.3455400E+05
5.7423500E-08	1.7365300E-09	-1.6856100E-09	7.2942300E+05
5.7459000E-08	1.7541100E-09	-1.7027000E-09	7.2335100E+05
5.7494600E-08	1.7742400E-09	-1.7223200E-09	7.1645000E+05
5.7530200E-08	1.7953100E-09	-1.7427500E-09	7.0887400E+05
5.7565800E-08	1.8155000E-09	-1.7621600E-09	7.0081900E+05
5.7601300E-08	1.8329400E-09	-1.7786800E-09	6.9250300E+05
5.7636900E-08	1.8458100E-09	-1.7906100E-09	6.8416200E+05
5.7672500E-08	1.8525100E-09	-1.7965100E-09	6.7602900E+05
5.7708100E-08	1.8518300E-09	-1.7953200E-09	6.6833000E+05
5.7743700E-08	1.8429900E-09	-1.7864400E-09	6.6126100E+05
5.7779200E-08	1.8257800E-09	-1.7698500E-09	6.5498400E+05
5.7814800E-08	1.8005900E-09	-1.7460300E-09	6.4961500E+05
5.7850400E-08	1.7684400E-09	-1.7160400E-09	6.4521200E+05
5.7886000E-08	1.7308500E-09	-1.6813900E-09	6.4177700E+05
5.7921500E-08	1.6898000E-09	-1.6439200E-09	6.3925400E+05
5.7957100E-08	1.6475600E-09	-1.6056800E-09	6.3753500E+05
5.7992700E-08	1.6065000E-09	-1.5687900E-09	6.3647000E+05
5.8028300E-08	1.5688500E-09	-1.5352000E-09	6.3587600E+05

5.8063800E-08	1.5366300E-09	-1.5065900E-09	6.3555300E+05
5.8099400E-08	1.5113900E-09	-1.4841800E-09	6.3530300E+05
5.8135000E-08	1.4940300E-09	-1.4686800E-09	6.3494600E+05
5.8170600E-08	1.4848600E-09	-1.4601900E-09	6.3433100E+05
5.8206200E-08	1.4835200E-09	-1.4582400E-09	6.3335500E+05
5.8241700E-08	1.4890100E-09	-1.4618600E-09	6.3196900E+05
5.8277300E-08	1.4998700E-09	-1.4696400E-09	6.3017900E+05
5.8312900E-08	1.5142500E-09	-1.4799300E-09	6.2805500E+05
5.8348500E-08	1.5301900E-09	-1.4909800E-09	6.2571900E+05
5.8384000E-08	1.5457300E-09	-1.5011500E-09	6.2334000E+05
5.8419600E-08	1.5591700E-09	-1.5090500E-09	6.2111500E+05
5.8455200E-08	1.5692300E-09	-1.5136800E-09	6.1925200E+05
5.8490800E-08	1.5750500E-09	-1.5145200E-09	6.1795400E+05
5.8526300E-08	1.5763500E-09	-1.5116000E-09	6.1740100E+05
5.8561900E-08	1.5734500E-09	-1.5054400E-09	6.1773000E+05
5.8597500E-08	1.5670900E-09	-1.4969900E-09	6.1901200E+05
5.8633100E-08	1.5584000E-09	-1.4875200E-09	6.2125300E+05
5.8668700E-08	1.5487500E-09	-1.4784600E-09	6.2438300E+05
5.8704200E-08	1.5395300E-09	-1.4712300E-09	6.2825800E+05
5.8739800E-08	1.5320300E-09	-1.4670800E-09	6.3267000E+05
5.8775400E-08	1.5272900E-09	-1.4669000E-09	6.3735200E+05
5.8811000E-08	1.5259600E-09	-1.4711700E-09	6.4200000E+05
5.8846500E-08	1.5282100E-09	-1.4798500E-09	6.4629800E+05
5.8882100E-08	1.5338200E-09	-1.4923700E-09	6.4993700E+05
5.8917700E-08	1.5420800E-09	-1.5077000E-09	6.5263500E+05
5.8953300E-08	1.5519400E-09	-1.5244300E-09	6.5416600E+05
5.8988800E-08	1.5621500E-09	-1.5408900E-09	6.5436900E+05
5.9024400E-08	1.5713500E-09	-1.5553500E-09	6.5316500E+05
5.9060000E-08	1.5782400E-09	-1.5661700E-09	6.5056500E+05
5.9095600E-08	1.5817400E-09	-1.5719900E-09	6.4666400E+05
5.9131100E-08	1.5810800E-09	-1.5718400E-09	6.4163600E+05
5.9166700E-08	1.5758600E-09	-1.5653000E-09	6.3571500E+05
5.9202300E-08	1.5662000E-09	-1.5524800E-09	6.2918300E+05
5.9237900E-08	1.5525900E-09	-1.5340900E-09	6.2234700E+05
5.9273500E-08	1.5359700E-09	-1.5113400E-09	6.1551000E+05
5.9309000E-08	1.5175100E-09	-1.4858600E-09	6.0894900E+05
5.9344600E-08	1.4986000E-09	-1.4595100E-09	6.0289600E+05
5.9380200E-08	1.4806900E-09	-1.4342700E-09	5.9752100E+05
5.9415800E-08	1.4651000E-09	-1.4120000E-09	5.9292100E+05
5.9451300E-08	1.4529200E-09	-1.3942900E-09	5.8911700E+05
5.9486900E-08	1.4449000E-09	-1.3823200E-09	5.8605300E+05
5.9522500E-08	1.4413800E-09	-1.3767200E-09	5.8360800E+05
5.9558100E-08	1.4422500E-09	-1.3775400E-09	5.8161300E+05
5.9593600E-08	1.4469700E-09	-1.3842100E-09	5.7986800E+05
5.9629200E-08	1.4545700E-09	-1.3956400E-09	5.7816600E+05
5.9664800E-08	1.4637900E-09	-1.4102800E-09	5.7631800E+05
5.9700400E-08	1.4731900E-09	-1.4262800E-09	5.7417000E+05
5.9736000E-08	1.4812900E-09	-1.4416600E-09	5.7161800E+05
5.9771600E-08	1.4866900E-09	-1.4544900E-09	5.6862600E+05
5.9807100E-08	1.4882100E-09	-1.4630800E-09	5.6522800E+05
5.9842700E-08	1.4850300E-09	-1.4661200E-09	5.6152500E+05

5.9878300E-08	1.4767600E-09	-1.4628300E-09	5.5767300E+05
5.9913900E-08	1.4634600E-09	-1.4529900E-09	5.5387600E+05
5.9949500E-08	1.4457400E-09	-1.4370300E-09	5.5036300E+05
5.9985100E-08	1.4246500E-09	-1.4159200E-09	5.4736400E+05
6.0020700E-08	1.4015800E-09	-1.3911700E-09	5.4509400E+05
6.0056200E-08	1.3782500E-09	-1.3646600E-09	5.4373000E+05
6.0091800E-08	1.3565000E-09	-1.3384700E-09	5.4339700E+05
6.0127400E-08	1.3381300E-09	-1.3147500E-09	5.4415100E+05
6.0163000E-08	1.3248100E-09	-1.2954500E-09	5.4597800E+05
6.0198600E-08	1.3178600E-09	-1.2822400E-09	5.4879400E+05
6.0234200E-08	1.3180800E-09	-1.2762600E-09	5.5245200E+05
6.0269700E-08	1.3257400E-09	-1.2780500E-09	5.5674800E+05
6.0305300E-08	1.3405100E-09	-1.2874500E-09	5.6144900E+05
6.0340900E-08	1.3613300E-09	-1.3036300E-09	5.6629600E+05
6.0376500E-08	1.3865800E-09	-1.3250800E-09	5.7103300E+05
6.0412100E-08	1.4141400E-09	-1.3497400E-09	5.7542100E+05
6.0447700E-08	1.4415300E-09	-1.3751800E-09	5.7925700E+05
6.0483200E-08	1.4661400E-09	-1.3988000E-09	5.8237600E+05
6.0518800E-08	1.4854900E-09	-1.4180500E-09	5.8467400E+05
6.0554400E-08	1.4973700E-09	-1.4307100E-09	5.8610100E+05
6.0590000E-08	1.5001900E-09	-1.4351100E-09	5.8666100E+05
6.0625600E-08	1.4931300E-09	-1.4303100E-09	5.8641400E+05
6.0661200E-08	1.4762700E-09	-1.4162600E-09	5.8545800E+05
6.0696700E-08	1.4506300E-09	-1.3938400E-09	5.8392600E+05
6.0732300E-08	1.4181600E-09	-1.3648200E-09	5.8196300E+05
6.0767900E-08	1.3816200E-09	-1.3317000E-09	5.7972600E+05
6.0803500E-08	1.3442800E-09	-1.2975400E-09	5.7735700E+05
6.0839100E-08	1.3097100E-09	-1.2656300E-09	5.7498600E+05
6.0874700E-08	1.2813600E-09	-1.2391500E-09	5.7271000E+05
6.0910300E-08	1.2622300E-09	-1.2209000E-09	5.7059700E+05
6.0945800E-08	1.2545600E-09	-1.2128800E-09	5.6867500E+05
6.0981400E-08	1.2595800E-09	-1.2161600E-09	5.6694300E+05
6.1017000E-08	1.2772500E-09	-1.2306400E-09	5.6536700E+05
6.1052600E-08	1.3062900E-09	-1.2551000E-09	5.6389100E+05
6.1088200E-08	1.3443400E-09	-1.2872300E-09	5.6244100E+05
6.1123800E-08	1.3880100E-09	-1.3238700E-09	5.6093600E+05
6.1159300E-08	1.4332700E-09	-1.3612800E-09	5.5929800E+05
6.1194900E-08	1.4759000E-09	-1.3955600E-09	5.5746200E+05
6.1230500E-08	1.5117900E-09	-1.4230000E-09	5.5538600E+05
6.1266100E-08	1.5374000E-09	-1.4404800E-09	5.5304900E+05
6.1301700E-08	1.5501100E-09	-1.4457600E-09	5.5045800E+05
6.1337300E-08	1.5484700E-09	-1.4377100E-09	5.4764900E+05
6.1372800E-08	1.5323100E-09	-1.4164000E-09	5.4468700E+05
6.1408400E-08	1.5027000E-09	-1.3831200E-09	5.4165800E+05
6.1444000E-08	1.4618900E-09	-1.3401600E-09	5.3866200E+05
6.1479600E-08	1.4130300E-09	-1.2906500E-09	5.3580800E+05
6.1515200E-08	1.3598600E-09	-1.2381800E-09	5.3320800E+05
6.1550800E-08	1.3063300E-09	-1.1864600E-09	5.3096100E+05
6.1586300E-08	1.2562400E-09	-1.1390300E-09	5.2915500E+05
6.1621900E-08	1.2129600E-09	-1.0988800E-09	5.2785900E+05
6.1657500E-08	1.1790600E-09	-1.0682700E-09	5.2711600E+05

6.1693100E-08	1.1563000E-09	-1.0485700E-09	5.2693900E+05
6.1728700E-08	1.1454900E-09	-1.0401900E-09	5.2731800E+05
6.1764300E-08	1.1463800E-09	-1.0426500E-09	5.2821700E+05
6.1799900E-08	1.1579900E-09	-1.0546800E-09	5.2958200E+05
6.1835400E-08	1.1786500E-09	-1.0744000E-09	5.3134400E+05
6.1871000E-08	1.2061900E-09	-1.0995500E-09	5.3342800E+05
6.1906600E-08	1.2382200E-09	-1.1276800E-09	5.3575400E+05
6.1942200E-08	1.2722800E-09	-1.1563700E-09	5.3824700E+05
6.1977800E-08	1.3060000E-09	-1.1834100E-09	5.4084600E+05
6.2013400E-08	1.3372800E-09	-1.2069300E-09	5.4349700E+05
6.2048900E-08	1.3644000E-09	-1.2254700E-09	5.4616400E+05
6.2084500E-08	1.3860300E-09	-1.2380600E-09	5.4882700E+05
6.2120100E-08	1.4012300E-09	-1.2442100E-09	5.5147800E+05
6.2155700E-08	1.4095400E-09	-1.2439100E-09	5.5412200E+05
6.2191300E-08	1.4109400E-09	-1.2375500E-09	5.5676800E+05
6.2226900E-08	1.4057500E-09	-1.2259200E-09	5.5942600E+05
6.2262400E-08	1.3947100E-09	-1.2101200E-09	5.6210400E+05
6.2298000E-08	1.3788900E-09	-1.1914800E-09	5.6479800E+05
6.2333600E-08	1.3596500E-09	-1.1715400E-09	5.6749400E+05
6.2369200E-08	1.3386000E-09	-1.1519000E-09	5.7015700E+05
6.2404800E-08	1.3174900E-09	-1.1342000E-09	5.7273000E+05
6.2440400E-08	1.2981300E-09	-1.1199300E-09	5.7514500E+05
6.2475900E-08	1.2822600E-09	-1.1104400E-09	5.7731300E+05
6.2511500E-08	1.2714300E-09	-1.1067300E-09	5.7913600E+05
6.2547100E-08	1.2668700E-09	-1.1094400E-09	5.8050900E+05
6.2582700E-08	1.2692900E-09	-1.1187000E-09	5.8133000E+05
6.2618300E-08	1.2789300E-09	-1.1341500E-09	5.8150400E+05
6.2653900E-08	1.2953900E-09	-1.1549200E-09	5.8095100E+05
6.2689500E-08	1.3176500E-09	-1.1796400E-09	5.7961700E+05
6.2725000E-08	1.3441800E-09	-1.2066100E-09	5.7748000E+05
6.2760600E-08	1.3730400E-09	-1.2338500E-09	5.7455500E+05
6.2796200E-08	1.4020300E-09	-1.2593500E-09	5.7089400E+05
6.2831800E-08	1.4289100E-09	-1.2811800E-09	5.6658400E+05
6.2867400E-08	1.4516100E-09	-1.2977400E-09	5.6174700E+05
6.2903000E-08	1.4684700E-09	-1.3078600E-09	5.5653500E+05
6.2938500E-08	1.4783400E-09	-1.3109600E-09	5.5111800E+05
6.2974100E-08	1.4807100E-09	-1.3070700E-09	5.4567400E+05
6.3009700E-08	1.4757900E-09	-1.2968100E-09	5.4037700E+05
6.3045300E-08	1.4644000E-09	-1.2813600E-09	5.3538500E+05
6.3080900E-08	1.4478700E-09	-1.2622700E-09	5.3083200E+05
6.3116500E-08	1.4279500E-09	-1.2413400E-09	5.2681300E+05
6.3152000E-08	1.4065200E-09	-1.2203900E-09	5.2338400E+05
6.3187600E-08	1.3853900E-09	-1.2010800E-09	5.2055500E+05
6.3223200E-08	1.3662000E-09	-1.1847800E-09	5.1828800E+05
6.3258800E-08	1.3501900E-09	-1.1724300E-09	5.1650400E+05
6.3294400E-08	1.3381100E-09	-1.1644800E-09	5.1509300E+05
6.3330000E-08	1.3302600E-09	-1.1609500E-09	5.1392400E+05
6.3365500E-08	1.3264800E-09	-1.1614000E-09	5.1285400E+05
6.3401100E-08	1.3262000E-09	-1.1651400E-09	5.1174600E+05
6.3436700E-08	1.3286800E-09	-1.1712700E-09	5.1048400E+05
6.3472300E-08	1.3330800E-09	-1.1788600E-09	5.0898100E+05

6.3507900E-08	1.3385200E-09	-1.1870700E-09	5.0719000E+05
6.3543500E-08	1.3443100E-09	-1.1952100E-09	5.0510600E+05
6.3579100E-08	1.3499800E-09	-1.2028100E-09	5.0277200E+05
6.3614600E-08	1.3552000E-09	-1.2096200E-09	5.0027400E+05
6.3650200E-08	1.3598700E-09	-1.2155500E-09	4.9773300E+05
6.3685800E-08	1.3641200E-09	-1.2206900E-09	4.9529100E+05
6.3721400E-08	1.3680400E-09	-1.2251700E-09	4.9310500E+05
6.3757000E-08	1.3717300E-09	-1.2290900E-09	4.9132800E+05
6.3792600E-08	1.3753000E-09	-1.2325300E-09	4.9009500E+05
6.3828100E-08	1.3787300E-09	-1.2355000E-09	4.8951200E+05
6.3863700E-08	1.3818900E-09	-1.2379000E-09	4.8964000E+05
6.3899300E-08	1.3845700E-09	-1.2396100E-09	4.9049900E+05
6.3934900E-08	1.3865500E-09	-1.2404900E-09	4.9205200E+05
6.3970500E-08	1.3876100E-09	-1.2404500E-09	4.9421500E+05
6.4006100E-08	1.3876100E-09	-1.2395200E-09	4.9685900E+05
6.4041600E-08	1.3865300E-09	-1.2378100E-09	4.9982300E+05
6.4077200E-08	1.3844800E-09	-1.2355900E-09	5.0292100E+05
6.4112800E-08	1.3816900E-09	-1.2332000E-09	5.0595800E+05
6.4148400E-08	1.3785000E-09	-1.2310600E-09	5.0874100E+05
6.4184000E-08	1.3753200E-09	-1.2295800E-09	5.1109200E+05
6.4219600E-08	1.3724600E-09	-1.2290900E-09	5.1286100E+05
6.4255100E-08	1.3701900E-09	-1.2297600E-09	5.1393300E+05
6.4290700E-08	1.3686800E-09	-1.2316200E-09	5.1423800E+05
6.4326300E-08	1.3678800E-09	-1.2344900E-09	5.1374700E+05
6.4361900E-08	1.3676000E-09	-1.2380100E-09	5.1247800E+05
6.4397500E-08	1.3675500E-09	-1.2417100E-09	5.1048800E+05
6.4433100E-08	1.3673800E-09	-1.2450500E-09	5.0787000E+05
6.4468700E-08	1.3667200E-09	-1.2475300E-09	5.0474100E+05
6.4504200E-08	1.3652500E-09	-1.2487600E-09	5.0123600E+05
6.4539800E-08	1.3628600E-09	-1.2485400E-09	4.9749900E+05
6.4575400E-08	1.3596000E-09	-1.2468900E-09	4.9366700E+05
6.4611000E-08	1.3556800E-09	-1.2440700E-09	4.8986700E+05
6.4646600E-08	1.3515000E-09	-1.2405600E-09	4.8620700E+05
6.4682200E-08	1.3476400E-09	-1.2370000E-09	4.8277600E+05
6.4717700E-08	1.3446500E-09	-1.2340900E-09	4.7963000E+05
6.4753300E-08	1.3430300E-09	-1.2325000E-09	4.7680000E+05
6.4788900E-08	1.3431900E-09	-1.2327400E-09	4.7429200E+05
6.4824500E-08	1.3453000E-09	-1.2351000E-09	4.7208300E+05
6.4860100E-08	1.3492300E-09	-1.2395900E-09	4.7013200E+05
6.4895700E-08	1.3545500E-09	-1.2458800E-09	4.6838900E+05
6.4931200E-08	1.3605700E-09	-1.2532900E-09	4.6679200E+05
6.4966800E-08	1.3663400E-09	-1.2609100E-09	4.6528400E+05
6.5002400E-08	1.3708000E-09	-1.2676300E-09	4.6381300E+05
6.5038000E-08	1.3728400E-09	-1.2722900E-09	4.6234000E+05
6.5073600E-08	1.3714500E-09	-1.2737800E-09	4.6084200E+05
6.5109200E-08	1.3658500E-09	-1.2712200E-09	4.5931800E+05
6.5144700E-08	1.3556400E-09	-1.2640200E-09	4.5778200E+05
6.5180300E-08	1.3408000E-09	-1.2520600E-09	4.5626800E+05
6.5215900E-08	1.3217800E-09	-1.2356600E-09	4.5482800E+05
6.5251500E-08	1.2995000E-09	-1.2156500E-09	4.5352400E+05
6.5287100E-08	1.2752700E-09	-1.1933000E-09	4.5242700E+05

6.5322700E-08	1.2507500E-09	-1.1702400E-09	4.5160700E+05
6.5358300E-08	1.2277400E-09	-1.1483000E-09	4.5112700E+05
6.5393800E-08	1.2080400E-09	-1.1293800E-09	4.5104300E+05
6.5429400E-08	1.1932900E-09	-1.1152300E-09	4.5139400E+05
6.5465000E-08	1.1847900E-09	-1.1072800E-09	4.5219900E+05
6.5500600E-08	1.1833400E-09	-1.1064900E-09	4.5345800E+05
6.5536200E-08	1.1891500E-09	-1.1131700E-09	4.5514600E+05
6.5571800E-08	1.2017900E-09	-1.1269600E-09	4.5722200E+05
6.5607300E-08	1.2201600E-09	-1.1468100E-09	4.5962900E+05
6.5642900E-08	1.2425600E-09	-1.1710300E-09	4.6229600E+05
6.5678500E-08	1.2668900E-09	-1.1974200E-09	4.6514500E+05
6.5714100E-08	1.2907700E-09	-1.2234700E-09	4.6809800E+05
6.5749700E-08	1.3117600E-09	-1.2466100E-09	4.7107500E+05
6.5785300E-08	1.3276400E-09	-1.2644200E-09	4.7400500E+05
6.5820800E-08	1.3366100E-09	-1.2749500E-09	4.7682900E+05
6.5856400E-08	1.3374700E-09	-1.2768700E-09	4.7949800E+05
6.5892000E-08	1.3298200E-09	-1.2696800E-09	4.8197600E+05
6.5927600E-08	1.3140600E-09	-1.2537600E-09	4.8424200E+05
6.5963200E-08	1.2914000E-09	-1.2304000E-09	4.8628700E+05
6.5998800E-08	1.2637600E-09	-1.2016600E-09	4.8811000E+05
6.6034300E-08	1.2336000E-09	-1.1701900E-09	4.8971500E+05
6.6069900E-08	1.2037300E-09	-1.1390100E-09	4.9110800E+05
6.6105500E-08	1.1769100E-09	-1.1111500E-09	4.9229400E+05
6.6141100E-08	1.1556600E-09	-1.0893800E-09	4.9327600E+05
6.6176700E-08	1.1420200E-09	-1.0759000E-09	4.9405300E+05
6.6212300E-08	1.1372600E-09	-1.0721100E-09	4.9461400E+05
6.6247800E-08	1.1417700E-09	-1.0784300E-09	4.9494800E+05
6.6283400E-08	1.1550200E-09	-1.0942800E-09	4.9504000E+05
6.6319000E-08	1.1755900E-09	-1.1181000E-09	4.9487400E+05
6.6354600E-08	1.2013900E-09	-1.1475500E-09	4.9443200E+05
6.6390200E-08	1.2297700E-09	-1.1796900E-09	4.9370600E+05
6.6425800E-08	1.2578100E-09	-1.2113300E-09	4.9268900E+05
6.6461400E-08	1.2826600E-09	-1.2393600E-09	4.9138400E+05
6.6496900E-08	1.3018800E-09	-1.2610300E-09	4.8980200E+05
6.6532500E-08	1.3135400E-09	-1.2742500E-09	4.8796500E+05
6.6568100E-08	1.3165100E-09	-1.2778100E-09	4.8590200E+05
6.6603700E-08	1.3105000E-09	-1.2714400E-09	4.8364500E+05
6.6639300E-08	1.2961200E-09	-1.2558200E-09	4.8123300E+05
6.6674900E-08	1.2747500E-09	-1.2325000E-09	4.7870600E+05
6.6710400E-08	1.2483300E-09	-1.2036700E-09	4.7609800E+05
6.6746000E-08	1.2192700E-09	-1.1719600E-09	4.7344300E+05
6.6781600E-08	1.1900600E-09	-1.1401100E-09	4.7076800E+05
6.6817200E-08	1.1630300E-09	-1.1107200E-09	4.6809400E+05
6.6852800E-08	1.1402200E-09	-1.0859600E-09	4.6543500E+05
6.6888400E-08	1.1231600E-09	-1.0674300E-09	4.6280700E+05
6.6923900E-08	1.1127100E-09	-1.0560100E-09	4.6022100E+05
6.6959500E-08	1.1090000E-09	-1.0518400E-09	4.5768700E+05
6.6995100E-08	1.1115800E-09	-1.0543300E-09	4.5522100E+05
6.7030700E-08	1.1194400E-09	-1.0623200E-09	4.5284300E+05
6.7066300E-08	1.1310800E-09	-1.0742000E-09	4.5058100E+05
6.7101900E-08	1.1447700E-09	-1.0881100E-09	4.4846900E+05

6.7137400E-08	1.1586900E-09	-1.1021500E-09	4.4655000E+05
6.7173000E-08	1.1710500E-09	-1.1145000E-09	4.4487200E+05
6.7208600E-08	1.1803100E-09	-1.1236500E-09	4.4348600E+05
6.7244200E-08	1.1852100E-09	-1.1284300E-09	4.4244200E+05
6.7279800E-08	1.1848400E-09	-1.1281000E-09	4.4178300E+05
6.7315400E-08	1.1788200E-09	-1.1224200E-09	4.4153700E+05
6.7351000E-08	1.1671800E-09	-1.1115800E-09	4.4171700E+05
6.7386500E-08	1.1503900E-09	-1.0961700E-09	4.4231300E+05
6.7422100E-08	1.1293500E-09	-1.0771700E-09	4.4329200E+05
6.7457700E-08	1.1053100E-09	-1.0558200E-09	4.4459500E+05
6.7493300E-08	1.0797800E-09	-1.0335500E-09	4.4613900E+05
6.7528900E-08	1.0544900E-09	-1.0119300E-09	4.4782000E+05
6.7564500E-08	1.0312400E-09	-9.9247800E-10	4.4952100E+05
6.7600000E-08	1.0117900E-09	-9.7665000E-10	4.5111700E+05
6.7635600E-08	9.9766700E-10	-9.6567800E-10	4.5248400E+05
6.7671200E-08	9.9013200E-10	-9.6048400E-10	4.5350500E+05
6.7706800E-08	9.8996800E-10	-9.6158200E-10	4.5408300E+05
6.7742400E-08	9.9736500E-10	-9.6900100E-10	4.5415000E+05
6.7778000E-08	1.0119400E-09	-9.8225800E-10	4.5366500E+05
6.7813500E-08	1.0326600E-09	-1.0003500E-09	4.5262600E+05
6.7849100E-08	1.0578600E-09	-1.0218300E-09	4.5106200E+05
6.7884700E-08	1.0854700E-09	-1.0448900E-09	4.4903300E+05
6.7920300E-08	1.1131000E-09	-1.0675100E-09	4.4663000E+05
6.7955900E-08	1.1382700E-09	-1.0876700E-09	4.4396100E+05
6.7991500E-08	1.1586900E-09	-1.1035400E-09	4.4114200E+05
6.8027000E-08	1.1724700E-09	-1.1136800E-09	4.3828700E+05
6.8062600E-08	1.1783400E-09	-1.1172200E-09	4.3550300E+05
6.8098200E-08	1.1757900E-09	-1.1139200E-09	4.3288000E+05
6.8133800E-08	1.1651400E-09	-1.1042600E-09	4.3048100E+05
6.8169400E-08	1.1474900E-09	-1.0893700E-09	4.2833900E+05
6.8205000E-08	1.1246000E-09	-1.0708900E-09	4.2645500E+05
6.8240600E-08	1.0987000E-09	-1.0508100E-09	4.2479600E+05
6.8276100E-08	1.0722400E-09	-1.0311800E-09	4.2330300E+05
6.8311700E-08	1.0475500E-09	-1.0138800E-09	4.2189400E+05
6.8347300E-08	1.0265900E-09	-1.0003800E-09	4.2047400E+05
6.8382900E-08	1.0107400E-09	-9.9155500E-10	4.1893800E+05
6.8418500E-08	1.0006100E-09	-9.8758500E-10	4.1718500E+05
6.8454100E-08	9.9608300E-10	-9.8791900E-10	4.1512800E+05
6.8489600E-08	9.9628800E-10	-9.9138300E-10	4.1270400E+05
6.8525200E-08	9.9975100E-10	-9.9634800E-10	4.0987600E+05
6.8560800E-08	1.0046600E-09	-1.0009600E-09	4.0663900E+05
6.8596400E-08	1.0091600E-09	-1.0034400E-09	4.0302000E+05
6.8632000E-08	1.0115800E-09	-1.0022900E-09	3.9908200E+05
6.8667600E-08	1.0106200E-09	-9.9657700E-10	3.9492400E+05
6.8703100E-08	1.0056700E-09	-9.8601200E-10	3.9067000E+05
6.8738700E-08	9.9673300E-10	-9.7106200E-10	3.8646100E+05
6.8774300E-08	9.8448600E-10	-9.5286800E-10	3.8244900E+05
6.8809900E-08	9.7014800E-10	-9.3309200E-10	3.7879000E+05
6.8845500E-08	9.5527600E-10	-9.1370600E-10	3.7562800E+05
6.8881100E-08	9.4155600E-10	-8.9673400E-10	3.7309100E+05
6.8916600E-08	9.3053000E-10	-8.8397400E-10	3.7127100E+05

6.8952200E-08	9.2338500E-10	-8.7676300E-10	3.7022700E+05
6.8987800E-08	9.2076200E-10	-8.7581000E-10	3.6997400E+05
6.9023400E-08	9.2273500E-10	-8.8109600E-10	3.7048800E+05
6.9059000E-08	9.2877100E-10	-8.9189300E-10	3.7169700E+05
6.9094600E-08	9.3780800E-10	-9.0686200E-10	3.7349000E+05
6.9130200E-08	9.4847700E-10	-9.2423100E-10	3.7571900E+05
6.9165700E-08	9.5920300E-10	-9.4201400E-10	3.7821700E+05
6.9201300E-08	9.6848100E-10	-9.5825800E-10	3.8080200E+05
6.9236900E-08	9.7512800E-10	-9.7126900E-10	3.8329600E+05
6.9272500E-08	9.7829900E-10	-9.7979400E-10	3.8553200E+05
6.9308100E-08	9.7762300E-10	-9.8313000E-10	3.8737200E+05
6.9343700E-08	9.7326300E-10	-9.8116300E-10	3.8872100E+05
6.9379200E-08	9.6581900E-10	-9.7433600E-10	3.8952900E+05
6.9414800E-08	9.5619000E-10	-9.6354400E-10	3.8980200E+05
6.9450400E-08	9.4544700E-10	-9.4999300E-10	3.8958900E+05
6.9486000E-08	9.3471000E-10	-9.3503800E-10	3.8898500E+05
6.9521600E-08	9.2499300E-10	-9.2002700E-10	3.8812400E+05
6.9557200E-08	9.1706800E-10	-9.0616400E-10	3.8715600E+05
6.9592700E-08	9.1143800E-10	-8.9441300E-10	3.8623700E+05
6.9628300E-08	9.0830800E-10	-8.8545400E-10	3.8551400E+05
6.9663900E-08	9.0767200E-10	-8.7968900E-10	3.8510300E+05
6.9699500E-08	9.0935800E-10	-8.7726600E-10	3.8507900E+05
6.9735100E-08	9.1302500E-10	-8.7811800E-10	3.8546800E+05
6.9770700E-08	9.1831200E-10	-8.8203100E-10	3.8624000E+05
6.9806200E-08	9.2492500E-10	-8.8868400E-10	3.8731200E+05
6.9841800E-08	9.3257800E-10	-8.9768400E-10	3.8855300E+05
6.9877400E-08	9.4101100E-10	-9.0857300E-10	3.8979500E+05
6.9913000E-08	9.4996900E-10	-9.2080000E-10	3.9085200E+05
6.9948600E-08	9.5910200E-10	-9.3371400E-10	3.9153600E+05
6.9984200E-08	9.6796300E-10	-9.4653900E-10	3.9167600E+05
7.0019800E-08	9.7597400E-10	-9.5838500E-10	3.9114100E+05
7.0055300E-08	9.8238800E-10	-9.6825700E-10	3.8984700E+05
7.0090900E-08	9.8634200E-10	-9.7511400E-10	3.8776700E+05
7.0126500E-08	9.8695800E-10	-9.7796400E-10	3.8493700E+05
7.0162100E-08	9.8344200E-10	-9.7597800E-10	3.8145800E+05
7.0197700E-08	9.7523100E-10	-9.6861500E-10	3.7748200E+05
7.0233300E-08	9.6212500E-10	-9.5573000E-10	3.7320200E+05
7.0268800E-08	9.4439600E-10	-9.3767100E-10	3.6883600E+05
7.0304400E-08	9.2282300E-10	-9.1531100E-10	3.6460600E+05
7.0340000E-08	8.9868200E-10	-8.9002900E-10	3.6072000E+05
7.0375600E-08	8.7372600E-10	-8.6361700E-10	3.5735500E+05
7.0411200E-08	8.4997300E-10	-8.3813600E-10	3.5463600E+05
7.0446800E-08	8.2951900E-10	-8.1572100E-10	3.5263500E+05
7.0482300E-08	8.1433800E-10	-7.9834600E-10	3.5136000E+05
7.0517900E-08	8.0599700E-10	-7.8760400E-10	3.5075900E+05
7.0553500E-08	8.0543900E-10	-7.8451600E-10	3.5072800E+05
7.0589100E-08	8.1291400E-10	-7.8939500E-10	3.5112500E+05
7.0624700E-08	8.2787000E-10	-8.0177300E-10	3.5178000E+05
7.0660300E-08	8.4891400E-10	-8.2042900E-10	3.5251800E+05
7.0695800E-08	8.7401100E-10	-8.4349500E-10	3.5317400E+05
7.0731400E-08	9.0067700E-10	-8.6863500E-10	3.5361300E+05

7.0767000E-08	9.2618700E-10	-8.9328400E-10	3.5373400E+05
7.0802600E-08	9.4790600E-10	-9.1491000E-10	3.5348700E+05
7.0838200E-08	9.6355000E-10	-9.3125600E-10	3.5287600E+05
7.0873800E-08	9.7136900E-10	-9.4056000E-10	3.5195100E+05
7.0909400E-08	9.7036300E-10	-9.4172800E-10	3.5081100E+05
7.0944900E-08	9.6039800E-10	-9.3441300E-10	3.4958800E+05
7.0980500E-08	9.4212500E-10	-9.1902200E-10	3.4843400E+05
7.1016100E-08	9.1689900E-10	-8.9664800E-10	3.4750500E+05
7.1051700E-08	8.8666700E-10	-8.6893900E-10	3.4694600E+05
7.1087300E-08	8.5371800E-10	-8.3793000E-10	3.4687700E+05
7.1122900E-08	8.2047000E-10	-8.0584800E-10	3.4737800E+05
7.1158400E-08	7.8925600E-10	-7.7492100E-10	3.4848300E+05
7.1194000E-08	7.6214000E-10	-7.4720600E-10	3.5017400E+05
7.1229600E-08	7.4080600E-10	-7.2445800E-10	3.5238000E+05
7.1265200E-08	7.2640500E-10	-7.0802600E-10	3.5497800E+05
7.1300800E-08	7.1954900E-10	-6.9880000E-10	3.5780800E+05
7.1336400E-08	7.2039300E-10	-6.9720100E-10	3.6067500E+05
7.1371900E-08	7.2857300E-10	-7.0317900E-10	3.6336800E+05
7.1407500E-08	7.4332000E-10	-7.1623700E-10	3.6567200E+05
7.1443100E-08	7.6354200E-10	-7.3547700E-10	3.6738800E+05
7.1478700E-08	7.8785700E-10	-7.5965400E-10	3.6835000E+05
7.1514300E-08	8.1471300E-10	-7.8722600E-10	3.6843800E+05
7.1549900E-08	8.4241300E-10	-8.1642500E-10	3.6758100E+05
7.1585400E-08	8.6917500E-10	-8.4533400E-10	3.6577400E+05
7.1621000E-08	8.9327400E-10	-8.7199500E-10	3.6307400E+05
7.1656600E-08	9.1307600E-10	-8.9452800E-10	3.5960000E+05
7.1692200E-08	9.2712700E-10	-9.1126300E-10	3.5552200E+05
7.1727800E-08	9.3434200E-10	-9.2088000E-10	3.5104900E+05
7.1763400E-08	9.3406800E-10	-9.2254500E-10	3.4641300E+05
7.1799000E-08	9.2615300E-10	-9.1602100E-10	3.4185300E+05
7.1834500E-08	9.1102000E-10	-9.0173100E-10	3.3759700E+05
7.1870100E-08	8.8974600E-10	-8.8077300E-10	3.3384400E+05
7.1905700E-08	8.6395900E-10	-8.5485300E-10	3.3073900E+05
7.1941300E-08	8.3568600E-10	-8.2615100E-10	3.2837000E+05
7.1976900E-08	8.0727300E-10	-7.9712800E-10	3.2676400E+05
7.2012500E-08	7.8111400E-10	-7.7029000E-10	3.2587700E+05
7.2048000E-08	7.5936500E-10	-7.4792800E-10	3.2560100E+05
7.2083600E-08	7.4379500E-10	-7.3185200E-10	3.2577500E+05
7.2119200E-08	7.3549300E-10	-7.2318800E-10	3.2620100E+05
7.2154800E-08	7.3474500E-10	-7.2224100E-10	3.2666200E+05
7.2190400E-08	7.4107300E-10	-7.2845000E-10	3.2694200E+05
7.2226000E-08	7.5311100E-10	-7.4044600E-10	3.2684700E+05
7.2261500E-08	7.6884900E-10	-7.5620600E-10	3.2621800E+05
7.2297100E-08	7.8590000E-10	-7.7328500E-10	3.2495200E+05
7.2332700E-08	8.0170200E-10	-7.8911500E-10	3.2300800E+05
7.2368300E-08	8.1381900E-10	-8.0131600E-10	3.2041500E+05
7.2403900E-08	8.2033400E-10	-8.0797900E-10	3.1726800E+05
7.2439500E-08	8.1999200E-10	-8.0790900E-10	3.1372000E+05
7.2475000E-08	8.1236000E-10	-8.0075800E-10	3.0996700E+05
7.2510600E-08	7.9796700E-10	-7.8706300E-10	3.0623700E+05
7.2546200E-08	7.7813700E-10	-7.6818500E-10	3.0276400E+05

7.2581800E-08	7.5489300E-10	-7.4614200E-10	2.9977000E+05
7.2617400E-08	7.3068900E-10	-7.2335800E-10	2.9744000E+05
7.2653000E-08	7.0813400E-10	-7.0236500E-10	2.9591000E+05
7.2688600E-08	6.8967700E-10	-6.8551600E-10	2.9525600E+05
7.2724100E-08	6.7731600E-10	-6.7470900E-10	2.9548700E+05
7.2759700E-08	6.7238300E-10	-6.7117200E-10	2.9653700E+05
7.2795300E-08	6.7542600E-10	-6.7535000E-10	2.9828400E+05
7.2830900E-08	6.8613400E-10	-6.8686700E-10	3.0055100E+05
7.2866500E-08	7.0341400E-10	-7.0458700E-10	3.0312400E+05
7.2902100E-08	7.2553300E-10	-7.2676400E-10	3.0577200E+05
7.2937600E-08	7.5034100E-10	-7.5125700E-10	3.0826700E+05
7.2973200E-08	7.7547900E-10	-7.7577100E-10	3.1039900E+05
7.3008800E-08	7.9864900E-10	-7.9810200E-10	3.1199600E+05
7.3044400E-08	8.1786500E-10	-8.1634300E-10	3.1294200E+05
7.3080000E-08	8.3162200E-10	-8.2907400E-10	3.1318400E+05
7.3115600E-08	8.3901500E-10	-8.3547700E-10	3.1273100E+05
7.3151100E-08	8.3974900E-10	-8.3535800E-10	3.1165300E+05
7.3186700E-08	8.3418000E-10	-8.2912000E-10	3.1007300E+05
7.3222300E-08	8.2319300E-10	-8.1768500E-10	3.0815800E+05
7.3257900E-08	8.0804700E-10	-8.0235500E-10	3.0609700E+05
7.3293500E-08	7.9024600E-10	-7.8465900E-10	3.0408800E+05
7.3329100E-08	7.7140800E-10	-7.6619700E-10	3.0231100E+05
7.3364600E-08	7.5305500E-10	-7.4849200E-10	3.0091600E+05
7.3400200E-08	7.3651300E-10	-7.3285500E-10	3.0001500E+05
7.3435800E-08	7.2285400E-10	-7.2030500E-10	2.9966400E+05
7.3471400E-08	7.1280100E-10	-7.1152000E-10	2.9985500E+05
7.3507000E-08	7.0672500E-10	-7.0682400E-10	3.0051900E+05
7.3542600E-08	7.0469500E-10	-7.0619000E-10	3.0153500E+05
7.3578100E-08	7.0647600E-10	-7.0928500E-10	3.0273400E+05
7.3613700E-08	7.1158500E-10	-7.1552500E-10	3.0392400E+05
7.3649300E-08	7.1935500E-10	-7.2413100E-10	3.0489500E+05
7.3684900E-08	7.2897800E-10	-7.3419900E-10	3.0544500E+05
7.3720500E-08	7.3958600E-10	-7.4476200E-10	3.0539900E+05
7.3756100E-08	7.5025500E-10	-7.5484200E-10	3.0462500E+05
7.3791700E-08	7.6003400E-10	-7.6349800E-10	3.0304800E+05
7.3827200E-08	7.6806200E-10	-7.6988900E-10	3.0065300E+05
7.3862800E-08	7.7354600E-10	-7.7333000E-10	2.9749300E+05
7.3898400E-08	7.7583700E-10	-7.7333100E-10	2.9368200E+05
7.3934000E-08	7.7451600E-10	-7.6964700E-10	2.8939500E+05
7.3969600E-08	7.6937600E-10	-7.6230300E-10	2.8484600E+05
7.4005200E-08	7.6049700E-10	-7.5163900E-10	2.8027900E+05
7.4040700E-08	7.4831600E-10	-7.3830700E-10	2.7594400E+05
7.4076300E-08	7.3360500E-10	-7.2324100E-10	2.7207700E+05
7.4111900E-08	7.1739800E-10	-7.0760600E-10	2.6888700E+05
7.4147500E-08	7.0094600E-10	-6.9271700E-10	2.6653500E+05
7.4183100E-08	6.8568200E-10	-6.7992600E-10	2.6512400E+05
7.4218700E-08	6.7301700E-10	-6.7049600E-10	2.6468300E+05
7.4254200E-08	6.6419600E-10	-6.6545400E-10	2.6517700E+05
7.4289800E-08	6.6017500E-10	-6.6545600E-10	2.6650200E+05
7.4325400E-08	6.6149700E-10	-6.7068100E-10	2.6850000E+05
7.4361000E-08	6.6814900E-10	-6.8076700E-10	2.7097200E+05

7.4396600E-08	6.7950700E-10	-6.9479300E-10	2.7369000E+05
7.4432200E-08	6.9445600E-10	-7.1132700E-10	2.7641600E+05
7.4467700E-08	7.1134700E-10	-7.2853900E-10	2.7892400E+05
7.4503300E-08	7.2814400E-10	-7.4435700E-10	2.8101900E+05
7.4538900E-08	7.4270200E-10	-7.5667900E-10	2.8255200E+05
7.4574500E-08	7.5296000E-10	-7.6359900E-10	2.8342900E+05
7.4610100E-08	7.5714500E-10	-7.6362200E-10	2.8361800E+05
7.4645700E-08	7.5397400E-10	-7.5586300E-10	2.8314800E+05
7.4681300E-08	7.4292200E-10	-7.4016900E-10	2.8211000E+05
7.4716800E-08	7.2418500E-10	-7.1718500E-10	2.8064200E+05
7.4752400E-08	6.9876000E-10	-6.8834000E-10	2.7891800E+05
7.4788000E-08	6.6846900E-10	-6.5574100E-10	2.7712900E+05
7.4823600E-08	6.3566300E-10	-6.2198700E-10	2.7546100E+05
7.4859200E-08	6.0304900E-10	-5.8992100E-10	2.7408800E+05
7.4894800E-08	5.7348400E-10	-5.6236400E-10	2.7314600E+05
7.4930300E-08	5.4962500E-10	-5.4183400E-10	2.7272600E+05
7.4965900E-08	5.3369200E-10	-5.3029000E-10	2.7286600E+05
7.5001500E-08	5.2723100E-10	-5.2891400E-10	2.7354400E+05
7.5037100E-08	5.3091700E-10	-5.3798200E-10	2.7468100E+05
7.5072700E-08	5.4449900E-10	-5.5682300E-10	2.7615200E+05
7.5108300E-08	5.6681800E-10	-5.8386300E-10	2.7779100E+05
7.5143800E-08	5.9587500E-10	-6.1676600E-10	2.7940800E+05
7.5179400E-08	6.2903200E-10	-6.5263500E-10	2.8080500E+05
7.5215000E-08	6.6325200E-10	-6.8828300E-10	2.8179500E+05
7.5250600E-08	6.9537400E-10	-7.2052600E-10	2.8221600E+05
7.5286200E-08	7.2244500E-10	-7.4648000E-10	2.8195000E+05
7.5321800E-08	7.4197100E-10	-7.6383000E-10	2.8092200E+05
7.5357300E-08	7.5213600E-10	-7.7103300E-10	2.7911500E+05
7.5392900E-08	7.5203000E-10	-7.6746800E-10	2.7656700E+05
7.5428500E-08	7.4165900E-10	-7.5348700E-10	2.7337100E+05
7.5464100E-08	7.2196500E-10	-7.3037600E-10	2.6966000E+05
7.5499700E-08	6.9479900E-10	-7.0023900E-10	2.6560100E+05
7.5535300E-08	6.6262200E-10	-6.6579600E-10	2.6137800E+05
7.5570900E-08	6.2835200E-10	-6.3012900E-10	2.5718000E+05
7.5606400E-08	5.9506500E-10	-5.9639000E-10	2.5318200E+05
7.5642000E-08	5.6563300E-10	-5.6751500E-10	2.4952900E+05
7.5677600E-08	5.4256900E-10	-5.4595000E-10	2.4633000E+05
7.5713200E-08	5.2770800E-10	-5.3342900E-10	2.4364900E+05
7.5748800E-08	5.2204700E-10	-5.3081800E-10	2.4149900E+05
7.5784400E-08	5.2569000E-10	-5.3804200E-10	2.3984400E+05
7.5819900E-08	5.3781500E-10	-5.5410200E-10	2.3860500E+05
7.5855500E-08	5.5681100E-10	-5.7718000E-10	2.3767200E+05
7.5891100E-08	5.8043500E-10	-6.0481700E-10	2.3691300E+05
7.5926700E-08	6.0601200E-10	-6.3414700E-10	2.3618300E+05
7.5962300E-08	6.3072000E-10	-6.6217000E-10	2.3535000E+05
7.5997900E-08	6.5187900E-10	-6.8602900E-10	2.3429500E+05
7.6033400E-08	6.6719800E-10	-7.0327300E-10	2.3292800E+05
7.6069000E-08	6.7495400E-10	-7.1206600E-10	2.3119600E+05
7.6104600E-08	6.7414900E-10	-7.1133600E-10	2.2908800E+05
7.6140200E-08	6.6458000E-10	-7.0085400E-10	2.2663900E+05
7.6175800E-08	6.4681300E-10	-6.8123600E-10	2.2391600E+05

7.6211400E-08	6.2212800E-10	-6.5386300E-10	2.2102300E+05
7.6246900E-08	5.9234900E-10	-6.2073900E-10	2.1808700E+05
7.6282500E-08	5.5966600E-10	-5.8430200E-10	2.1524800E+05
7.6318100E-08	5.2646300E-10	-5.4721800E-10	2.1264200E+05
7.6353700E-08	4.9508000E-10	-5.1216100E-10	2.1039200E+05
7.6389300E-08	4.6766100E-10	-4.8160700E-10	2.0859900E+05
7.6424900E-08	4.4598400E-10	-4.5765500E-10	2.0732700E+05
7.6460500E-08	4.3138000E-10	-4.4188900E-10	2.0660000E+05
7.6496000E-08	4.2466600E-10	-4.3528400E-10	2.0640400E+05
7.6531600E-08	4.2608800E-10	-4.3816800E-10	2.0667900E+05
7.6567200E-08	4.3535700E-10	-4.5021100E-10	2.0732600E+05
7.6602800E-08	4.5171200E-10	-4.7046700E-10	2.0821500E+05
7.6638400E-08	4.7394000E-10	-4.9745100E-10	2.0919300E+05
7.6674000E-08	5.0047300E-10	-5.2923500E-10	2.1009900E+05
7.6709500E-08	5.2948600E-10	-5.6357000E-10	2.1077300E+05
7.6745100E-08	5.5896900E-10	-5.9803600E-10	2.1106800E+05
7.6780700E-08	5.8687600E-10	-6.3019000E-10	2.1086500E+05
7.6816300E-08	6.1124500E-10	-6.5772400E-10	2.1008100E+05
7.6851900E-08	6.3026700E-10	-6.7862700E-10	2.0867800E+05
7.6887500E-08	6.4244600E-10	-6.9133000E-10	2.0666300E+05
7.6923000E-08	6.4677600E-10	-6.9483900E-10	2.0408900E+05
7.6958600E-08	6.4275400E-10	-6.8882800E-10	2.0104200E+05
7.6994200E-08	6.3049000E-10	-6.7369100E-10	1.9764500E+05
7.7029800E-08	6.1074800E-10	-6.5052600E-10	1.9404400E+05
7.7065400E-08	5.8486400E-10	-6.2107500E-10	1.9039400E+05
7.7101000E-08	5.5471900E-10	-5.8760000E-10	1.8684000E+05
7.7136500E-08	5.2257400E-10	-5.5269400E-10	1.8350900E+05
7.7172100E-08	4.9085800E-10	-5.1905000E-10	1.8049800E+05
7.7207700E-08	4.6196200E-10	-4.8921300E-10	1.7786400E+05
7.7243300E-08	4.3797200E-10	-4.6533300E-10	1.7562300E+05
7.7278900E-08	4.2049200E-10	-4.4896300E-10	1.7374100E+05
7.7314500E-08	4.1046900E-10	-4.4089700E-10	1.7213800E+05
7.7350100E-08	4.0804600E-10	-4.4108600E-10	1.7070100E+05
7.7385600E-08	4.1261000E-10	-4.4864600E-10	1.6929100E+05
7.7421200E-08	4.2282800E-10	-4.6196600E-10	1.6776100E+05
7.7456800E-08	4.3676200E-10	-4.7889600E-10	1.6596500E+05
7.7492400E-08	4.5215800E-10	-4.9698300E-10	1.6377700E+05
7.7528000E-08	4.6672900E-10	-5.1375500E-10	1.6110100E+05
7.7563600E-08	4.7835300E-10	-5.2699100E-10	1.5788400E+05
7.7599100E-08	4.8530700E-10	-5.3497100E-10	1.5412100E+05
7.7634700E-08	4.8654900E-10	-5.3664600E-10	1.4985900E+05
7.7670300E-08	4.8178500E-10	-5.3173900E-10	1.4518800E+05
7.7705900E-08	4.7137000E-10	-5.2073700E-10	1.4024300E+05
7.7741500E-08	4.5637400E-10	-5.0478600E-10	1.3519000E+05
7.7777100E-08	4.3839800E-10	-4.8552600E-10	1.3020800E+05
7.7812600E-08	4.1918300E-10	-4.6486600E-10	1.2548100E+05
7.7848200E-08	4.0057700E-10	-4.4473000E-10	1.2118000E+05
7.7883800E-08	3.8424400E-10	-4.2682600E-10	1.1744300E+05
7.7919400E-08	3.7133000E-10	-4.1244600E-10	1.1437300E+05
7.7955000E-08	3.6247600E-10	-4.0233500E-10	1.1202400E+05
7.7990600E-08	3.5777800E-10	-3.9665300E-10	1.1040400E+05

7.8026100E-08	3.5671600E-10	-3.9500000E-10	1.0947000E+05
7.8061700E-08	3.5833600E-10	-3.9650700E-10	1.0913300E+05
7.8097300E-08	3.6139600E-10	-3.9999400E-10	1.0927100E+05
7.8132900E-08	3.6454700E-10	-4.0414900E-10	1.0974000E+05
7.8168500E-08	3.6652900E-10	-4.0770200E-10	1.1038200E+05
7.8204100E-08	3.6631100E-10	-4.0958100E-10	1.1103900E+05
7.8239700E-08	3.6326600E-10	-4.0902100E-10	1.1156600E+05
7.8275200E-08	3.5717000E-10	-4.0562100E-10	1.1184200E+05
7.8310800E-08	3.4816300E-10	-3.9934500E-10	1.1177900E+05
7.8346400E-08	3.3675300E-10	-3.9047000E-10	1.1132300E+05
7.8382000E-08	3.2367200E-10	-3.7950900E-10	1.1045500E+05
7.8417600E-08	3.0975500E-10	-3.6710800E-10	1.0919300E+05
7.8453200E-08	2.9584800E-10	-3.5394300E-10	1.0758500E+05
7.8488700E-08	2.8262700E-10	-3.4064300E-10	1.0570300E+05
7.8524300E-08	2.7058700E-10	-3.2773400E-10	1.0363200E+05
7.8559900E-08	2.6004500E-10	-3.1561000E-10	1.0146500E+05
7.8595500E-08	2.5111000E-10	-3.0454300E-10	9.9294500E+04
7.8631100E-08	2.4371900E-10	-2.9471500E-10	9.7200200E+04
7.8666700E-08	2.3771500E-10	-2.8624700E-10	9.5249200E+04
7.8702200E-08	2.3295200E-10	-2.7924300E-10	9.3489400E+04
7.8737800E-08	2.2930400E-10	-2.7381300E-10	9.1944400E+04
7.8773400E-08	2.2670700E-10	-2.7007600E-10	9.0614800E+04
7.8809000E-08	2.2515300E-10	-2.6813100E-10	8.9480600E+04
7.8844600E-08	2.2468800E-10	-2.6803200E-10	8.8502500E+04
7.8880200E-08	2.2535200E-10	-2.6972800E-10	8.7626600E+04
7.8915700E-08	2.2705200E-10	-2.7301300E-10	8.6791400E+04
7.8951300E-08	2.2959600E-10	-2.7748500E-10	8.5932900E+04
7.8986900E-08	2.3263700E-10	-2.8254500E-10	8.4991100E+04
7.9022500E-08	2.3561900E-10	-2.8740900E-10	8.3917400E+04
7.9058100E-08	2.3781100E-10	-2.9116300E-10	8.2678100E+04
7.9093700E-08	2.3839300E-10	-2.9284800E-10	8.1257600E+04
7.9129300E-08	2.3658700E-10	-2.9158300E-10	7.9663000E+04
7.9164800E-08	2.3170300E-10	-2.8667900E-10	7.7920300E+04
7.9200400E-08	2.2324700E-10	-2.7773600E-10	7.6073100E+04
7.9236000E-08	2.1111000E-10	-2.6473500E-10	7.4182500E+04
7.9271600E-08	1.9555300E-10	-2.4807600E-10	7.2319200E+04
7.9307200E-08	1.7719900E-10	-2.2856800E-10	7.0557500E+04
7.9342800E-08	1.5705300E-10	-2.0736500E-10	6.8969100E+04
7.9378300E-08	1.3638500E-10	-1.8585300E-10	6.7615700E+04
7.9413900E-08	1.1657500E-10	-1.6549900E-10	6.6543900E+04
7.9449500E-08	9.8971700E-11	-1.4768600E-10	6.5781300E+04
7.9485100E-08	8.4739800E-11	-1.3354900E-10	6.5333000E+04
7.9520700E-08	7.4714000E-11	-1.2384100E-10	6.5181900E+04
7.9556300E-08	6.9227100E-11	-1.1882900E-10	6.5291700E+04
7.9591800E-08	6.8144900E-11	-1.1826500E-10	6.5612400E+04
7.9627400E-08	7.0851900E-11	-1.2140500E-10	6.6086500E+04
7.9663000E-08	7.6278400E-11	-1.2710000E-10	6.6652900E+04
7.9698600E-08	8.3059200E-11	-1.3393100E-10	6.7257900E+04
7.9734200E-08	8.9710900E-11	-1.4036800E-10	6.7863200E+04
7.9769800E-08	9.4773100E-11	-1.4496100E-10	6.8445800E+04
7.9805300E-08	9.6994800E-11	-1.4650800E-10	6.9003400E+04

7.9840900E-08	9.5497400E-11	-1.4418400E-10	6.9559600E+04
7.9876500E-08	8.9822000E-11	-1.3762800E-10	7.0159300E+04
7.9912100E-08	8.0017300E-11	-1.2697600E-10	7.0860700E+04
7.9947700E-08	6.6583300E-11	-1.1283000E-10	7.1733800E+04
7.9983300E-08	5.0386800E-11	-9.6175200E-11	7.2853100E+04
8.0018900E-08	3.2608100E-11	-7.8260200E-11	7.4283500E+04
8.0054400E-08	1.4556500E-11	-6.0445600E-11	7.6071500E+04
8.0090000E-08	-2.4710300E-12	-4.4057400E-11	7.8243700E+04
8.0125600E-08	-1.7309000E-11	-3.0258600E-11	8.0800700E+04
8.0161200E-08	-2.9049200E-11	-1.9944800E-11	8.3710300E+04
8.0196800E-08	-3.7043900E-11	-1.3665200E-11	8.6908100E+04
8.0232400E-08	-4.0989900E-11	-1.1604700E-11	9.0303800E+04
8.0267900E-08	-4.0917200E-11	-1.3604200E-11	9.3787500E+04
8.0303500E-08	-3.7132000E-11	-1.9192600E-11	9.7239100E+04
8.0339100E-08	-3.0147600E-11	-2.7655600E-11	1.0053700E+05
8.0374700E-08	-2.0655000E-11	-3.8111000E-11	1.0356300E+05
8.0410300E-08	-9.4759200E-12	-4.9582400E-11	1.0621900E+05
8.0445900E-08	2.5157900E-12	-6.1072900E-11	1.0842600E+05
8.0481400E-08	1.4420100E-11	-7.1619900E-11	1.1014100E+05
8.0517000E-08	2.5358700E-11	-8.0339200E-11	1.1135000E+05
8.0552600E-08	3.4462400E-11	-8.6465900E-11	1.1207500E+05
8.0588200E-08	4.0909100E-11	-8.9390900E-11	1.1235800E+05
8.0623800E-08	4.4011500E-11	-8.8693100E-11	1.1226600E+05
8.0659400E-08	4.3219200E-11	-8.4169400E-11	1.1188200E+05
8.0694900E-08	3.8204800E-11	-7.5864600E-11	1.1129300E+05
8.0730500E-08	2.8875500E-11	-6.4098600E-11	1.1058000E+05
8.0766100E-08	1.5477300E-11	-4.9485200E-11	1.0981000E+05
8.0801700E-08	-1.3407500E-12	-3.2918500E-11	1.0903400E+05
8.0837300E-08	-2.0632000E-11	-1.5539900E-11	1.0828000E+05
8.0872900E-08	-4.1113100E-11	1.3172700E-12	1.0754800E+05
8.0908400E-08	-6.1197100E-11	1.6226900E-11	1.0681300E+05
8.0944000E-08	-7.9246600E-11	2.7803100E-11	1.0602900E+05
8.0979600E-08	-9.3668300E-11	3.4831200E-11	1.0513300E+05
8.1015200E-08	-1.0302900E-10	3.6414300E-11	1.0404700E+05
8.1050800E-08	-1.0632500E-10	3.2105300E-11	1.0269600E+05
8.1086400E-08	-1.0309900E-10	2.1997900E-11	1.0100900E+05
8.1122000E-08	-9.3530100E-11	6.7703800E-12	9.8925100E+04
8.1157500E-08	-7.8470500E-11	-1.2325100E-11	9.6407600E+04
8.1193100E-08	-5.9380300E-11	-3.3552100E-11	9.3444500E+04
8.1228700E-08	-3.8253200E-11	-5.4839400E-11	9.0052400E+04
8.1264300E-08	-1.7395400E-11	-7.3973600E-11	8.6277500E+04
8.1299900E-08	8.2013300E-13	-8.8827700E-11	8.2191600E+04
8.1335500E-08	1.4177600E-11	-9.7605800E-11	7.7887400E+04
8.1371000E-08	2.0923800E-11	-9.9052100E-11	7.3474300E+04
8.1406600E-08	1.9961500E-11	-9.2613500E-11	6.9069200E+04
8.1442200E-08	1.0984200E-11	-7.8538500E-11	6.4786200E+04
8.1477800E-08	-5.4278600E-12	-5.7899400E-11	6.0726700E+04
8.1513400E-08	-2.7868600E-11	-3.2506900E-11	5.6973200E+04
8.1549000E-08	-5.4210400E-11	-4.7391500E-12	5.3583600E+04
8.1584500E-08	-8.1855500E-11	2.2674300E-11	5.0584900E+04
8.1620100E-08	-1.0798200E-10	4.6927900E-11	4.7971100E+04

8.1655700E-08	-1.2982100E-10	6.5441000E-11	4.5705400E+04
8.1691300E-08	-1.4502700E-10	7.6131100E-11	4.3726400E+04
8.1726900E-08	-1.5190900E-10	7.7639600E-11	4.1954600E+04
8.1762500E-08	-1.4958500E-10	6.9486300E-11	4.0300800E+04
8.1798000E-08	-1.3815000E-10	5.2133100E-11	3.8675900E+04
8.1833600E-08	-1.1863000E-10	2.6955800E-11	3.7001100E+04
8.1869200E-08	-9.2858200E-11	-3.8931000E-12	3.5215500E+04
8.1904800E-08	-6.3345100E-11	-3.7696700E-11	3.3282700E+04
8.1940400E-08	-3.2976600E-11	-7.1447100E-11	3.1195100E+04
8.1976000E-08	-4.6511500E-12	-1.0214500E-10	2.8976000E+04
8.2011600E-08	1.8937300E-11	-1.2710300E-10	2.6677800E+04
8.2047100E-08	3.5594600E-11	-1.4421000E-10	2.4374600E+04
8.2082700E-08	4.3902100E-11	-1.5212700E-10	2.2156200E+04
8.2118300E-08	4.3313500E-11	-1.5041400E-10	2.0123000E+04
8.2153900E-08	3.4157300E-11	-1.3955400E-10	1.8372200E+04
8.2189500E-08	1.7614700E-11	-1.2087200E-10	1.6984900E+04
8.2225100E-08	-4.4199600E-12	-9.6383400E-11	1.6022300E+04
8.2260600E-08	-2.9604700E-11	-6.8568800E-11	1.5516000E+04
8.2296200E-08	-5.5387000E-11	-4.0109400E-11	1.5461300E+04
8.2331800E-08	-7.9270800E-11	-1.3618600E-11	1.5820600E+04
8.2367400E-08	-9.9097700E-11	8.6147700E-12	1.6522000E+04
8.2403000E-08	-1.1325200E-10	2.4833100E-11	1.7462700E+04
8.2438600E-08	-1.2071500E-10	3.3946000E-11	1.8518000E+04
8.2474100E-08	-1.2121100E-10	3.5600300E-11	1.9551300E+04
8.2509700E-08	-1.1521200E-10	3.0157600E-11	2.0425700E+04
8.2545300E-08	-1.0372600E-10	1.8599700E-11	2.1012600E+04
8.2580900E-08	-8.8194800E-11	2.3861100E-12	2.1201400E+04
8.2616500E-08	-7.0336600E-11	-1.6726300E-11	2.0912700E+04
8.2652100E-08	-5.1913400E-11	-3.6882300E-11	2.0102500E+04
8.2687600E-08	-3.4526600E-11	-5.6324600E-11	1.8766100E+04
8.2723200E-08	-1.9519800E-11	-7.3549700E-11	1.6940900E+04
8.2758800E-08	-7.8518900E-12	-8.7418000E-11	1.4700600E+04
8.2794400E-08	-3.2759100E-15	-9.7224700E-11	1.2150600E+04
8.2830000E-08	3.9627400E-12	-1.0271500E-10	9.4235400E+03
8.2865600E-08	4.3387000E-12	-1.0404900E-10	6.6674300E+03
8.2901200E-08	1.7515600E-12	-1.0172900E-10	4.0317800E+03
8.2936700E-08	-3.0076100E-12	-9.6508800E-11	1.6599500E+03
8.2972300E-08	-9.0823900E-12	-8.9293500E-11	-3.2070000E+02
8.3007900E-08	-1.5590500E-11	-8.1034500E-11	-1.8074300E+03
8.3043500E-08	-2.1782600E-11	-7.2645500E-11	-2.7269700E+03
8.3079100E-08	-2.7089300E-11	-6.4934200E-11	-3.0424900E+03
8.3114700E-08	-3.1087200E-11	-5.8559200E-11	-2.7581500E+03
8.3150200E-08	-3.3520100E-11	-5.4003800E-11	-1.9145200E+03
8.3185800E-08	-3.4333000E-11	-5.1575300E-11	-5.8213700E+02
8.3221400E-08	-3.3543300E-11	-5.1417300E-11	1.1414600E+03
8.3257000E-08	-3.1270300E-11	-5.3520800E-11	3.1387200E+03
8.3292600E-08	-2.7717800E-11	-5.7742300E-11	5.2844400E+03
8.3328200E-08	-2.3074100E-11	-6.3830400E-11	7.4567200E+03
8.3363700E-08	-1.7543400E-11	-7.1440600E-11	9.5450100E+03
8.3399300E-08	-1.1352100E-11	-8.0153200E-11	1.1460300E+04
8.3434900E-08	-4.7224500E-12	-8.9489700E-11	1.3143200E+04

8.3470500E-08	2.1112500E-12	-9.8933900E-11	1.4565900E+04
8.3506100E-08	8.8979000E-12	-1.0795900E-10	1.5734600E+04
8.3541700E-08	1.5386700E-11	-1.1606300E-10	1.6690400E+04
8.3577200E-08	2.1311400E-11	-1.2280600E-10	1.7502300E+04
8.3612800E-08	2.6429400E-11	-1.2784900E-10	1.8259100E+04
8.3648400E-08	3.0600400E-11	-1.3101300E-10	1.9062200E+04
8.3684000E-08	3.3740700E-11	-1.3230900E-10	2.0014900E+04
8.3719600E-08	3.5863600E-11	-1.3195300E-10	2.1209700E+04
8.3755200E-08	3.7172600E-11	-1.3037700E-10	2.2719200E+04
8.3790800E-08	3.7994800E-11	-1.2821100E-10	2.4588800E+04
8.3826300E-08	3.8791400E-11	-1.2623200E-10	2.6829900E+04
8.3861900E-08	4.0080200E-11	-1.2528100E-10	2.9419700E+04
8.3897500E-08	4.2383800E-11	-1.2616800E-10	3.2304900E+04
8.3933100E-08	4.6202200E-11	-1.2957600E-10	3.5403700E+04
8.3968700E-08	5.1851200E-11	-1.3593500E-10	3.8613600E+04
8.4004300E-08	5.9398500E-11	-1.4533900E-10	4.1821400E+04
8.4039800E-08	6.8632800E-11	-1.5749200E-10	4.4913500E+04
8.4075400E-08	7.9022900E-11	-1.7167600E-10	4.7787200E+04
8.4111000E-08	8.9726400E-11	-1.8679400E-10	5.0359300E+04
8.4146600E-08	9.9664600E-11	-2.0145500E-10	5.2571700E+04
8.4182200E-08	1.0764400E-10	-2.1411900E-10	5.4397100E+04
8.4217800E-08	1.1247300E-10	-2.2326100E-10	5.5841400E+04
8.4253300E-08	1.1312900E-10	-2.2756400E-10	5.6942400E+04
8.4288900E-08	1.0895400E-10	-2.2611500E-10	5.7765900E+04
8.4324500E-08	9.9748400E-11	-2.1855000E-10	5.8397900E+04
8.4360100E-08	8.5874600E-11	-2.0515600E-10	5.8936400E+04
8.4395700E-08	6.8311700E-11	-1.8689900E-10	5.9479600E+04
8.4431300E-08	4.8526800E-11	-1.6536900E-10	6.0116500E+04
8.4466800E-08	2.8420900E-11	-1.4263700E-10	6.0922000E+04
8.4502400E-08	1.0102500E-11	-1.2105400E-10	6.1948300E+04
8.4538000E-08	-4.3565000E-12	-1.0297700E-10	6.3215300E+04
8.4573600E-08	-1.3152800E-11	-9.0501400E-11	6.4710300E+04
8.4609200E-08	-1.5033200E-11	-8.5191700E-11	6.6391300E+04
8.4644800E-08	-9.4364000E-12	-8.7878500E-11	6.8191500E+04
8.4680400E-08	3.3582600E-12	-9.8520400E-11	7.0022400E+04
8.4715900E-08	2.2206500E-11	-1.1617000E-10	7.1780900E+04
8.4751500E-08	4.5245000E-11	-1.3906100E-10	7.3361500E+04
8.4787100E-08	7.0044200E-11	-1.6479100E-10	7.4665100E+04
8.4822700E-08	9.3827400E-11	-1.9057300E-10	7.5608200E+04
8.4858300E-08	1.1385400E-10	-2.1355500E-10	7.6133600E+04
8.4893900E-08	1.2773500E-10	-2.3114900E-10	7.6212800E+04
8.4929400E-08	1.3366500E-10	-2.4132600E-10	7.5846200E+04
8.4965000E-08	1.3065900E-10	-2.4284600E-10	7.5067900E+04
8.5000600E-08	1.1871800E-10	-2.3539600E-10	7.3942700E+04
8.5036200E-08	9.8787200E-11	-2.1961500E-10	7.2556800E+04
8.5071800E-08	7.2635900E-11	-1.9700300E-10	7.1010300E+04
8.5107400E-08	4.2691100E-11	-1.6973400E-10	6.9410300E+04
8.5142900E-08	1.1729200E-11	-1.4040000E-10	6.7859200E+04
8.5178500E-08	-1.7447400E-11	-1.1170000E-10	6.6446700E+04
8.5214100E-08	-4.2309000E-11	-8.6150900E-11	6.5242500E+04
8.5249700E-08	-6.0840900E-11	-6.5820300E-11	6.4287100E+04

8.5285300E-08	-7.1769300E-11	-5.2129000E-11	6.3587700E+04
8.5320900E-08	-7.4664600E-11	-4.5748400E-11	6.3120500E+04
8.5356400E-08	-6.9908200E-11	-4.6596900E-11	6.2831900E+04
8.5392000E-08	-5.8564600E-11	-5.3919600E-11	6.2640900E+04
8.5427600E-08	-4.2228200E-11	-6.6436800E-11	6.2447800E+04
8.5463200E-08	-2.2814700E-11	-8.2547100E-11	6.2144300E+04
8.5498800E-08	-2.2732800E-12	-1.0054800E-10	6.1622900E+04
8.5534400E-08	1.7609400E-11	-1.1883400E-10	6.0787600E+04
8.5570000E-08	3.5376000E-11	-1.3605400E-10	5.9565800E+04
8.5605500E-08	5.0030600E-11	-1.5121400E-10	5.7913800E+04
8.5641100E-08	6.1065500E-11	-1.6372200E-10	5.5820600E+04
8.5676700E-08	6.8427200E-11	-1.7336400E-10	5.3313500E+04
8.5712300E-08	7.2411100E-11	-1.8023400E-10	5.0457900E+04
8.5747900E-08	7.3537300E-11	-1.8463300E-10	4.7351600E+04
8.5783500E-08	7.2462300E-11	-1.8696000E-10	4.4116900E+04
8.5819000E-08	6.9838900E-11	-1.8762800E-10	4.0891400E+04
8.5854600E-08	6.6195200E-11	-1.8699100E-10	3.7817000E+04
8.5890200E-08	6.1924300E-11	-1.8530900E-10	3.5028400E+04
8.5925800E-08	5.7290900E-11	-1.8276300E-10	3.2643000E+04
8.5961400E-08	5.2481200E-11	-1.7949200E-10	3.0748500E+04
8.5997000E-08	4.7661500E-11	-1.7564300E-10	2.9393000E+04
8.6032500E-08	4.3000300E-11	-1.7143500E-10	2.8585100E+04
8.6068100E-08	3.8777400E-11	-1.6718900E-10	2.8294500E+04
8.6103700E-08	3.5417500E-11	-1.6334300E-10	2.8454100E+04
8.6139300E-08	3.3411200E-11	-1.6042900E-10	2.8965400E+04
8.6174900E-08	3.3311000E-11	-1.5901500E-10	2.9705300E+04
8.6210500E-08	3.5641300E-11	-1.5960700E-10	3.0539600E+04
8.6246000E-08	4.0754400E-11	-1.6255800E-10	3.1339100E+04
8.6281600E-08	4.8719300E-11	-1.6797500E-10	3.1988900E+04
8.6317200E-08	5.9281100E-11	-1.7563500E-10	3.2398300E+04
8.6352800E-08	7.1807400E-11	-1.8496900E-10	3.2510000E+04
8.6388400E-08	8.5277900E-11	-1.9507400E-10	3.2305700E+04
8.6424000E-08	9.8357500E-11	-2.0478900E-10	3.1810000E+04
8.6459600E-08	1.0953500E-10	-2.1282700E-10	3.1089200E+04
8.6495100E-08	1.1733500E-10	-2.1792600E-10	3.0245000E+04
8.6530700E-08	1.2048600E-10	-2.1902000E-10	2.9404100E+04
8.6566300E-08	1.1806100E-10	-2.1540500E-10	2.8705200E+04
8.6601900E-08	1.0968100E-10	-2.0685600E-10	2.8289700E+04
8.6637500E-08	9.5610700E-11	-1.9370200E-10	2.8289400E+04
8.6673100E-08	7.6751800E-11	-1.7682400E-10	2.8809600E+04
8.6708600E-08	5.4578600E-11	-1.5757800E-10	2.9916800E+04
8.6744200E-08	3.0999600E-11	-1.3766600E-10	3.1634300E+04
8.6779800E-08	8.1561700E-12	-1.1894900E-10	3.3938200E+04
8.6815400E-08	-1.1796600E-11	-1.0322900E-10	3.6757000E+04
8.6851000E-08	-2.6967400E-11	-9.2050500E-11	3.9976100E+04
8.6886600E-08	-3.5937600E-11	-8.6507200E-11	4.3444600E+04
8.6922100E-08	-3.7874900E-11	-8.7121000E-11	4.6987700E+04
8.6957700E-08	-3.2674000E-11	-9.3783000E-11	5.0422700E+04
8.6993300E-08	-2.0994900E-11	-1.0576400E-10	5.3576500E+04
8.7028900E-08	-4.0849600E-12	-1.2180200E-10	5.6297800E+04
8.7064500E-08	1.6341800E-11	-1.4025000E-10	5.8466900E+04

8.7100100E-08	3.8228100E-11	-1.5926200E-10	6.0008800E+04
8.7135600E-08	5.9459600E-11	-1.7698500E-10	6.0902100E+04
8.7171200E-08	7.8105700E-11	-1.9176500E-10	6.1180200E+04
8.7206800E-08	9.2579000E-11	-2.0229700E-10	6.0926400E+04
8.7242400E-08	1.0175800E-10	-2.0774400E-10	6.0267300E+04
8.7278000E-08	1.0509900E-10	-2.0779100E-10	5.9362200E+04
8.7313600E-08	1.0265900E-10	-2.0264600E-10	5.8391000E+04
8.7349200E-08	9.5004300E-11	-1.9298500E-10	5.7540800E+04
8.7384700E-08	8.3155000E-11	-1.7986000E-10	5.6990600E+04
8.7420300E-08	6.8450300E-11	-1.6458700E-10	5.6892700E+04
8.7455900E-08	5.2373700E-11	-1.4860100E-10	5.7359400E+04
8.7491500E-08	3.6439000E-11	-1.3333600E-10	5.8457500E+04
8.7527100E-08	2.2083400E-11	-1.2011600E-10	6.0202800E+04
8.7562700E-08	1.0521900E-11	-1.1005700E-10	6.2555700E+04
8.7598200E-08	2.6633300E-12	-1.0399400E-10	6.5424500E+04
8.7633800E-08	-8.8437700E-13	-1.0244500E-10	6.8672300E+04
8.7669400E-08	1.2235400E-13	-1.0558200E-10	7.2125800E+04
8.7705000E-08	5.5694000E-12	-1.1322600E-10	7.5590600E+04
8.7740600E-08	1.4994600E-11	-1.2485800E-10	7.8868200E+04
8.7776200E-08	2.7651700E-11	-1.3965000E-10	8.1767200E+04
8.7811700E-08	4.2538200E-11	-1.5651200E-10	8.4117700E+04
8.7847300E-08	5.8387400E-11	-1.7414900E-10	8.5786000E+04
8.7882900E-08	7.3874500E-11	-1.9114000E-10	8.6683300E+04
8.7918500E-08	8.7667200E-11	-2.0604700E-10	8.6772500E+04
8.7954100E-08	9.8439900E-11	-2.1752600E-10	8.6071000E+04
8.7989700E-08	1.0507900E-10	-2.2443900E-10	8.4643600E+04
8.8025200E-08	1.0678400E-10	-2.2598100E-10	8.2597900E+04
8.8060800E-08	1.0316200E-10	-2.2177800E-10	8.0078200E+04
8.8096400E-08	9.4290800E-11	-2.1195400E-10	7.7251300E+04
8.8132000E-08	8.0708300E-11	-1.9716400E-10	7.4291300E+04
8.8167600E-08	6.3477200E-11	-1.7856600E-10	7.1366000E+04
8.8203200E-08	4.4073300E-11	-1.5774000E-10	6.8625400E+04
8.8238700E-08	2.4246900E-11	-1.3654700E-10	6.6194900E+04
8.8274300E-08	5.8924600E-12	-1.1695300E-10	6.4165000E+04
8.8309900E-08	-9.2020400E-12	-1.0082900E-10	6.2586100E+04
8.8345500E-08	-1.9529100E-11	-8.9731800E-11	6.1470200E+04
8.8381100E-08	-2.4010000E-11	-8.4726500E-11	6.0790900E+04
8.8416700E-08	-2.2224200E-11	-8.6247700E-11	6.0490000E+04
8.8452300E-08	-1.4440200E-11	-9.4027800E-11	6.0488200E+04
8.8487800E-08	-1.5574100E-12	-1.0710300E-10	6.0692900E+04
8.8523400E-08	1.4867500E-11	-1.2390600E-10	6.1006400E+04
8.8559000E-08	3.2832300E-11	-1.4242900E-10	6.1334000E+04
8.8594600E-08	5.0140700E-11	-1.6044700E-10	6.1593600E+04
8.8630200E-08	6.4606900E-11	-1.7576500E-10	6.1724000E+04
8.8665800E-08	7.4310600E-11	-1.8647400E-10	6.1688800E+04
8.8701300E-08	7.7871600E-11	-1.9117300E-10	6.1473500E+04
8.8736900E-08	7.4606100E-11	-1.8913200E-10	6.1084200E+04
8.8772500E-08	6.4553200E-11	-1.8038000E-10	6.0543700E+04
8.8808100E-08	4.8561100E-11	-1.6569300E-10	5.9888700E+04
8.8843700E-08	2.8173900E-11	-1.4650400E-10	5.9163500E+04
8.8879300E-08	5.3945800E-12	-1.2474700E-10	5.8411200E+04

8.8914800E-08	-1.7472800E-11	-1.0261500E-10	5.7665700E+04
8.8950400E-08	-3.8125100E-11	-8.2317900E-11	5.6947400E+04
8.8986000E-08	-5.4540100E-11	-6.5838300E-11	5.6261900E+04
8.9021600E-08	-6.5145100E-11	-5.4713800E-11	5.5601500E+04
8.9057200E-08	-6.9038700E-11	-4.9876500E-11	5.4942900E+04
8.9092800E-08	-6.6064500E-11	-5.1575500E-11	5.4247600E+04
8.9128300E-08	-5.6794500E-11	-5.9377900E-11	5.3469200E+04
8.9163900E-08	-4.2448900E-11	-7.2234700E-11	5.2563100E+04
8.9199500E-08	-2.4727100E-11	-8.8624500E-11	5.1489300E+04
8.9235100E-08	-5.6491000E-12	-1.0673700E-10	5.0218600E+04
8.9270700E-08	1.2714800E-11	-1.2467400E-10	4.8738300E+04
8.9306300E-08	2.8489500E-11	-1.4064700E-10	4.7052000E+04
8.9341900E-08	4.0138400E-11	-1.5316500E-10	4.5181200E+04
8.9377400E-08	4.6622800E-11	-1.6116600E-10	4.3168300E+04
8.9413000E-08	4.7526100E-11	-1.6409400E-10	4.1073700E+04
8.9448600E-08	4.3032100E-11	-1.6192800E-10	3.8966400E+04
8.9484200E-08	3.3844900E-11	-1.5516600E-10	3.6917900E+04
8.9519800E-08	2.1166700E-11	-1.4474100E-10	3.5000300E+04
8.9555400E-08	6.5535600E-12	-1.3191900E-10	3.3281500E+04
8.9590900E-08	-8.3344600E-12	-1.1816600E-10	3.1817000E+04
8.9626500E-08	-2.1830100E-11	-1.0501900E-10	3.0641500E+04
8.9662100E-08	-3.2386200E-11	-9.3946900E-11	2.9768800E+04
8.9697700E-08	-3.8767700E-11	-8.6227700E-11	2.9192400E+04
8.9733300E-08	-4.0137200E-11	-8.2848300E-11	2.8882300E+04
8.9768900E-08	-3.6091900E-11	-8.4424300E-11	2.8791300E+04
8.9804400E-08	-2.6706600E-11	-9.1152600E-11	2.8857800E+04
8.9840000E-08	-1.2526400E-11	-1.0279300E-10	2.9006600E+04
8.9875600E-08	5.5040700E-12	-1.1867300E-10	2.9158400E+04
8.9911200E-08	2.6082400E-11	-1.3772600E-10	2.9239200E+04
8.9946800E-08	4.7659500E-11	-1.5856900E-10	2.9186300E+04
8.9982400E-08	6.8569100E-11	-1.7959600E-10	2.8952400E+04
9.0017900E-08	8.7129800E-11	-1.9909000E-10	2.8507900E+04
9.0053500E-08	1.0179200E-10	-2.1536400E-10	2.7844400E+04
9.0089100E-08	1.1125000E-10	-2.2690000E-10	2.6980900E+04
9.0124700E-08	1.1456600E-10	-2.3248500E-10	2.5963100E+04
9.0160300E-08	1.1123400E-10	-2.3133000E-10	2.4857800E+04
9.0195900E-08	1.0123800E-10	-2.2315800E-10	2.3745200E+04
9.0231500E-08	8.5096800E-11	-2.0825700E-10	2.2714400E+04
9.0267000E-08	6.3813100E-11	-1.8748000E-10	2.1860700E+04
9.0302600E-08	3.8787200E-11	-1.6219100E-10	2.1277500E+04
9.0338200E-08	1.1750000E-11	-1.3416300E-10	2.1046200E+04
9.0373800E-08	-1.5380200E-11	-1.0543500E-10	2.1229100E+04
9.0409400E-08	-4.0685600E-11	-7.8137500E-11	2.1863700E+04
9.0445000E-08	-6.2393800E-11	-5.4304200E-11	2.2959600E+04
9.0480500E-08	-7.8999700E-11	-3.5680700E-11	2.4501500E+04
9.0516100E-08	-8.9413700E-11	-2.3576500E-11	2.6449000E+04
9.0551700E-08	-9.3084300E-11	-1.8745000E-11	2.8735300E+04
9.0587300E-08	-8.9983700E-11	-2.1318000E-11	3.1268400E+04
9.0622900E-08	-8.0619100E-11	-3.0804000E-11	3.3942700E+04
9.0658500E-08	-6.5994900E-11	-4.6140300E-11	3.6652100E+04
9.0694000E-08	-4.7484300E-11	-6.5805500E-11	3.9293000E+04

```
9.0729600E-08  -2.6721300E-11  -8.7971200E-11   4.1766300E+04
9.0765200E-08  -5.4668100E-12  -1.1067700E-10   4.3987900E+04
9.0800800E-08   1.4587900E-11  -1.3200500E-10   4.5897000E+04
9.0836400E-08   3.1942600E-11  -1.5024400E-10   4.7461300E+04
9.0872000E-08   4.5359800E-11  -1.6404000E-10   4.8677100E+04
9.0907500E-08   5.3983000E-11  -1.7248300E-10   4.9565400E+04
9.0943100E-08   5.7394900E-11  -1.7516000E-10   5.0168300E+04
9.0978700E-08   5.5604200E-11  -1.7215800E-10   5.0546400E+04
9.1014300E-08   4.9011900E-11  -1.6403000E-10   5.0775100E+04
9.1049900E-08   3.8353000E-11  -1.5170300E-10   5.0936900E+04
9.1085500E-08   2.4669200E-11  -1.3638600E-10   5.1109000E+04
9.1121100E-08   9.1436500E-12  -1.1945100E-10   5.1355100E+04
9.1156600E-08  -6.9672400E-12  -1.0230600E-10   5.1722200E+04
9.1192200E-08  -2.2383400E-11  -8.6287400E-11   5.2238300E+04
9.1227800E-08  -3.5964000E-11  -7.2565700E-11   5.2909200E+04
9.1263400E-08  -4.6740300E-11  -6.2063200E-11   5.3715400E+04
9.1299000E-08  -5.3993300E-11  -5.5403700E-11   5.4613400E+04
9.1334600E-08  -5.7298800E-11  -5.2882600E-11   5.5540100E+04
9.1370100E-08  -5.6528600E-11  -5.4461200E-11   5.6421300E+04
9.1405700E-08  -5.1929700E-11  -5.9782900E-11   5.7183100E+04
9.1441300E-08  -4.4063000E-11  -6.8207800E-11   5.7752200E+04
9.1476900E-08  -3.3710400E-11  -7.8865100E-11   5.8060100E+04
9.1512500E-08  -2.1928900E-11  -9.0719500E-11   5.8052900E+04
9.1548100E-08  -9.9263700E-12  -1.0264800E-10   5.7698900E+04
9.1583600E-08   1.0924300E-12  -1.1353200E-10   5.6991600E+04
9.1619200E-08   9.9884700E-12  -1.2235100E-10   5.5948000E+04
9.1654800E-08   1.5771000E-11  -1.2826200E-10   5.4602500E+04
9.1690400E-08   1.7734600E-11  -1.3068700E-10   5.3006700E+04
9.1726000E-08   1.5577900E-11  -1.2936700E-10   5.1228200E+04
9.1761600E-08   9.4026500E-12  -1.2440400E-10   4.9344800E+04
9.1797100E-08  -2.8617300E-13  -1.1626200E-10   4.7436800E+04
9.1832700E-08  -1.2572500E-11  -1.0573600E-10   4.5574700E+04
9.1868300E-08  -2.6207000E-11  -9.3878700E-11   4.3813400E+04
9.1903900E-08  -3.9774200E-11  -8.1906500E-11   4.2194500E+04
9.1939500E-08  -5.1822800E-11  -7.1076500E-11   4.0741300E+04
9.1975100E-08  -6.0997500E-11  -6.2544000E-11   3.9451200E+04
9.2010700E-08  -6.6235500E-11  -5.7239300E-11   3.8297600E+04
9.2046200E-08  -6.6881400E-11  -5.5752700E-11   3.7233400E+04
9.2081800E-08  -6.2765700E-11  -5.8263400E-11   3.6191800E+04
9.2117400E-08  -5.4265900E-11  -6.4504100E-11   3.5094600E+04
9.2153000E-08  -4.2253100E-11  -7.3784300E-11   3.3862100E+04
9.2188600E-08  -2.7992700E-11  -8.5065100E-11   3.2413700E+04
9.2224200E-08  -1.3011600E-11  -9.7070100E-11   3.0671600E+04
9.2259700E-08   1.0475400E-12  -1.0843700E-10   2.8572100E+04
9.2295300E-08   1.2641300E-11  -1.1787300E-10   2.6075100E+04
9.2330900E-08   2.0502400E-11  -1.2430100E-10   2.3163100E+04
9.2366500E-08   2.3772100E-11  -1.2698200E-10   1.9839300E+04
9.2402100E-08   2.2074800E-11  -1.2559900E-10   1.6128600E+04
9.2437700E-08   1.5564800E-11  -1.2027300E-10   1.2078600E+04
9.2473200E-08   4.8618200E-12  -1.1153200E-10   7.7535700E+03
9.2508800E-08  -9.0319000E-12  -1.0024300E-10   3.2282400E+03
```

```
9.2544400E-08  -2.4845100E-11  -8.7480800E-11  -1.4165500E+03
9.2580000E-08  -4.1222700E-11  -7.4389300E-11  -6.1060100E+03
9.2615600E-08  -5.6888700E-11  -6.2049100E-11  -1.0781000E+04
9.2651200E-08  -7.0751600E-11  -5.1353300E-11  -1.5396600E+04
9.2686700E-08  -8.2023500E-11  -4.2922600E-11  -1.9927200E+04
9.2722300E-08  -9.0276600E-11  -3.7063700E-11  -2.4377100E+04
9.2757900E-08  -9.5400800E-11  -3.3769800E-11  -2.8783500E+04
9.2793500E-08  -9.7583100E-11  -3.2769300E-11  -3.3214200E+04
9.2829100E-08  -9.7242200E-11  -3.3595400E-11  -3.7764800E+04
9.2864700E-08  -9.4925300E-11  -3.5678100E-11  -4.2553400E+04
9.2900300E-08  -9.1196500E-11  -3.8447300E-11  -4.7717300E+04
9.2935800E-08  -8.6582700E-11  -4.1403100E-11  -5.3408900E+04
9.2971400E-08  -8.1536700E-11  -4.4161000E-11  -5.9788700E+04
9.3007000E-08  -7.6384900E-11  -4.6483000E-11  -6.7014800E+04
9.3042600E-08  -7.1367800E-11  -4.8268500E-11  -7.5231600E+04
9.3078200E-08  -6.6696000E-11  -4.9524300E-11  -8.4566200E+04
9.3113800E-08  -6.2537400E-11  -5.0313700E-11  -9.5128100E+04
9.3149300E-08  -5.9066800E-11  -5.0714500E-11  -1.0700200E+05
9.3184900E-08  -5.6512000E-11  -5.0779800E-11  -1.2024300E+05
9.3220500E-08  -5.5174800E-11  -5.0516900E-11  -1.3487900E+05
9.3256100E-08  -5.5334800E-11  -4.9882000E-11  -1.5090900E+05
9.3291700E-08  -5.7211800E-11  -4.8807900E-11  -1.6831100E+05
9.3327300E-08  -6.0987000E-11  -4.7234700E-11  -1.8704800E+05
9.3362800E-08  -6.6670000E-11  -4.5136000E-11  -2.0706900E+05
9.3398400E-08  -7.4061600E-11  -4.2551000E-11  -2.2831800E+05
9.3434000E-08  -8.2805700E-11  -3.9605400E-11  -2.5073200E+05
9.3469600E-08  -9.2366900E-11  -3.6504900E-11  -2.7425300E+05
9.3505200E-08  -1.0207300E-10  -3.3488200E-11  -2.9883500E+05
9.3540800E-08  -1.1126500E-10  -3.0779700E-11  -3.2444100E+05
9.3576300E-08  -1.1936200E-10  -2.8539700E-11  -3.5104300E+05
9.3611900E-08  -1.2599500E-10  -2.6788500E-11  -3.7861900E+05
9.3647500E-08  -1.3106700E-10  -2.5357500E-11  -4.0715900E+05
9.3683100E-08  -1.3481900E-10  -2.3876400E-11  -4.3666100E+05
9.3718700E-08  -1.3785600E-10  -2.1772500E-11  -4.6712400E+05
9.3754300E-08  -1.4105900E-10  -1.8316100E-11  -4.9854700E+05
9.3789900E-08  -1.4553300E-10  -1.2699900E-11  -5.3092000E+05
9.3825400E-08  -1.5244700E-10  -4.1472400E-12  -5.6423000E+05
9.3861000E-08  -1.6286400E-10   7.9878700E-12  -5.9845800E+05
9.3896600E-08  -1.7762800E-10   2.4121100E-11  -6.3357900E+05
9.3932200E-08  -1.9722400E-10   4.4360400E-11  -6.6956000E+05
9.3967800E-08  -2.2169200E-10   6.8466700E-11  -7.0635800E+05
9.4003400E-08  -2.5060000E-10   9.5861500E-11  -7.4393200E+05
9.4038900E-08  -2.8313700E-10   1.2569300E-10  -7.8224000E+05
9.4074500E-08  -3.1815300E-10   1.5693700E-10  -8.2124400E+05
9.4110100E-08  -3.5433100E-10   1.8852200E-10  -8.6090900E+05
9.4145700E-08  -3.9040300E-10   2.1947700E-10  -9.0120000E+05
9.4181300E-08  -4.2528400E-10   2.4906600E-10  -9.4208400E+05
9.4216900E-08  -4.5821100E-10   2.7689600E-10  -9.8352500E+05
9.4252400E-08  -4.8888200E-10   3.0298000E-10  -1.0254800E+06
9.4288000E-08  -5.1750600E-10   3.2775400E-10  -1.0679100E+06
9.4323600E-08  -5.4473700E-10   3.5203400E-10  -1.1107200E+06
```

9.4359200E-08	-5.7162700E-10	3.7691900E-10	-1.1538300E+06
9.4394800E-08	-5.9949900E-10	4.0367200E-10	-1.1970900E+06
9.4430400E-08	-6.2972700E-10	4.3356500E-10	-1.2403400E+06
9.4465900E-08	-6.6359800E-10	4.6773900E-10	-1.2834000E+06
9.4501500E-08	-7.0217900E-10	5.0706700E-10	-1.3260300E+06
9.4537100E-08	-7.4613300E-10	5.5207700E-10	-1.3679700E+06
9.4572700E-08	-7.9568100E-10	6.0290000E-10	-1.4089500E+06
9.4608300E-08	-8.5063900E-10	6.5927700E-10	-1.4486900E+06
9.4643900E-08	-9.1040000E-10	7.2061800E-10	-1.4869400E+06
9.4679500E-08	-9.7402700E-10	7.8609300E-10	-1.5234300E+06
9.4715000E-08	-1.0404400E-09	8.5474300E-10	-1.5579500E+06
9.4750600E-08	-1.1085300E-09	9.2560200E-10	-1.5903500E+06
9.4786200E-08	-1.1773000E-09	9.9780200E-10	-1.6205000E+06
9.4821800E-08	-1.2459300E-09	1.0706600E-09	-1.6483600E+06
9.4857400E-08	-1.3138700E-09	1.1437300E-09	-1.6739400E+06
9.4893000E-08	-1.3809000E-09	1.2168100E-09	-1.6973300E+06
9.4928500E-08	-1.4470400E-09	1.2899300E-09	-1.7186300E+06
9.4964100E-08	-1.5125900E-09	1.3632900E-09	-1.7380400E+06
9.4999700E-08	-1.5779700E-09	1.4372200E-09	-1.7557500E+06
9.5035300E-08	-1.6436700E-09	1.5121100E-09	-1.7719900E+06
9.5070900E-08	-1.7102300E-09	1.5883700E-09	-1.7869800E+06
9.5106500E-08	-1.7781300E-09	1.6663700E-09	-1.8009200E+06
9.5142000E-08	-1.8477600E-09	1.7464300E-09	-1.8140000E+06
9.5177600E-08	-1.9194600E-09	1.8288100E-09	-1.8263800E+06
9.5213200E-08	-1.9935300E-09	1.9137800E-09	-1.8381500E+06
9.5248800E-08	-2.0701900E-09	2.0015600E-09	-1.8493800E+06
9.5284400E-08	-2.1496900E-09	2.0923800E-09	-1.8600700E+06
9.5320000E-08	-2.2322800E-09	2.1864700E-09	-1.8702000E+06
9.5355500E-08	-2.3181700E-09	2.2840000E-09	-1.8797100E+06
9.5391100E-08	-2.4075400E-09	2.3851100E-09	-1.8885100E+06
9.5426700E-08	-2.5005000E-09	2.4897900E-09	-1.8965000E+06
9.5462300E-08	-2.5969500E-09	2.5978400E-09	-1.9036000E+06
9.5497900E-08	-2.6965400E-09	2.7088100E-09	-1.9097300E+06
9.5533500E-08	-2.7987000E-09	2.8219800E-09	-1.9148300E+06
9.5569000E-08	-2.9024900E-09	2.9363000E-09	-1.9189000E+06
9.5604600E-08	-3.0067100E-09	3.0505100E-09	-1.9219700E+06
9.5640200E-08	-3.1099600E-09	3.1631400E-09	-1.9241000E+06
9.5675800E-08	-3.2107200E-09	3.2726800E-09	-1.9254300E+06
9.5711400E-08	-3.3074600E-09	3.3776900E-09	-1.9261100E+06
9.5747000E-08	-3.3989300E-09	3.4769700E-09	-1.9263300E+06
9.5782600E-08	-3.4841900E-09	3.5697000E-09	-1.9263000E+06
9.5818100E-08	-3.5627900E-09	3.6555500E-09	-1.9262300E+06
9.5853700E-08	-3.6348400E-09	3.7347200E-09	-1.9263300E+06
9.5889300E-08	-3.7009900E-09	3.8079900E-09	-1.9267900E+06
9.5924900E-08	-3.7624700E-09	3.8766000E-09	-1.9277600E+06
9.5960500E-08	-3.8209200E-09	3.9421800E-09	-1.9293700E+06
9.5996100E-08	-3.8781600E-09	4.0065400E-09	-1.9316700E+06
9.6031600E-08	-3.9360700E-09	4.0715000E-09	-1.9346800E+06
9.6067200E-08	-3.9963800E-09	4.1387000E-09	-1.9383700E+06
9.6102800E-08	-4.0604400E-09	4.2093700E-09	-1.9426500E+06
9.6138400E-08	-4.1291000E-09	4.2842900E-09	-1.9474300E+06

9.6174000E-08	-4.2026200E-09	4.3636500E-09	-1.9525700E+06
9.6209600E-08	-4.2806400E-09	4.4470600E-09	-1.9579200E+06
9.6245100E-08	-4.3623200E-09	4.5336600E-09	-1.9633400E+06
9.6280700E-08	-4.4464100E-09	4.6221800E-09	-1.9687200E+06
9.6316300E-08	-4.5313700E-09	4.7111400E-09	-1.9739700E+06
9.6351900E-08	-4.6156900E-09	4.7990000E-09	-1.9790300E+06
9.6387500E-08	-4.6979400E-09	4.8843700E-09	-1.9838800E+06
9.6423100E-08	-4.7770400E-09	4.9661300E-09	-1.9885500E+06
9.6458600E-08	-4.8522600E-09	5.0435100E-09	-1.9931000E+06
9.6494200E-08	-4.9233800E-09	5.1162200E-09	-1.9976300E+06
9.6529800E-08	-4.9906300E-09	5.1843700E-09	-2.0022400E+06
9.6565400E-08	-5.0546200E-09	5.2485100E-09	-2.0070600E+06
9.6601000E-08	-5.1162600E-09	5.3094700E-09	-2.0121900E+06
9.6636600E-08	-5.1766900E-09	5.3683000E-09	-2.0177300E+06
9.6672200E-08	-5.2370400E-09	5.4261700E-09	-2.0237500E+06
9.6707700E-08	-5.2984200E-09	5.4842400E-09	-2.0302900E+06
9.6743300E-08	-5.3618000E-09	5.5435900E-09	-2.0373400E+06
9.6778900E-08	-5.4279000E-09	5.6051300E-09	-2.0448600E+06
9.6814500E-08	-5.4972300E-09	5.6696100E-09	-2.0527900E+06
9.6850100E-08	-5.5700600E-09	5.7375400E-09	-2.0610400E+06
9.6885700E-08	-5.6464100E-09	5.8092100E-09	-2.0695100E+06
9.6921200E-08	-5.7261600E-09	5.8847000E-09	-2.0781000E+06
9.6956800E-08	-5.8090300E-09	5.9639000E-09	-2.0867100E+06
9.6992400E-08	-5.8946100E-09	6.0464900E-09	-2.0952900E+06
9.7028000E-08	-5.9824300E-09	6.1320000E-09	-2.1038100E+06
9.7063600E-08	-6.0719600E-09	6.2198100E-09	-2.1122700E+06
9.7099200E-08	-6.1626600E-09	6.3091700E-09	-2.1207300E+06
9.7134700E-08	-6.2539000E-09	6.3992500E-09	-2.1292600E+06
9.7170300E-08	-6.3451000E-09	6.4891700E-09	-2.1380000E+06
9.7205900E-08	-6.4356800E-09	6.5781000E-09	-2.1470800E+06
9.7241500E-08	-6.5251400E-09	6.6652700E-09	-2.1566700E+06
9.7277100E-08	-6.6130800E-09	6.7501000E-09	-2.1669300E+06
9.7312700E-08	-6.6992500E-09	6.8322500E-09	-2.1780000E+06
9.7348200E-08	-6.7835900E-09	6.9116500E-09	-2.1900000E+06
9.7383800E-08	-6.8662800E-09	6.9885500E-09	-2.2030300E+06
9.7419400E-08	-6.9477000E-09	7.0635100E-09	-2.2171200E+06
9.7455000E-08	-7.0283900E-09	7.1373400E-09	-2.2322900E+06
9.7490600E-08	-7.1090700E-09	7.2110000E-09	-2.2484700E+06
9.7526200E-08	-7.1904900E-09	7.2855500E-09	-2.2655800E+06
9.7561800E-08	-7.2733200E-09	7.3619200E-09	-2.2835000E+06
9.7597300E-08	-7.3580500E-09	7.4408300E-09	-2.3020600E+06
9.7632900E-08	-7.4449000E-09	7.5226600E-09	-2.3211200E+06
9.7668500E-08	-7.5337300E-09	7.6073300E-09	-2.3405000E+06
9.7704100E-08	-7.6239700E-09	7.6942700E-09	-2.3600700E+06
9.7739700E-08	-7.7146500E-09	7.7824400E-09	-2.3797000E+06
9.7775300E-08	-7.8044600E-09	7.8703700E-09	-2.3993100E+06
9.7810800E-08	-7.8918300E-09	7.9563100E-09	-2.4188800E+06
9.7846400E-08	-7.9751000E-09	8.0384000E-09	-2.4384100E+06
9.7882000E-08	-8.0526600E-09	8.1148400E-09	-2.4579600E+06
9.7917600E-08	-8.1231300E-09	8.1840800E-09	-2.4776400E+06
9.7953200E-08	-8.1855400E-09	8.2450200E-09	-2.4975800E+06

9.7988800E-08	-8.2394400E-09	8.2971100E-09	-2.5179500E+06
9.8024300E-08	-8.2849400E-09	8.3404400E-09	-2.5389200E+06
9.8059900E-08	-8.3227300E-09	8.3757000E-09	-2.5606500E+06
9.8095500E-08	-8.3540300E-09	8.4041500E-09	-2.5832900E+06
9.8131100E-08	-8.3804700E-09	8.4274700E-09	-2.6069500E+06
9.8166700E-08	-8.4039000E-09	8.4475500E-09	-2.6317200E+06
9.8202300E-08	-8.4262000E-09	8.4663000E-09	-2.6576000E+06
9.8237800E-08	-8.4491100E-09	8.4854600E-09	-2.6845800E+06
9.8273400E-08	-8.4739600E-09	8.5063300E-09	-2.7125600E+06
9.8309000E-08	-8.5016000E-09	8.5297100E-09	-2.7414300E+06
9.8344600E-08	-8.5322900E-09	8.5557600E-09	-2.7710200E+06
9.8380200E-08	-8.5656800E-09	8.5840100E-09	-2.8011400E+06
9.8415800E-08	-8.6007800E-09	8.6133900E-09	-2.8315700E+06
9.8451400E-08	-8.6361800E-09	8.6424300E-09	-2.8621200E+06
9.8486900E-08	-8.6701400E-09	8.6693400E-09	-2.8925800E+06
9.8522500E-08	-8.7008200E-09	8.6923100E-09	-2.9227900E+06
9.8558100E-08	-8.7264700E-09	8.7096900E-09	-2.9526100E+06
9.8593700E-08	-8.7456400E-09	8.7201900E-09	-2.9819400E+06
9.8629300E-08	-8.7573700E-09	8.7230700E-09	-3.0107200E+06
9.8664900E-08	-8.7612700E-09	8.7182000E-09	-3.0389100E+06
9.8700400E-08	-8.7576100E-09	8.7061400E-09	-3.0665100E+06
9.8736000E-08	-8.7472800E-09	8.6880800E-09	-3.0935300E+06
9.8771600E-08	-8.7317400E-09	8.6657100E-09	-3.1200000E+06
9.8807200E-08	-8.7128800E-09	8.6410800E-09	-3.1459200E+06
9.8842800E-08	-8.6928100E-09	8.6164200E-09	-3.1712800E+06
9.8878400E-08	-8.6737200E-09	8.5938400E-09	-3.1960600E+06
9.8913900E-08	-8.6576000E-09	8.5752200E-09	-3.2202000E+06
9.8949500E-08	-8.6460800E-09	8.5619200E-09	-3.2435800E+06
9.8985100E-08	-8.6403300E-09	8.5547300E-09	-3.2660900E+06
9.9020700E-08	-8.6409000E-09	8.5537800E-09	-3.2875500E+06
9.9056300E-08	-8.6476800E-09	8.5585100E-09	-3.3078100E+06
9.9091900E-08	-8.6599700E-09	8.5677700E-09	-3.3266700E+06
9.9127400E-08	-8.6765400E-09	8.5799500E-09	-3.3439500E+06
9.9163000E-08	-8.6957600E-09	8.5931200E-09	-3.3595100E+06
9.9198600E-08	-8.7158000E-09	8.6053000E-09	-3.3732200E+06
9.9234200E-08	-8.7348000E-09	8.6145900E-09	-3.3849900E+06
9.9269800E-08	-8.7510100E-09	8.6194300E-09	-3.3948000E+06
9.9305400E-08	-8.7630700E-09	8.6187500E-09	-3.4026600E+06
9.9341000E-08	-8.7700500E-09	8.6120200E-09	-3.4086500E+06
9.9376500E-08	-8.7715700E-09	8.5993700E-09	-3.4129000E+06
9.9412100E-08	-8.7678000E-09	8.5815300E-09	-3.4155600E+06
9.9447700E-08	-8.7594300E-09	8.5597600E-09	-3.4168400E+06
9.9483300E-08	-8.7476300E-09	8.5357200E-09	-3.4169500E+06
9.9518900E-08	-8.7338900E-09	8.5112900E-09	-3.4161100E+06
9.9554500E-08	-8.7198300E-09	8.4883700E-09	-3.4145200E+06
9.9590000E-08	-8.7071000E-09	8.4687400E-09	-3.4123800E+06
9.9625600E-08	-8.6971600E-09	8.4538400E-09	-3.4098400E+06
9.9661200E-08	-8.6911600E-09	8.4446300E-09	-3.4070000E+06
9.9696800E-08	-8.6898600E-09	8.4415500E-09	-3.4039300E+06
9.9732400E-08	-8.6935000E-09	8.4444600E-09	-3.4006500E+06
9.9768000E-08	-8.7018900E-09	8.4527100E-09	-3.3971200E+06

```
9.9803500E-08  -8.7143600E-09  8.4651300E-09  -3.3932700E+06
9.9839100E-08  -8.7298700E-09  8.4802500E-09  -3.3890400E+06
9.9874700E-08  -8.7471400E-09  8.4964200E-09  -3.3843000E+06
9.9910300E-08  -8.7647900E-09  8.5119500E-09  -3.3789600E+06
9.9945900E-08  -8.7814400E-09  8.5252600E-09  -3.3729300E+06
9.9981500E-08  -8.7958100E-09  8.5350600E-09  -3.3661700E+06
1.0001700E-07  -8.8069300E-09  8.5403800E-09  -3.3586700E+06
1.0005300E-07  -8.8140700E-09  8.5406700E-09  -3.3504600E+06
1.0008800E-07  -8.8168400E-09  8.5357900E-09  -3.3416400E+06
1.0012400E-07  -8.8151900E-09  8.5259600E-09  -3.3323400E+06
1.0015900E-07  -8.8093500E-09  8.5117600E-09  -3.3227600E+06
1.0019500E-07  -8.7997500E-09  8.4939400E-09  -3.3131100E+06
1.0023100E-07  -8.7870000E-09  8.4734200E-09  -3.3036400E+06
1.0026600E-07  -8.7718500E-09  8.4511700E-09  -3.2946000E+06
1.0030200E-07  -8.7550200E-09  8.4281000E-09  -3.2862100E+06
1.0033700E-07  -8.7372400E-09  8.4050600E-09  -3.2786800E+06
1.0037300E-07  -8.7192400E-09  8.3827700E-09  -3.2721500E+06
1.0040800E-07  -8.7016500E-09  8.3618400E-09  -3.2667300E+06
1.0044400E-07  -8.6850700E-09  8.3427200E-09  -3.2624600E+06
1.0048000E-07  -8.6700400E-09  8.3257700E-09  -3.2592800E+06
1.0051500E-07  -8.6570300E-09  8.3112400E-09  -3.2571200E+06
1.0055100E-07  -8.6463900E-09  8.2992800E-09  -3.2557900E+06
1.0058600E-07  -8.6383900E-09  8.2899100E-09  -3.2551200E+06
1.0062200E-07  -8.6331000E-09  8.2830300E-09  -3.2548500E+06
1.0065800E-07  -8.6304000E-09  8.2783600E-09  -3.2547500E+06
1.0069300E-07  -8.6299300E-09  8.2754500E-09  -3.2545800E+06
1.0072900E-07  -8.6310500E-09  8.2736200E-09  -3.2541200E+06
1.0076400E-07  -8.6328700E-09  8.2720100E-09  -3.2532000E+06
1.0080000E-07  -8.6343100E-09  8.2696000E-09  -3.2516900E+06
1.0083500E-07  -8.6341200E-09  8.2652800E-09  -3.2495200E+06
1.0087100E-07  -8.6310200E-09  8.2579300E-09  -3.2466700E+06
1.0090700E-07  -8.6239000E-09  8.2465500E-09  -3.2431900E+06
1.0094200E-07  -8.6118200E-09  8.2303700E-09  -3.2391700E+06
1.0097800E-07  -8.5942500E-09  8.2090000E-09  -3.2347500E+06
1.0101300E-07  -8.5711500E-09  8.1824500E-09  -3.2300900E+06
1.0104900E-07  -8.5429900E-09  8.1512400E-09  -3.2253500E+06
1.0108500E-07  -8.5107300E-09  8.1163500E-09  -3.2207000E+06
1.0112000E-07  -8.4758200E-09  8.0791600E-09  -3.2162800E+06
1.0115600E-07  -8.4399900E-09  8.0413300E-09  -3.2122100E+06
1.0119100E-07  -8.4051000E-09  8.0046400E-09  -3.2085500E+06
1.0122700E-07  -8.3729800E-09  7.9707900E-09  -3.2053400E+06
1.0126200E-07  -8.3451900E-09  7.9412600E-09  -3.2025400E+06
1.0129800E-07  -8.3228400E-09  7.9170600E-09  -3.2000900E+06
1.0133400E-07  -8.3064700E-09  7.8987100E-09  -3.1979000E+06
1.0136900E-07  -8.2960400E-09  7.8861100E-09  -3.1958300E+06
1.0140500E-07  -8.2908500E-09  7.8785700E-09  -3.1937300E+06
1.0144000E-07  -8.2897100E-09  7.8749200E-09  -3.1914600E+06
1.0147600E-07  -8.2910200E-09  7.8735700E-09  -3.1889000E+06
1.0151200E-07  -8.2929500E-09  7.8727600E-09  -3.1859300E+06
1.0154700E-07  -8.2936900E-09  7.8707000E-09  -3.1824900E+06
1.0158300E-07  -8.2916400E-09  7.8658000E-09  -3.1785300E+06
```

1.0161800E-07	-8.2855100E-09	7.8567700E-09	-3.1740700E+06
1.0165400E-07	-8.2744600E-09	7.8427900E-09	-3.1691600E+06
1.0168900E-07	-8.2582000E-09	7.8235600E-09	-3.1638700E+06
1.0172500E-07	-8.2369500E-09	7.7992900E-09	-3.1583100E+06
1.0176100E-07	-8.2113800E-09	7.7706400E-09	-3.1526200E+06
1.0179600E-07	-8.1824800E-09	7.7386400E-09	-3.1469100E+06
1.0183200E-07	-8.1515000E-09	7.7045900E-09	-3.1413000E+06
1.0186700E-07	-8.1197200E-09	7.6698400E-09	-3.1358800E+06
1.0190300E-07	-8.0883800E-09	7.6357400E-09	-3.1307300E+06
1.0193900E-07	-8.0585700E-09	7.6034500E-09	-3.1258600E+06
1.0197400E-07	-8.0310900E-09	7.5739000E-09	-3.1212700E+06
1.0201000E-07	-8.0064400E-09	7.5476700E-09	-3.1169200E+06
1.0204500E-07	-7.9848300E-09	7.5250100E-09	-3.1127400E+06
1.0208100E-07	-7.9661500E-09	7.5058000E-09	-3.1086400E+06
1.0211700E-07	-7.9499800E-09	7.4896200E-09	-3.1045100E+06
1.0215200E-07	-7.9357600E-09	7.4757300E-09	-3.1002700E+06
1.0218800E-07	-7.9227100E-09	7.4632300E-09	-3.0958200E+06
1.0222300E-07	-7.9099900E-09	7.4510600E-09	-3.0911200E+06
1.0225900E-07	-7.8967000E-09	7.4381200E-09	-3.0861400E+06
1.0229400E-07	-7.8819900E-09	7.4233500E-09	-3.0808800E+06
1.0233000E-07	-7.8651000E-09	7.4058200E-09	-3.0754000E+06
1.0236600E-07	-7.8453900E-09	7.3848000E-09	-3.0697700E+06
1.0240100E-07	-7.8224500E-09	7.3598200E-09	-3.0640900E+06
1.0243700E-07	-7.7961300E-09	7.3307500E-09	-3.0584800E+06
1.0247200E-07	-7.7665200E-09	7.2978200E-09	-3.0530500E+06
1.0250800E-07	-7.7340100E-09	7.2616000E-09	-3.0479000E+06
1.0254400E-07	-7.6993000E-09	7.2230100E-09	-3.0431200E+06
1.0257900E-07	-7.6633600E-09	7.1832400E-09	-3.0387600E+06
1.0261500E-07	-7.6273100E-09	7.1437000E-09	-3.0348400E+06
1.0265000E-07	-7.5924100E-09	7.1058300E-09	-3.0313500E+06
1.0268600E-07	-7.5599200E-09	7.0710500E-09	-3.0282200E+06
1.0272100E-07	-7.5309500E-09	7.0405700E-09	-3.0253600E+06
1.0275700E-07	-7.5063900E-09	7.0152900E-09	-3.0226700E+06
1.0279300E-07	-7.4867600E-09	6.9956800E-09	-3.0200000E+06
1.0282800E-07	-7.4721400E-09	6.9817300E-09	-3.0172400E+06
1.0286400E-07	-7.4621600E-09	6.9729200E-09	-3.0142500E+06
1.0289900E-07	-7.4559500E-09	6.9682600E-09	-3.0109100E+06
1.0293500E-07	-7.4523000E-09	6.9663700E-09	-3.0071400E+06
1.0297100E-07	-7.4496800E-09	6.9656100E-09	-3.0029000E+06
1.0300600E-07	-7.4464600E-09	6.9642800E-09	-2.9981400E+06
1.0304200E-07	-7.4410400E-09	6.9607200E-09	-2.9928900E+06
1.0307700E-07	-7.4320600E-09	6.9535700E-09	-2.9871900E+06
1.0311300E-07	-7.4185500E-09	6.9418500E-09	-2.9810900E+06
1.0314800E-07	-7.4000100E-09	6.9251200E-09	-2.9746900E+06
1.0318400E-07	-7.3765500E-09	6.9034900E-09	-2.9680900E+06
1.0322000E-07	-7.3488500E-09	6.8776500E-09	-2.9613600E+06
1.0325500E-07	-7.3180800E-09	6.8487600E-09	-2.9545900E+06
1.0329100E-07	-7.2857800E-09	6.8183600E-09	-2.9478500E+06
1.0332600E-07	-7.2537200E-09	6.7881600E-09	-2.9412000E+06
1.0336200E-07	-7.2236900E-09	6.7598900E-09	-2.9346400E+06
1.0339800E-07	-7.1972800E-09	6.7350900E-09	-2.9281700E+06

1.0343300E-07	-7.1757100E-09	6.7149700E-09	-2.9217700E+06
1.0346900E-07	-7.1597000E-09	6.7002500E-09	-2.9153700E+06
1.0350400E-07	-7.1494100E-09	6.6911000E-09	-2.9089200E+06
1.0354000E-07	-7.1443600E-09	6.6871700E-09	-2.9023400E+06
1.0357500E-07	-7.1435900E-09	6.6875700E-09	-2.8955600E+06
1.0361100E-07	-7.1456900E-09	6.6910100E-09	-2.8885200E+06
1.0364700E-07	-7.1489800E-09	6.6959500E-09	-2.8811500E+06
1.0368200E-07	-7.1517000E-09	6.7007400E-09	-2.8734500E+06
1.0371800E-07	-7.1522400E-09	6.7037700E-09	-2.8653900E+06
1.0375300E-07	-7.1491700E-09	6.7036900E-09	-2.8570100E+06
1.0378900E-07	-7.1415100E-09	6.6994500E-09	-2.8483300E+06
1.0382500E-07	-7.1287600E-09	6.6904200E-09	-2.8394200E+06
1.0386000E-07	-7.1108600E-09	6.6764300E-09	-2.8303500E+06
1.0389600E-07	-7.0882700E-09	6.6577200E-09	-2.8212000E+06
1.0393100E-07	-7.0617900E-09	6.6349500E-09	-2.8120500E+06
1.0396700E-07	-7.0325500E-09	6.6090600E-09	-2.8029500E+06
1.0400200E-07	-7.0018100E-09	6.5812100E-09	-2.7939600E+06
1.0403800E-07	-6.9708900E-09	6.5526500E-09	-2.7850900E+06
1.0407400E-07	-6.9410500E-09	6.5246600E-09	-2.7763400E+06
1.0410900E-07	-6.9133600E-09	6.4983900E-09	-2.7676600E+06
1.0414500E-07	-6.8887200E-09	6.4748400E-09	-2.7590000E+06
1.0418000E-07	-6.8677300E-09	6.4547800E-09	-2.7502700E+06
1.0421600E-07	-6.8507100E-09	6.4386900E-09	-2.7413800E+06
1.0425200E-07	-6.8377200E-09	6.4267800E-09	-2.7322500E+06
1.0428700E-07	-6.8285500E-09	6.4189300E-09	-2.7227900E+06
1.0432300E-07	-6.8227300E-09	6.4147300E-09	-2.7129400E+06
1.0435800E-07	-6.8195800E-09	6.4134700E-09	-2.7026700E+06
1.0439400E-07	-6.8182000E-09	6.4141900E-09	-2.6919800E+06
1.0442900E-07	-6.8175800E-09	6.4157400E-09	-2.6809300E+06
1.0446500E-07	-6.8165600E-09	6.4168300E-09	-2.6696100E+06
1.0450100E-07	-6.8139300E-09	6.4161000E-09	-2.6581200E+06
1.0453600E-07	-6.8085100E-09	6.4122900E-09	-2.6466200E+06
1.0457200E-07	-6.7992400E-09	6.4042400E-09	-2.6352800E+06
1.0460700E-07	-6.7852500E-09	6.3911200E-09	-2.6242700E+06
1.0464300E-07	-6.7659800E-09	6.3724400E-09	-2.6137600E+06
1.0467900E-07	-6.7413200E-09	6.3481900E-09	-2.6039000E+06
1.0471400E-07	-6.7115400E-09	6.3188400E-09	-2.5947900E+06
1.0475000E-07	-6.6774400E-09	6.2853600E-09	-2.5865200E+06
1.0478500E-07	-6.6403000E-09	6.2491500E-09	-2.5791100E+06
1.0482100E-07	-6.6017700E-09	6.2119900E-09	-2.5725700E+06
1.0485600E-07	-6.5637800E-09	6.1758200E-09	-2.5668200E+06
1.0489200E-07	-6.5283900E-09	6.1426100E-09	-2.5617700E+06
1.0492800E-07	-6.4975300E-09	6.1141400E-09	-2.5572900E+06
1.0496300E-07	-6.4728500E-09	6.0918000E-09	-2.5532300E+06
1.0499900E-07	-6.4555100E-09	6.0764500E-09	-2.5494000E+06
1.0503400E-07	-6.4460500E-09	6.0683100E-09	-2.5456500E+06
1.0507000E-07	-6.4442500E-09	6.0668500E-09	-2.5418000E+06
1.0510600E-07	-6.4491200E-09	6.0708700E-09	-2.5377200E+06
1.0514100E-07	-6.4589900E-09	6.0785800E-09	-2.5332700E+06
1.0517700E-07	-6.4716500E-09	6.0877300E-09	-2.5283800E+06
1.0521200E-07	-6.4844900E-09	6.0959000E-09	-2.5230000E+06

1.0524800E-07	-6.4948400E-09	6.1006500E-09	-2.5171200E+06
1.0528300E-07	-6.5002100E-09	6.0998800E-09	-2.5107800E+06
1.0531900E-07	-6.4985400E-09	6.0919600E-09	-2.5040600E+06
1.0535500E-07	-6.4884400E-09	6.0759900E-09	-2.4970600E+06
1.0539000E-07	-6.4693300E-09	6.0518600E-09	-2.4899100E+06
1.0542600E-07	-6.4415900E-09	6.0203000E-09	-2.4827700E+06
1.0546100E-07	-6.4064300E-09	5.9828100E-09	-2.4757900E+06
1.0549700E-07	-6.3658500E-09	5.9415100E-09	-2.4691300E+06
1.0553300E-07	-6.3224000E-09	5.8989500E-09	-2.4629300E+06
1.0556800E-07	-6.2789800E-09	5.8578400E-09	-2.4573000E+06
1.0560400E-07	-6.2384700E-09	5.8207900E-09	-2.4523500E+06
1.0563900E-07	-6.2034600E-09	5.7900100E-09	-2.4481300E+06
1.0567500E-07	-6.1760400E-09	5.7671300E-09	-2.4446600E+06
1.0571000E-07	-6.1575300E-09	5.7530300E-09	-2.4419300E+06
1.0574600E-07	-6.1483600E-09	5.7477400E-09	-2.4398800E+06
1.0578200E-07	-6.1481200E-09	5.7504800E-09	-2.4384400E+06
1.0581700E-07	-6.1555000E-09	5.7597000E-09	-2.4375000E+06
1.0585300E-07	-6.1685600E-09	5.7733300E-09	-2.4369300E+06
1.0588800E-07	-6.1848700E-09	5.7889000E-09	-2.4366200E+06
1.0592400E-07	-6.2018000E-09	5.8038900E-09	-2.4364200E+06
1.0596000E-07	-6.2168100E-09	5.8159000E-09	-2.4362100E+06
1.0599500E-07	-6.2276700E-09	5.8229500E-09	-2.4358700E+06
1.0603100E-07	-6.2326800E-09	5.8235900E-09	-2.4353200E+06
1.0606600E-07	-6.2308300E-09	5.8170800E-09	-2.4344700E+06
1.0610200E-07	-6.2218100E-09	5.8034100E-09	-2.4332900E+06
1.0613700E-07	-6.2060600E-09	5.7832300E-09	-2.4317500E+06
1.0617300E-07	-6.1846400E-09	5.7578100E-09	-2.4298500E+06
1.0620900E-07	-6.1590800E-09	5.7288200E-09	-2.4276400E+06
1.0624400E-07	-6.1311600E-09	5.6981900E-09	-2.4251500E+06
1.0628000E-07	-6.1027600E-09	5.6678700E-09	-2.4224700E+06
1.0631500E-07	-6.0756100E-09	5.6396400E-09	-2.4196800E+06
1.0635100E-07	-6.0511600E-09	5.6149900E-09	-2.4168800E+06
1.0638700E-07	-6.0304000E-09	5.5949400E-09	-2.4141700E+06
1.0642200E-07	-6.0138600E-09	5.5800000E-09	-2.4116700E+06
1.0645800E-07	-6.0015800E-09	5.5702000E-09	-2.4094800E+06
1.0649300E-07	-5.9931400E-09	5.5650800E-09	-2.4076800E+06
1.0652900E-07	-5.9878000E-09	5.5638200E-09	-2.4063700E+06
1.0656400E-07	-5.9845700E-09	5.5653400E-09	-2.4055800E+06
1.0660000E-07	-5.9824300E-09	5.5684400E-09	-2.4053700E+06
1.0663600E-07	-5.9803700E-09	5.5719500E-09	-2.4057200E+06
1.0667100E-07	-5.9775500E-09	5.5748300E-09	-2.4066200E+06
1.0670700E-07	-5.9733900E-09	5.5762500E-09	-2.4080000E+06
1.0674200E-07	-5.9675600E-09	5.5756700E-09	-2.4097700E+06
1.0677800E-07	-5.9600500E-09	5.5728300E-09	-2.4118000E+06
1.0681400E-07	-5.9510700E-09	5.5677500E-09	-2.4139600E+06
1.0684900E-07	-5.9409900E-09	5.5606900E-09	-2.4160700E+06
1.0688500E-07	-5.9303100E-09	5.5520600E-09	-2.4179800E+06
1.0692000E-07	-5.9195400E-09	5.5423900E-09	-2.4195200E+06
1.0695600E-07	-5.9091100E-09	5.5322100E-09	-2.4205400E+06
1.0699200E-07	-5.8993700E-09	5.5220200E-09	-2.4208800E+06
1.0702700E-07	-5.8904600E-09	5.5122200E-09	-2.4204500E+06

```
1.0706300E-07  -5.8824100E-09  5.5031300E-09  -2.4191700E+06
1.0709800E-07  -5.8751000E-09  5.4949200E-09  -2.4169900E+06
1.0713400E-07  -5.8683200E-09  5.4877000E-09  -2.4139100E+06
1.0716900E-07  -5.8617700E-09  5.4814600E-09  -2.4099500E+06
1.0720500E-07  -5.8551800E-09  5.4761600E-09  -2.4051800E+06
1.0724100E-07  -5.8483200E-09  5.4717000E-09  -2.3996800E+06
1.0727600E-07  -5.8410000E-09  5.4679900E-09  -2.3935500E+06
1.0731200E-07  -5.8331500E-09  5.4649200E-09  -2.3869300E+06
1.0734700E-07  -5.8247500E-09  5.4623600E-09  -2.3799300E+06
1.0738300E-07  -5.8158100E-09  5.4601200E-09  -2.3727000E+06
1.0741900E-07  -5.8063600E-09  5.4580100E-09  -2.3653500E+06
1.0745400E-07  -5.7964100E-09  5.4557400E-09  -2.3580000E+06
1.0749000E-07  -5.7859200E-09  5.4529700E-09  -2.3507700E+06
1.0752500E-07  -5.7747600E-09  5.4492900E-09  -2.3437500E+06
1.0756100E-07  -5.7627500E-09  5.4442900E-09  -2.3370000E+06
1.0759600E-07  -5.7496800E-09  5.4375700E-09  -2.3305800E+06
1.0763200E-07  -5.7353900E-09  5.4288300E-09  -2.3244900E+06
1.0766800E-07  -5.7197600E-09  5.4179200E-09  -2.3187500E+06
1.0770300E-07  -5.7028300E-09  5.4049200E-09  -2.3133000E+06
1.0773900E-07  -5.6848400E-09  5.3901400E-09  -2.3080900E+06
1.0777400E-07  -5.6662000E-09  5.3741600E-09  -2.3030200E+06
1.0781000E-07  -5.6475900E-09  5.3578200E-09  -2.2979800E+06
1.0784600E-07  -5.6298000E-09  5.3421200E-09  -2.2928300E+06
1.0788100E-07  -5.6137100E-09  5.3281100E-09  -2.2874500E+06
1.0791700E-07  -5.6001500E-09  5.3168400E-09  -2.2816900E+06
1.0795200E-07  -5.5898200E-09  5.3091300E-09  -2.2754300E+06
1.0798800E-07  -5.5831500E-09  5.3055100E-09  -2.2685700E+06
1.0802300E-07  -5.5801900E-09  5.3060600E-09  -2.2610400E+06
1.0805900E-07  -5.5805500E-09  5.3104100E-09  -2.2528100E+06
1.0809500E-07  -5.5834300E-09  5.3177200E-09  -2.2438800E+06
1.0813000E-07  -5.5876600E-09  5.3267200E-09  -2.2343200E+06
1.0816600E-07  -5.5918000E-09  5.3358400E-09  -2.2242300E+06
1.0820100E-07  -5.5943000E-09  5.3434000E-09  -2.2137400E+06
1.0823700E-07  -5.5936800E-09  5.3477900E-09  -2.2030300E+06
1.0827300E-07  -5.5887400E-09  5.3476700E-09  -2.1923000E+06
1.0830800E-07  -5.5787200E-09  5.3421500E-09  -2.1817300E+06
1.0834400E-07  -5.5634000E-09  5.3309100E-09  -2.1715500E+06
1.0837900E-07  -5.5431700E-09  5.3142800E-09  -2.1619300E+06
1.0841500E-07  -5.5189800E-09  5.2931900E-09  -2.1530400E+06
1.0845000E-07  -5.4923100E-09  5.2691300E-09  -2.1450100E+06
1.0848600E-07  -5.4649900E-09  5.2439400E-09  -2.1379300E+06
1.0852200E-07  -5.4389100E-09  5.2196200E-09  -2.1318400E+06
1.0855700E-07  -5.4159100E-09  5.1980900E-09  -2.1267400E+06
1.0859300E-07  -5.3974900E-09  5.1809600E-09  -2.1226000E+06
1.0862800E-07  -5.3846800E-09  5.1693600E-09  -2.1193200E+06
1.0866400E-07  -5.3778600E-09  5.1637900E-09  -2.1167800E+06
1.0870000E-07  -5.3767800E-09  5.1640800E-09  -2.1148500E+06
1.0873500E-07  -5.3805800E-09  5.1694300E-09  -2.1133500E+06
1.0877100E-07  -5.3879000E-09  5.1785100E-09  -2.1121200E+06
1.0880600E-07  -5.3970400E-09  5.1896300E-09  -2.1110100E+06
1.0884200E-07  -5.4062300E-09  5.2009800E-09  -2.1098600E+06
```

1.0887700E-07	-5.4137800E-09	5.2108000E-09	-2.1085600E+06
1.0891300E-07	-5.4182900E-09	5.2176200E-09	-2.1070100E+06
1.0894900E-07	-5.4188200E-09	5.2203800E-09	-2.1051600E+06
1.0898400E-07	-5.4149400E-09	5.2185700E-09	-2.1029900E+06
1.0902000E-07	-5.4067600E-09	5.2122000E-09	-2.1005400E+06
1.0905500E-07	-5.3948500E-09	5.2017800E-09	-2.0978700E+06
1.0909100E-07	-5.3801400E-09	5.1882100E-09	-2.0950700E+06
1.0912700E-07	-5.3637600E-09	5.1726400E-09	-2.0922600E+06
1.0916200E-07	-5.3469300E-09	5.1563000E-09	-2.0895700E+06
1.0919800E-07	-5.3307100E-09	5.1403500E-09	-2.0871300E+06
1.0923300E-07	-5.3159900E-09	5.1258000E-09	-2.0850800E+06
1.0926900E-07	-5.3033600E-09	5.1133600E-09	-2.0835100E+06
1.0930400E-07	-5.2931500E-09	5.1034500E-09	-2.0825100E+06
1.0934000E-07	-5.2853400E-09	5.0962200E-09	-2.0821000E+06
1.0937600E-07	-5.2797800E-09	5.0915900E-09	-2.0823000E+06
1.0941100E-07	-5.2761900E-09	5.0892800E-09	-2.0830700E+06
1.0944700E-07	-5.2742000E-09	5.0889800E-09	-2.0843200E+06
1.0948200E-07	-5.2734900E-09	5.0903200E-09	-2.0859600E+06
1.0951800E-07	-5.2738200E-09	5.0929400E-09	-2.0878600E+06
1.0955400E-07	-5.2749800E-09	5.0965500E-09	-2.0898900E+06
1.0958900E-07	-5.2768200E-09	5.1008500E-09	-2.0919200E+06
1.0962500E-07	-5.2791600E-09	5.1055100E-09	-2.0938500E+06
1.0966000E-07	-5.2817700E-09	5.1101800E-09	-2.0955800E+06
1.0969600E-07	-5.2842700E-09	5.1143900E-09	-2.0970700E+06
1.0973100E-07	-5.2861700E-09	5.1175900E-09	-2.0982900E+06
1.0976700E-07	-5.2868400E-09	5.1191300E-09	-2.0992800E+06
1.0980300E-07	-5.2855900E-09	5.1183500E-09	-2.1000900E+06
1.0983800E-07	-5.2817300E-09	5.1146500E-09	-2.1007900E+06
1.0987400E-07	-5.2747200E-09	5.1075700E-09	-2.1015000E+06
1.0990900E-07	-5.2642700E-09	5.0969400E-09	-2.1023000E+06
1.0994500E-07	-5.2504000E-09	5.0829300E-09	-2.1032900E+06
1.0998100E-07	-5.2335700E-09	5.0661000E-09	-2.1045600E+06
1.1001600E-07	-5.2146900E-09	5.0474200E-09	-2.1061500E+06
1.1005200E-07	-5.1949800E-09	5.0281900E-09	-2.1081100E+06
1.1008700E-07	-5.1759500E-09	5.0099400E-09	-2.1104100E+06
1.1012300E-07	-5.1592000E-09	4.9942600E-09	-2.1130200E+06
1.1015800E-07	-5.1462000E-09	4.9826100E-09	-2.1158800E+06
1.1019400E-07	-5.1381500E-09	4.9761200E-09	-2.1188900E+06
1.1023000E-07	-5.1357600E-09	4.9754700E-09	-2.1219500E+06
1.1026500E-07	-5.1391200E-09	4.9807400E-09	-2.1249600E+06
1.1030100E-07	-5.1477000E-09	4.9913600E-09	-2.1277900E+06
1.1033600E-07	-5.1603300E-09	5.0061700E-09	-2.1303500E+06
1.1037200E-07	-5.1753500E-09	5.0235100E-09	-2.1325400E+06
1.1040800E-07	-5.1907600E-09	5.0414200E-09	-2.1343200E+06
1.1044300E-07	-5.2044400E-09	5.0578000E-09	-2.1356200E+06
1.1047900E-07	-5.2144300E-09	5.0707300E-09	-2.1364400E+06
1.1051400E-07	-5.2191700E-09	5.0785900E-09	-2.1367600E+06
1.1055000E-07	-5.2176200E-09	5.0803300E-09	-2.1366000E+06
1.1058500E-07	-5.2094300E-09	5.0755300E-09	-2.1360100E+06
1.1062100E-07	-5.1949500E-09	5.0644200E-09	-2.1350200E+06
1.1065700E-07	-5.1752100E-09	5.0478900E-09	-2.1336900E+06

1.1069200E-07	-5.1517700E-09	5.0273600E-09	-2.1320700E+06
1.1072800E-07	-5.1265000E-09	5.0045700E-09	-2.1301900E+06
1.1076300E-07	-5.1014300E-09	4.9814400E-09	-2.1281000E+06
1.1079900E-07	-5.0785000E-09	4.9597900E-09	-2.1258300E+06
1.1083500E-07	-5.0593300E-09	4.9412400E-09	-2.1233900E+06
1.1087000E-07	-5.0450700E-09	4.9269700E-09	-2.1207800E+06
1.1090600E-07	-5.0363400E-09	4.9177000E-09	-2.1179900E+06
1.1094100E-07	-5.0332000E-09	4.9136100E-09	-2.1150000E+06
1.1097700E-07	-5.0351400E-09	4.9143800E-09	-2.1118000E+06
1.1101200E-07	-5.0412300E-09	4.9193000E-09	-2.1083600E+06
1.1104800E-07	-5.0502100E-09	4.9273000E-09	-2.1046500E+06
1.1108400E-07	-5.0606300E-09	4.9371500E-09	-2.1006500E+06
1.1111900E-07	-5.0710400E-09	4.9475500E-09	-2.0963600E+06
1.1115500E-07	-5.0801100E-09	4.9572700E-09	-2.0917800E+06
1.1119000E-07	-5.0866700E-09	4.9651900E-09	-2.0869000E+06
1.1122600E-07	-5.0898200E-09	4.9704500E-09	-2.0817500E+06
1.1126200E-07	-5.0889800E-09	4.9723900E-09	-2.0763400E+06
1.1129700E-07	-5.0838600E-09	4.9706400E-09	-2.0707100E+06
1.1133300E-07	-5.0744300E-09	4.9650700E-09	-2.0648900E+06
1.1136800E-07	-5.0609300E-09	4.9557800E-09	-2.0589000E+06
1.1140400E-07	-5.0438400E-09	4.9431100E-09	-2.0527800E+06
1.1144000E-07	-5.0238000E-09	4.9275700E-09	-2.0465500E+06
1.1147500E-07	-5.0016200E-09	4.9098400E-09	-2.0402300E+06
1.1151100E-07	-4.9782600E-09	4.8907500E-09	-2.0338300E+06
1.1154600E-07	-4.9547500E-09	4.8712100E-09	-2.0273500E+06
1.1158200E-07	-4.9322300E-09	4.8522200E-09	-2.0208000E+06
1.1161700E-07	-4.9117800E-09	4.8347600E-09	-2.0141800E+06
1.1165300E-07	-4.8944500E-09	4.8197500E-09	-2.0074800E+06
1.1168900E-07	-4.8811300E-09	4.8079600E-09	-2.0007100E+06
1.1172400E-07	-4.8724900E-09	4.7999400E-09	-1.9938900E+06
1.1176000E-07	-4.8688500E-09	4.7959600E-09	-1.9870200E+06
1.1179500E-07	-4.8701600E-09	4.7959300E-09	-1.9801300E+06
1.1183100E-07	-4.8759300E-09	4.7994100E-09	-1.9732800E+06
1.1186700E-07	-4.8852900E-09	4.8056200E-09	-1.9664900E+06
1.1190200E-07	-4.8969900E-09	4.8135000E-09	-1.9598300E+06
1.1193800E-07	-4.9095500E-09	4.8218200E-09	-1.9533800E+06
1.1197300E-07	-4.9214100E-09	4.8293100E-09	-1.9471800E+06
1.1200900E-07	-4.9310300E-09	4.8348000E-09	-1.9413200E+06
1.1204400E-07	-4.9371400E-09	4.8373900E-09	-1.9358400E+06
1.1208000E-07	-4.9389200E-09	4.8365300E-09	-1.9308100E+06
1.1211600E-07	-4.9359900E-09	4.8321600E-09	-1.9262400E+06
1.1215100E-07	-4.9286100E-09	4.8246500E-09	-1.9221600E+06
1.1218700E-07	-4.9175400E-09	4.8148500E-09	-1.9185600E+06
1.1222200E-07	-4.9040200E-09	4.8039000E-09	-1.9154100E+06
1.1225800E-07	-4.8896100E-09	4.7931600E-09	-1.9126900E+06
1.1229400E-07	-4.8759500E-09	4.7839400E-09	-1.9103300E+06
1.1232900E-07	-4.8645700E-09	4.7773800E-09	-1.9082500E+06
1.1236500E-07	-4.8566800E-09	4.7742400E-09	-1.9063900E+06
1.1240000E-07	-4.8529700E-09	4.7747900E-09	-1.9046700E+06
1.1243600E-07	-4.8535700E-09	4.7787400E-09	-1.9030200E+06
1.1247100E-07	-4.8579500E-09	4.7852700E-09	-1.9013800E+06

1.1250700E-07	-4.8650600E-09	4.7931500E-09	-1.8997000E+06
1.1254300E-07	-4.8734300E-09	4.8008300E-09	-1.8979600E+06
1.1257800E-07	-4.8813800E-09	4.8067300E-09	-1.8961500E+06
1.1261400E-07	-4.8872400E-09	4.8093700E-09	-1.8943100E+06
1.1264900E-07	-4.8896300E-09	4.8076600E-09	-1.8924700E+06
1.1268500E-07	-4.8876000E-09	4.8009800E-09	-1.8907000E+06
1.1272100E-07	-4.8807600E-09	4.7893100E-09	-1.8890800E+06
1.1275600E-07	-4.8693800E-09	4.7732500E-09	-1.8877000E+06
1.1279200E-07	-4.8543200E-09	4.7539300E-09	-1.8866700E+06
1.1282700E-07	-4.8369000E-09	4.7328500E-09	-1.8860600E+06
1.1286300E-07	-4.8187600E-09	4.7117400E-09	-1.8859500E+06
1.1289800E-07	-4.8016100E-09	4.6923200E-09	-1.8864100E+06
1.1293400E-07	-4.7870600E-09	4.6760800E-09	-1.8874600E+06
1.1297000E-07	-4.7763600E-09	4.6641400E-09	-1.8891000E+06
1.1300500E-07	-4.7703000E-09	4.6571000E-09	-1.8912900E+06
1.1304100E-07	-4.7691300E-09	4.6550400E-09	-1.8939700E+06
1.1307600E-07	-4.7725500E-09	4.6575100E-09	-1.8970400E+06
1.1311200E-07	-4.7797700E-09	4.6636200E-09	-1.9003600E+06
1.1314800E-07	-4.7896900E-09	4.6722000E-09	-1.9038100E+06
1.1318300E-07	-4.8009600E-09	4.6819600E-09	-1.9072400E+06
1.1321900E-07	-4.8122100E-09	4.6916000E-09	-1.9104900E+06
1.1325400E-07	-4.8222100E-09	4.7000100E-09	-1.9134400E+06
1.1329000E-07	-4.8299600E-09	4.7063300E-09	-1.9159900E+06
1.1332500E-07	-4.8347800E-09	4.7100000E-09	-1.9180400E+06
1.1336100E-07	-4.8363000E-09	4.7107700E-09	-1.9195600E+06
1.1339700E-07	-4.8345100E-09	4.7086600E-09	-1.9205200E+06
1.1343200E-07	-4.8296400E-09	4.7039000E-09	-1.9209500E+06
1.1346800E-07	-4.8221200E-09	4.6968300E-09	-1.9209100E+06
1.1350300E-07	-4.8124600E-09	4.6878700E-09	-1.9204800E+06
1.1353900E-07	-4.8012400E-09	4.6774000E-09	-1.9197500E+06
1.1357500E-07	-4.7890500E-09	4.6658100E-09	-1.9188500E+06
1.1361000E-07	-4.7763900E-09	4.6534500E-09	-1.9178900E+06
1.1364600E-07	-4.7637600E-09	4.6406400E-09	-1.9169900E+06
1.1368100E-07	-4.7516500E-09	4.6277400E-09	-1.9162500E+06
1.1371700E-07	-4.7405100E-09	4.6151500E-09	-1.9157500E+06
1.1375200E-07	-4.7307800E-09	4.6033600E-09	-1.9155500E+06
1.1378800E-07	-4.7229500E-09	4.5929200E-09	-1.9156900E+06
1.1382400E-07	-4.7174700E-09	4.5844000E-09	-1.9161400E+06
1.1385900E-07	-4.7147300E-09	4.5783900E-09	-1.9169000E+06
1.1389500E-07	-4.7149900E-09	4.5753300E-09	-1.9179000E+06
1.1393000E-07	-4.7183600E-09	4.5755300E-09	-1.9190500E+06
1.1396600E-07	-4.7246900E-09	4.5790100E-09	-1.9202600E+06
1.1400200E-07	-4.7335700E-09	4.5854700E-09	-1.9214200E+06
1.1403700E-07	-4.7442900E-09	4.5942700E-09	-1.9224200E+06
1.1407300E-07	-4.7558800E-09	4.6044600E-09	-1.9231600E+06
1.1410800E-07	-4.7671800E-09	4.6148100E-09	-1.9235500E+06
1.1414400E-07	-4.7769200E-09	4.6239500E-09	-1.9235200E+06
1.1417900E-07	-4.7838400E-09	4.6305000E-09	-1.9230300E+06
1.1421500E-07	-4.7868400E-09	4.6332000E-09	-1.9220700E+06
1.1425100E-07	-4.7851400E-09	4.6311000E-09	-1.9206300E+06
1.1428600E-07	-4.7783200E-09	4.6236700E-09	-1.9187600E+06

1.1432200E-07	-4.7664400E-09	4.6109000E-09	-1.9165100E+06
1.1435700E-07	-4.7500500E-09	4.5932800E-09	-1.9139400E+06
1.1439300E-07	-4.7301600E-09	4.5718500E-09	-1.9111200E+06
1.1442900E-07	-4.7081500E-09	4.5480500E-09	-1.9081200E+06
1.1446400E-07	-4.6856600E-09	4.5236000E-09	-1.9050000E+06
1.1450000E-07	-4.6643900E-09	4.5003300E-09	-1.9018000E+06
1.1453500E-07	-4.6459200E-09	4.4799800E-09	-1.8985600E+06
1.1457100E-07	-4.6315800E-09	4.4639800E-09	-1.8952700E+06
1.1460600E-07	-4.6223000E-09	4.4533200E-09	-1.8919400E+06
1.1464200E-07	-4.6184600E-09	4.4484300E-09	-1.8885300E+06
1.1467800E-07	-4.6198400E-09	4.4491300E-09	-1.8850200E+06
1.1471300E-07	-4.6257000E-09	4.4546000E-09	-1.8813600E+06
1.1474900E-07	-4.6348300E-09	4.4635200E-09	-1.8775000E+06
1.1478400E-07	-4.6456100E-09	4.4741800E-09	-1.8734100E+06
1.1482000E-07	-4.6562900E-09	4.4846900E-09	-1.8690600E+06
1.1485600E-07	-4.6651200E-09	4.4931300E-09	-1.8644400E+06
1.1489100E-07	-4.6705200E-09	4.4978400E-09	-1.8595500E+06
1.1492700E-07	-4.6712900E-09	4.4975500E-09	-1.8544000E+06
1.1496200E-07	-4.6667300E-09	4.4915000E-09	-1.8490300E+06
1.1499800E-07	-4.6566900E-09	4.4795900E-09	-1.8435000E+06
1.1503300E-07	-4.6416400E-09	4.4623100E-09	-1.8378400E+06
1.1506900E-07	-4.6225300E-09	4.4407500E-09	-1.8321200E+06
1.1510500E-07	-4.6007800E-09	4.4164100E-09	-1.8264000E+06
1.1514000E-07	-4.5780400E-09	4.3910900E-09	-1.8207200E+06
1.1517600E-07	-4.5560900E-09	4.3666800E-09	-1.8151100E+06
1.1521100E-07	-4.5366100E-09	4.3449400E-09	-1.8096000E+06
1.1524700E-07	-4.5210100E-09	4.3273400E-09	-1.8042000E+06
1.1528300E-07	-4.5102800E-09	4.3148800E-09	-1.7989000E+06
1.1531800E-07	-4.5049200E-09	4.3080500E-09	-1.7936900E+06
1.1535400E-07	-4.5049200E-09	4.3067400E-09	-1.7885600E+06
1.1538900E-07	-4.5096800E-09	4.3103200E-09	-1.7834600E+06
1.1542500E-07	-4.5182100E-09	4.3176800E-09	-1.7783900E+06
1.1546000E-07	-4.5291500E-09	4.3273900E-09	-1.7733300E+06
1.1549600E-07	-4.5409700E-09	4.3378400E-09	-1.7682500E+06
1.1553200E-07	-4.5521000E-09	4.3474500E-09	-1.7631800E+06
1.1556700E-07	-4.5611600E-09	4.3548000E-09	-1.7581200E+06
1.1560300E-07	-4.5670100E-09	4.3587800E-09	-1.7531000E+06
1.1563800E-07	-4.5689200E-09	4.3587200E-09	-1.7481500E+06
1.1567400E-07	-4.5666100E-09	4.3543600E-09	-1.7433300E+06
1.1571000E-07	-4.5602400E-09	4.3459700E-09	-1.7387000E+06
1.1574500E-07	-4.5504000E-09	4.3341700E-09	-1.7343000E+06
1.1578100E-07	-4.5380300E-09	4.3199400E-09	-1.7302000E+06
1.1581600E-07	-4.5242600E-09	4.3044300E-09	-1.7264600E+06
1.1585200E-07	-4.5103000E-09	4.2888500E-09	-1.7231200E+06
1.1588800E-07	-4.4973200E-09	4.2743500E-09	-1.7202200E+06
1.1592300E-07	-4.4863000E-09	4.2618500E-09	-1.7177700E+06
1.1595900E-07	-4.4779400E-09	4.2519800E-09	-1.7157700E+06
1.1599400E-07	-4.4725700E-09	4.2450700E-09	-1.7142000E+06
1.1603000E-07	-4.4702200E-09	4.2410600E-09	-1.7130400E+06
1.1606500E-07	-4.4705500E-09	4.2395900E-09	-1.7122100E+06
1.1610100E-07	-4.4729800E-09	4.2400900E-09	-1.7116500E+06

1.1613700E-07	-4.4767600E-09	4.2418000E-09	-1.7112900E+06
1.1617200E-07	-4.4810900E-09	4.2439500E-09	-1.7110400E+06
1.1620800E-07	-4.4852000E-09	4.2458300E-09	-1.7108400E+06
1.1624300E-07	-4.4885000E-09	4.2468700E-09	-1.7106100E+06
1.1627900E-07	-4.4905800E-09	4.2467000E-09	-1.7103200E+06
1.1631500E-07	-4.4912500E-09	4.2451700E-09	-1.7099400E+06
1.1635000E-07	-4.4905800E-09	4.2423500E-09	-1.7094700E+06
1.1638600E-07	-4.4887300E-09	4.2384500E-09	-1.7089400E+06
1.1642100E-07	-4.4860400E-09	4.2337600E-09	-1.7083800E+06
1.1645700E-07	-4.4828100E-09	4.2285900E-09	-1.7078400E+06
1.1649200E-07	-4.4793000E-09	4.2231700E-09	-1.7074000E+06
1.1652800E-07	-4.4756100E-09	4.2175900E-09	-1.7071100E+06
1.1656400E-07	-4.4717000E-09	4.2118000E-09	-1.7070300E+06
1.1659900E-07	-4.4673700E-09	4.2056100E-09	-1.7071800E+06
1.1663500E-07	-4.4623100E-09	4.1986800E-09	-1.7075900E+06
1.1667000E-07	-4.4561700E-09	4.1906700E-09	-1.7082400E+06
1.1670600E-07	-4.4486400E-09	4.1812700E-09	-1.7091200E+06
1.1674200E-07	-4.4395300E-09	4.1703100E-09	-1.7101600E+06
1.1677700E-07	-4.4288900E-09	4.1578200E-09	-1.7112900E+06
1.1681300E-07	-4.4170100E-09	4.1440600E-09	-1.7124400E+06
1.1684800E-07	-4.4044400E-09	4.1295700E-09	-1.7135000E+06
1.1688400E-07	-4.3919200E-09	4.1150600E-09	-1.7144000E+06
1.1691900E-07	-4.3803100E-09	4.1014000E-09	-1.7150500E+06
1.1695500E-07	-4.3705000E-09	4.0894500E-09	-1.7154000E+06
1.1699100E-07	-4.3632000E-09	4.0799700E-09	-1.7154000E+06
1.1702600E-07	-4.3588700E-09	4.0734800E-09	-1.7150400E+06
1.1706200E-07	-4.3576600E-09	4.0702100E-09	-1.7143100E+06
1.1709700E-07	-4.3593000E-09	4.0700000E-09	-1.7132500E+06
1.1713300E-07	-4.3631300E-09	4.0723100E-09	-1.7119200E+06
1.1716900E-07	-4.3681700E-09	4.0763100E-09	-1.7103700E+06
1.1720400E-07	-4.3732700E-09	4.0809500E-09	-1.7086900E+06
1.1724000E-07	-4.3772300E-09	4.0851000E-09	-1.7069600E+06
1.1727500E-07	-4.3789800E-09	4.0877100E-09	-1.7052600E+06
1.1731100E-07	-4.3777600E-09	4.0879700E-09	-1.7036300E+06
1.1734600E-07	-4.3732100E-09	4.0853700E-09	-1.7021400E+06
1.1738200E-07	-4.3654500E-09	4.0798400E-09	-1.7007900E+06
1.1741800E-07	-4.3550600E-09	4.0716700E-09	-1.6995800E+06
1.1745300E-07	-4.3429900E-09	4.0615400E-09	-1.6984700E+06
1.1748900E-07	-4.3304800E-09	4.0503500E-09	-1.6974000E+06
1.1752400E-07	-4.3188300E-09	4.0391000E-09	-1.6963000E+06
1.1756000E-07	-4.3092200E-09	4.0287500E-09	-1.6950500E+06
1.1759600E-07	-4.3025400E-09	4.0200500E-09	-1.6935600E+06
1.1763100E-07	-4.2992700E-09	4.0134300E-09	-1.6917300E+06
1.1766700E-07	-4.2993800E-09	4.0089500E-09	-1.6894800E+06
1.1770200E-07	-4.3022900E-09	4.0062700E-09	-1.6867300E+06
1.1773800E-07	-4.3070700E-09	4.0047300E-09	-1.6834700E+06
1.1777300E-07	-4.3124300E-09	4.0034600E-09	-1.6796700E+06
1.1780900E-07	-4.3170000E-09	4.0015100E-09	-1.6754000E+06
1.1784500E-07	-4.3194400E-09	3.9979900E-09	-1.6707100E+06
1.1788000E-07	-4.3187300E-09	3.9922400E-09	-1.6657000E+06
1.1791600E-07	-4.3142000E-09	3.9839300E-09	-1.6605200E+06

1.1795100E-07	-4.3057000E-09	3.9731100E-09	-1.6552900E+06
1.1798700E-07	-4.2935600E-09	3.9602100E-09	-1.6501700E+06
1.1802300E-07	-4.2786100E-09	3.9459700E-09	-1.6452800E+06
1.1805800E-07	-4.2619500E-09	3.9313700E-09	-1.6407300E+06
1.1809400E-07	-4.2449100E-09	3.9174600E-09	-1.6366000E+06
1.1812900E-07	-4.2288200E-09	3.9052200E-09	-1.6329400E+06
1.1816500E-07	-4.2148300E-09	3.8954200E-09	-1.6297200E+06
1.1820000E-07	-4.2037900E-09	3.8885300E-09	-1.6269100E+06
1.1823600E-07	-4.1961800E-09	3.8846500E-09	-1.6244000E+06
1.1827200E-07	-4.1920300E-09	3.8835100E-09	-1.6220600E+06
1.1830700E-07	-4.1909900E-09	3.8845000E-09	-1.6197600E+06
1.1834300E-07	-4.1923500E-09	3.8867500E-09	-1.6173300E+06
1.1837800E-07	-4.1952100E-09	3.8892800E-09	-1.6146200E+06
1.1841400E-07	-4.1985500E-09	3.8910500E-09	-1.6115000E+06
1.1845000E-07	-4.2013600E-09	3.8911300E-09	-1.6078600E+06
1.1848500E-07	-4.2027600E-09	3.8887800E-09	-1.6036600E+06
1.1852100E-07	-4.2020700E-09	3.8834700E-09	-1.5988600E+06
1.1855600E-07	-4.1988200E-09	3.8749600E-09	-1.5935100E+06
1.1859200E-07	-4.1927800E-09	3.8632400E-09	-1.5876800E+06
1.1862700E-07	-4.1838900E-09	3.8485300E-09	-1.5814600E+06
1.1866300E-07	-4.1723300E-09	3.8312100E-09	-1.5750100E+06
1.1869900E-07	-4.1583300E-09	3.8117900E-09	-1.5684500E+06
1.1873400E-07	-4.1422500E-09	3.7908500E-09	-1.5619500E+06
1.1877000E-07	-4.1245500E-09	3.7690000E-09	-1.5556300E+06
1.1880500E-07	-4.1057000E-09	3.7468900E-09	-1.5496100E+06
1.1884100E-07	-4.0862800E-09	3.7251900E-09	-1.5439900E+06
1.1887700E-07	-4.0669400E-09	3.7045700E-09	-1.5388100E+06
1.1891200E-07	-4.0484000E-09	3.6857500E-09	-1.5340900E+06
1.1894800E-07	-4.0314800E-09	3.6694100E-09	-1.5298200E+06
1.1898300E-07	-4.0169800E-09	3.6562300E-09	-1.5259500E+06
1.1901900E-07	-4.0057100E-09	3.6467600E-09	-1.5224200E+06
1.1905400E-07	-3.9983300E-09	3.6414400E-09	-1.5191400E+06
1.1909000E-07	-3.9953100E-09	3.6404400E-09	-1.5160300E+06
1.1912600E-07	-3.9967800E-09	3.6436400E-09	-1.5130100E+06
1.1916100E-07	-4.0025000E-09	3.6505600E-09	-1.5099900E+06
1.1919700E-07	-4.0118100E-09	3.6603400E-09	-1.5069200E+06
1.1923200E-07	-4.0236000E-09	3.6717800E-09	-1.5037700E+06
1.1926800E-07	-4.0364700E-09	3.6834100E-09	-1.5005200E+06
1.1930400E-07	-4.0487300E-09	3.6935900E-09	-1.4971900E+06
1.1933900E-07	-4.0586700E-09	3.7007300E-09	-1.4938300E+06
1.1937500E-07	-4.0647100E-09	3.7034100E-09	-1.4904800E+06
1.1941000E-07	-4.0656000E-09	3.7005700E-09	-1.4872300E+06
1.1944600E-07	-4.0605800E-09	3.6916900E-09	-1.4841500E+06
1.1948100E-07	-4.0495500E-09	3.6768600E-09	-1.4813100E+06
1.1951700E-07	-4.0330700E-09	3.6568100E-09	-1.4787600E+06
1.1955300E-07	-4.0123500E-09	3.6328800E-09	-1.4765700E+06
1.1958800E-07	-3.9891600E-09	3.6068800E-09	-1.4747300E+06
1.1962400E-07	-3.9655800E-09	3.5809300E-09	-1.4732400E+06
1.1965900E-07	-3.9438000E-09	3.5571800E-09	-1.4720600E+06
1.1969500E-07	-3.9258800E-09	3.5376200E-09	-1.4711500E+06
1.1973100E-07	-3.9134400E-09	3.5237700E-09	-1.4704300E+06

1.1976600E-07	-3.9074900E-09	3.5165700E-09	-1.4698000E+06
1.1980200E-07	-3.9082900E-09	3.5161900E-09	-1.4692000E+06
1.1983700E-07	-3.9153000E-09	3.5220500E-09	-1.4685300E+06
1.1987300E-07	-3.9272400E-09	3.5328400E-09	-1.4677600E+06
1.1990800E-07	-3.9422400E-09	3.5467100E-09	-1.4668200E+06
1.1994400E-07	-3.9580700E-09	3.5614400E-09	-1.4657200E+06
1.1998000E-07	-3.9724300E-09	3.5747500E-09	-1.4644800E+06
1.2001500E-07	-3.9831800E-09	3.5845300E-09	-1.4631200E+06
1.2005100E-07	-3.9886200E-09	3.5891200E-09	-1.4617200E+06
1.2008600E-07	-3.9877000E-09	3.5874400E-09	-1.4603600E+06
1.2012200E-07	-3.9801400E-09	3.5791500E-09	-1.4591000E+06
1.2015800E-07	-3.9663800E-09	3.5646500E-09	-1.4580200E+06
1.2019300E-07	-3.9475600E-09	3.5450100E-09	-1.4571900E+06
1.2022900E-07	-3.9253300E-09	3.5217900E-09	-1.4566500E+06
1.2026400E-07	-3.9016100E-09	3.4968800E-09	-1.4564100E+06
1.2030000E-07	-3.8784000E-09	3.4722100E-09	-1.4564500E+06
1.2033500E-07	-3.8574400E-09	3.4495700E-09	-1.4567300E+06
1.2037100E-07	-3.8401000E-09	3.4303600E-09	-1.4571800E+06
1.2040700E-07	-3.8271900E-09	3.4155000E-09	-1.4577200E+06
1.2044200E-07	-3.8189000E-09	3.4053100E-09	-1.4582300E+06
1.2047800E-07	-3.8148300E-09	3.3995500E-09	-1.4586000E+06
1.2051300E-07	-3.8141200E-09	3.3975000E-09	-1.4587400E+06
1.2054900E-07	-3.8155800E-09	3.3980800E-09	-1.4585400E+06
1.2058500E-07	-3.8178400E-09	3.4000200E-09	-1.4579500E+06
1.2062000E-07	-3.8196000E-09	3.4020300E-09	-1.4569100E+06
1.2065600E-07	-3.8197700E-09	3.4029800E-09	-1.4554200E+06
1.2069100E-07	-3.8175900E-09	3.4019800E-09	-1.4534800E+06
1.2072700E-07	-3.8126700E-09	3.3985000E-09	-1.4511600E+06
1.2076300E-07	-3.8050600E-09	3.3923500E-09	-1.4485100E+06
1.2079800E-07	-3.7951100E-09	3.3836800E-09	-1.4456200E+06
1.2083400E-07	-3.7834300E-09	3.3728800E-09	-1.4425900E+06
1.2086900E-07	-3.7707800E-09	3.3605300E-09	-1.4395100E+06
1.2090500E-07	-3.7579300E-09	3.3472800E-09	-1.4364800E+06
1.2094000E-07	-3.7455300E-09	3.3337600E-09	-1.4335600E+06
1.2097600E-07	-3.7341100E-09	3.3205400E-09	-1.4308300E+06
1.2101200E-07	-3.7239900E-09	3.3080400E-09	-1.4283000E+06
1.2104700E-07	-3.7153200E-09	3.2966100E-09	-1.4259800E+06
1.2108300E-07	-3.7080700E-09	3.2864400E-09	-1.4238400E+06
1.2111800E-07	-3.7021200E-09	3.2776700E-09	-1.4218500E+06
1.2115400E-07	-3.6973200E-09	3.2703700E-09	-1.4199300E+06
1.2119000E-07	-3.6935100E-09	3.2645900E-09	-1.4179900E+06
1.2122500E-07	-3.6905700E-09	3.2603500E-09	-1.4159500E+06
1.2126100E-07	-3.6884000E-09	3.2576600E-09	-1.4137000E+06
1.2129600E-07	-3.6869600E-09	3.2564600E-09	-1.4111700E+06
1.2133200E-07	-3.6861800E-09	3.2566200E-09	-1.4082800E+06
1.2136700E-07	-3.6859200E-09	3.2578900E-09	-1.4049700E+06
1.2140300E-07	-3.6859800E-09	3.2598900E-09	-1.4012300E+06
1.2143900E-07	-3.6860200E-09	3.2620900E-09	-1.3970600E+06
1.2147400E-07	-3.6856000E-09	3.2638500E-09	-1.3924900E+06
1.2151000E-07	-3.6842000E-09	3.2644700E-09	-1.3875900E+06
1.2154500E-07	-3.6812600E-09	3.2632500E-09	-1.3824500E+06

1.2158100E-07	-3.6762900E-09	3.2595700E-09	-1.3771600E+06
1.2161700E-07	-3.6689200E-09	3.2530000E-09	-1.3718400E+06
1.2165200E-07	-3.6589600E-09	3.2433500E-09	-1.3666000E+06
1.2168800E-07	-3.6465300E-09	3.2307500E-09	-1.3615300E+06
1.2172300E-07	-3.6320000E-09	3.2156300E-09	-1.3567200E+06
1.2175900E-07	-3.6160200E-09	3.1987200E-09	-1.3522100E+06
1.2179400E-07	-3.5994300E-09	3.1809500E-09	-1.3480400E+06
1.2183000E-07	-3.5832300E-09	3.1634200E-09	-1.3442000E+06
1.2186600E-07	-3.5683800E-09	3.1471700E-09	-1.3406400E+06
1.2190100E-07	-3.5557400E-09	3.1331500E-09	-1.3373000E+06
1.2193700E-07	-3.5459500E-09	3.1220300E-09	-1.3341100E+06
1.2197200E-07	-3.5393000E-09	3.1141600E-09	-1.3309700E+06
1.2200800E-07	-3.5357000E-09	3.1095100E-09	-1.3277800E+06
1.2204400E-07	-3.5347200E-09	3.1076300E-09	-1.3244500E+06
1.2207900E-07	-3.5355700E-09	3.1077400E-09	-1.3209100E+06
1.2211500E-07	-3.5372200E-09	3.1088100E-09	-1.3171000E+06
1.2215000E-07	-3.5384900E-09	3.1097000E-09	-1.3130100E+06
1.2218600E-07	-3.5382900E-09	3.1092500E-09	-1.3086100E+06
1.2222100E-07	-3.5356100E-09	3.1064800E-09	-1.3039400E+06
1.2225700E-07	-3.5297600E-09	3.1006900E-09	-1.2990500E+06
1.2229300E-07	-3.5204200E-09	3.0915400E-09	-1.2940000E+06
1.2232800E-07	-3.5076700E-09	3.0791100E-09	-1.2888700E+06
1.2236400E-07	-3.4920200E-09	3.0638800E-09	-1.2837600E+06
1.2239900E-07	-3.4743200E-09	3.0466900E-09	-1.2787500E+06
1.2243500E-07	-3.4556700E-09	3.0286100E-09	-1.2739300E+06
1.2247100E-07	-3.4373400E-09	3.0108800E-09	-1.2693700E+06
1.2250600E-07	-3.4205400E-09	2.9946800E-09	-1.2651100E+06
1.2254200E-07	-3.4063700E-09	2.9810700E-09	-1.2612000E+06
1.2257700E-07	-3.3956500E-09	2.9708200E-09	-1.2576300E+06
1.2261300E-07	-3.3888400E-09	2.9643600E-09	-1.2543900E+06
1.2264800E-07	-3.3860100E-09	2.9616900E-09	-1.2514200E+06
1.2268400E-07	-3.3868000E-09	2.9624400E-09	-1.2486700E+06
1.2272000E-07	-3.3905200E-09	2.9658800E-09	-1.2460600E+06
1.2275500E-07	-3.3962000E-09	2.9710100E-09	-1.2435000E+06
1.2279100E-07	-3.4026900E-09	2.9767000E-09	-1.2409100E+06
1.2282600E-07	-3.4087900E-09	2.9817600E-09	-1.2382100E+06
1.2286200E-07	-3.4134100E-09	2.9851400E-09	-1.2353300E+06
1.2289800E-07	-3.4156200E-09	2.9859800E-09	-1.2322500E+06
1.2293300E-07	-3.4147600E-09	2.9837100E-09	-1.2289400E+06
1.2296900E-07	-3.4105200E-09	2.9781100E-09	-1.2254300E+06
1.2300400E-07	-3.4028900E-09	2.9693100E-09	-1.2217400E+06
1.2304000E-07	-3.3922400E-09	2.9577600E-09	-1.2179600E+06
1.2307500E-07	-3.3791700E-09	2.9441600E-09	-1.2141700E+06
1.2311100E-07	-3.3644800E-09	2.9294000E-09	-1.2104800E+06
1.2314700E-07	-3.3490900E-09	2.9144400E-09	-1.2070000E+06
1.2318200E-07	-3.3339400E-09	2.9002200E-09	-1.2038400E+06
1.2321800E-07	-3.3199100E-09	2.8875600E-09	-1.2011000E+06
1.2325300E-07	-3.3077400E-09	2.8771100E-09	-1.1988500E+06
1.2328900E-07	-3.2979800E-09	2.8692800E-09	-1.1971500E+06
1.2332500E-07	-3.2909600E-09	2.8642400E-09	-1.1960100E+06
1.2336000E-07	-3.2868300E-09	2.8619000E-09	-1.1954300E+06

1.2339600E-07	-3.2854700E-09	2.8619700E-09	-1.1953500E+06
1.2343100E-07	-3.2865900E-09	2.8639800E-09	-1.1956900E+06
1.2346700E-07	-3.2897700E-09	2.8673300E-09	-1.1963300E+06
1.2350200E-07	-3.2944300E-09	2.8713500E-09	-1.1971500E+06
1.2353800E-07	-3.2999400E-09	2.8753800E-09	-1.1980100E+06
1.2357400E-07	-3.3056300E-09	2.8787700E-09	-1.1987700E+06
1.2360900E-07	-3.3108000E-09	2.8809600E-09	-1.1992900E+06
1.2364500E-07	-3.3147700E-09	2.8814600E-09	-1.1994700E+06
1.2368000E-07	-3.3169300E-09	2.8799000E-09	-1.1992200E+06
1.2371600E-07	-3.3167200E-09	2.8760300E-09	-1.1985000E+06
1.2375200E-07	-3.3136700E-09	2.8696900E-09	-1.1973100E+06
1.2378700E-07	-3.3074700E-09	2.8608900E-09	-1.1956700E+06
1.2382300E-07	-3.2979600E-09	2.8497200E-09	-1.1936400E+06
1.2385800E-07	-3.2851300E-09	2.8364100E-09	-1.1913300E+06
1.2389400E-07	-3.2692200E-09	2.8213000E-09	-1.1888400E+06
1.2392900E-07	-3.2506800E-09	2.8048900E-09	-1.1862900E+06
1.2396500E-07	-3.2301500E-09	2.7877300E-09	-1.1838000E+06
1.2400100E-07	-3.2084900E-09	2.7705100E-09	-1.1814800E+06
1.2403600E-07	-3.1867100E-09	2.7539500E-09	-1.1794000E+06
1.2407200E-07	-3.1658800E-09	2.7387900E-09	-1.1776100E+06
1.2410700E-07	-3.1470900E-09	2.7257400E-09	-1.1761500E+06
1.2414300E-07	-3.1313700E-09	2.7154000E-09	-1.1749800E+06
1.2417900E-07	-3.1195100E-09	2.7081800E-09	-1.1740700E+06
1.2421400E-07	-3.1120500E-09	2.7043000E-09	-1.1733300E+06
1.2425000E-07	-3.1091600E-09	2.7036800E-09	-1.1726900E+06
1.2428500E-07	-3.1106300E-09	2.7059500E-09	-1.1720300E+06
1.2432100E-07	-3.1158300E-09	2.7104500E-09	-1.1712500E+06
1.2435600E-07	-3.1237400E-09	2.7162700E-09	-1.1702700E+06
1.2439200E-07	-3.1330700E-09	2.7222900E-09	-1.1690000E+06
1.2442800E-07	-3.1423100E-09	2.7273100E-09	-1.1674000E+06
1.2446300E-07	-3.1499100E-09	2.7301500E-09	-1.1654300E+06
1.2449900E-07	-3.1544600E-09	2.7297700E-09	-1.1631100E+06
1.2453400E-07	-3.1547700E-09	2.7253900E-09	-1.1604500E+06
1.2457000E-07	-3.1500500E-09	2.7165900E-09	-1.1574900E+06
1.2460600E-07	-3.1400100E-09	2.7033900E-09	-1.1543100E+06
1.2464100E-07	-3.1248900E-09	2.6862400E-09	-1.1509700E+06
1.2467700E-07	-3.1054400E-09	2.6660300E-09	-1.1475500E+06
1.2471200E-07	-3.0828900E-09	2.6440100E-09	-1.1440900E+06
1.2474800E-07	-3.0587900E-09	2.6216400E-09	-1.1406600E+06
1.2478300E-07	-3.0348300E-09	2.6004900E-09	-1.1372900E+06
1.2481900E-07	-3.0127600E-09	2.5820700E-09	-1.1339900E+06
1.2485500E-07	-2.9940900E-09	2.5676400E-09	-1.1307700E+06
1.2489000E-07	-2.9800000E-09	2.5581200E-09	-1.1276100E+06
1.2492600E-07	-2.9712000E-09	2.5539500E-09	-1.1244700E+06
1.2496100E-07	-2.9678400E-09	2.5550600E-09	-1.1213100E+06
1.2499700E-07	-2.9695600E-09	2.5608800E-09	-1.1181000E+06
1.2503300E-07	-2.9754900E-09	2.5704100E-09	-1.1148000E+06
1.2506800E-07	-2.9843800E-09	2.5823000E-09	-1.1113800E+06
1.2510400E-07	-2.9947300E-09	2.5950100E-09	-1.1078300E+06
1.2513900E-07	-3.0049800E-09	2.6069600E-09	-1.1041300E+06
1.2517500E-07	-3.0136900E-09	2.6167200E-09	-1.1003000E+06

1.2521100E-07	-3.0196300E-09	2.6230800E-09	-1.0963700E+06
1.2524600E-07	-3.0219000E-09	2.6252100E-09	-1.0923700E+06
1.2528200E-07	-3.0200500E-09	2.6227100E-09	-1.0883600E+06
1.2531700E-07	-3.0140200E-09	2.6155600E-09	-1.0843900E+06
1.2535300E-07	-3.0041500E-09	2.6041900E-09	-1.0805000E+06
1.2538800E-07	-2.9911100E-09	2.5893200E-09	-1.0767600E+06
1.2542400E-07	-2.9757200E-09	2.5719000E-09	-1.0732100E+06
1.2546000E-07	-2.9589500E-09	2.5530100E-09	-1.0698600E+06
1.2549500E-07	-2.9417700E-09	2.5337300E-09	-1.0667400E+06
1.2553100E-07	-2.9250500E-09	2.5150800E-09	-1.0638300E+06
1.2556600E-07	-2.9095100E-09	2.4979400E-09	-1.0611100E+06
1.2560200E-07	-2.8957000E-09	2.4829800E-09	-1.0585300E+06
1.2563800E-07	-2.8839600E-09	2.4706700E-09	-1.0560300E+06
1.2567300E-07	-2.8744600E-09	2.4612500E-09	-1.0535300E+06
1.2570900E-07	-2.8672200E-09	2.4547700E-09	-1.0509600E+06
1.2574400E-07	-2.8620900E-09	2.4510700E-09	-1.0482300E+06
1.2578000E-07	-2.8588400E-09	2.4498500E-09	-1.0452900E+06
1.2581500E-07	-2.8571500E-09	2.4506400E-09	-1.0420700E+06
1.2585100E-07	-2.8566100E-09	2.4528600E-09	-1.0385600E+06
1.2588700E-07	-2.8567100E-09	2.4558200E-09	-1.0347500E+06
1.2592200E-07	-2.8569000E-09	2.4587600E-09	-1.0306800E+06
1.2595800E-07	-2.8565500E-09	2.4608800E-09	-1.0263900E+06
1.2599300E-07	-2.8550100E-09	2.4614000E-09	-1.0219900E+06
1.2602900E-07	-2.8516700E-09	2.4596200E-09	-1.0175700E+06
1.2606500E-07	-2.8459700E-09	2.4549500E-09	-1.0132500E+06
1.2610000E-07	-2.8375100E-09	2.4470300E-09	-1.0091300E+06
1.2613600E-07	-2.8261200E-09	2.4358000E-09	-1.0053200E+06
1.2617100E-07	-2.8119100E-09	2.4214700E-09	-1.0018900E+06
1.2620700E-07	-2.7953000E-09	2.4046300E-09	-9.9891000E+05
1.2624200E-07	-2.7770200E-09	2.3861500E-09	-9.9638100E+05
1.2627800E-07	-2.7580800E-09	2.3671800E-09	-9.9429100E+05
1.2631400E-07	-2.7397100E-09	2.3490100E-09	-9.9258900E+05
1.2634900E-07	-2.7231800E-09	2.3329500E-09	-9.9119400E+05
1.2638500E-07	-2.7096800E-09	2.3202000E-09	-9.9000700E+05
1.2642000E-07	-2.7002300E-09	2.3116700E-09	-9.8891600E+05
1.2645600E-07	-2.6954400E-09	2.3078700E-09	-9.8780800E+05
1.2649200E-07	-2.6954400E-09	2.3088500E-09	-9.8657800E+05
1.2652700E-07	-2.6998800E-09	2.3141200E-09	-9.8514200E+05
1.2656300E-07	-2.7078900E-09	2.3227200E-09	-9.8344000E+05
1.2659800E-07	-2.7181400E-09	2.3332700E-09	-9.8144000E+05
1.2663400E-07	-2.7290400E-09	2.3441500E-09	-9.7914000E+05
1.2666900E-07	-2.7388500E-09	2.3536700E-09	-9.7656800E+05
1.2670500E-07	-2.7459500E-09	2.3602800E-09	-9.7378100E+05
1.2674100E-07	-2.7489900E-09	2.3627100E-09	-9.7085400E+05
1.2677600E-07	-2.7470700E-09	2.3601900E-09	-9.6787600E+05
1.2681200E-07	-2.7398700E-09	2.3525100E-09	-9.6494300E+05
1.2684700E-07	-2.7277000E-09	2.3400300E-09	-9.6214800E+05
1.2688300E-07	-2.7114500E-09	2.3236900E-09	-9.5957400E+05
1.2691900E-07	-2.6924700E-09	2.3048500E-09	-9.5729500E+05
1.2695400E-07	-2.6724400E-09	2.2851600E-09	-9.5535900E+05
1.2699000E-07	-2.6531300E-09	2.2663300E-09	-9.5379200E+05

1.2702500E-07	-2.6362300E-09	2.2499700E-09	-9.5259500E+05
1.2706100E-07	-2.6230800E-09	2.2373700E-09	-9.5174800E+05
1.2709600E-07	-2.6145900E-09	2.2293800E-09	-9.5120500E+05
1.2713200E-07	-2.6110700E-09	2.2263100E-09	-9.5090600E+05
1.2716800E-07	-2.6122800E-09	2.2279400E-09	-9.5077000E+05
1.2720300E-07	-2.6174300E-09	2.2335500E-09	-9.5071300E+05
1.2723900E-07	-2.6253200E-09	2.2420300E-09	-9.5064700E+05
1.2727400E-07	-2.6344600E-09	2.2520100E-09	-9.5048900E+05
1.2731000E-07	-2.6433200E-09	2.2620300E-09	-9.5016400E+05
1.2734600E-07	-2.6504600E-09	2.2707100E-09	-9.4961200E+05
1.2738100E-07	-2.6546700E-09	2.2768600E-09	-9.4879500E+05
1.2741700E-07	-2.6551400E-09	2.2796100E-09	-9.4769100E+05
1.2745200E-07	-2.6514700E-09	2.2784400E-09	-9.4630600E+05
1.2748800E-07	-2.6437000E-09	2.2732000E-09	-9.4466700E+05
1.2752300E-07	-2.6322400E-09	2.2640900E-09	-9.4281800E+05
1.2755900E-07	-2.6178100E-09	2.2515800E-09	-9.4081400E+05
1.2759500E-07	-2.6013400E-09	2.2363700E-09	-9.3872500E+05
1.2763000E-07	-2.5838000E-09	2.2192700E-09	-9.3661800E+05
1.2766600E-07	-2.5661800E-09	2.2011700E-09	-9.3456100E+05
1.2770100E-07	-2.5493400E-09	2.1829300E-09	-9.3260700E+05
1.2773700E-07	-2.5340000E-09	2.1653700E-09	-9.3079400E+05
1.2777300E-07	-2.5206700E-09	2.1492300E-09	-9.2913900E+05
1.2780800E-07	-2.5097000E-09	2.1351300E-09	-9.2763800E+05
1.2784400E-07	-2.5012500E-09	2.1235800E-09	-9.2626700E+05
1.2787900E-07	-2.4953500E-09	2.1149500E-09	-9.2498400E+05
1.2791500E-07	-2.4918600E-09	2.1094300E-09	-9.2373000E+05
1.2795000E-07	-2.4905400E-09	2.1070500E-09	-9.2243900E+05
1.2798600E-07	-2.4910600E-09	2.1076200E-09	-9.2104500E+05
1.2802200E-07	-2.4929600E-09	2.1107200E-09	-9.1948800E+05
1.2805700E-07	-2.4956600E-09	2.1157000E-09	-9.1772200E+05
1.2809300E-07	-2.4985000E-09	2.1216800E-09	-9.1571700E+05
1.2812800E-07	-2.5007300E-09	2.1276300E-09	-9.1346200E+05
1.2816400E-07	-2.5015500E-09	2.1324000E-09	-9.1097000E+05
1.2820000E-07	-2.5001700E-09	2.1348600E-09	-9.0827300E+05
1.2823500E-07	-2.4958900E-09	2.1339800E-09	-9.0542200E+05
1.2827100E-07	-2.4882000E-09	2.1289800E-09	-9.0247800E+05
1.2830600E-07	-2.4768500E-09	2.1194600E-09	-8.9950400E+05
1.2834200E-07	-2.4619500E-09	2.1054400E-09	-8.9656800E+05
1.2837700E-07	-2.4439600E-09	2.0874600E-09	-8.9372800E+05
1.2841300E-07	-2.4237500E-09	2.0665300E-09	-8.9103200E+05
1.2844900E-07	-2.4025400E-09	2.0440800E-09	-8.8850800E+05
1.2848400E-07	-2.3818000E-09	2.0218300E-09	-8.8616700E+05
1.2852000E-07	-2.3631300E-09	2.0016400E-09	-8.8400200E+05
1.2855500E-07	-2.3480600E-09	1.9852800E-09	-8.8198500E+05
1.2859100E-07	-2.3379000E-09	1.9742200E-09	-8.8007700E+05
1.2862700E-07	-2.3335800E-09	1.9694800E-09	-8.7822300E+05
1.2866200E-07	-2.3354900E-09	1.9714800E-09	-8.7636300E+05
1.2869800E-07	-2.3433900E-09	1.9799200E-09	-8.7443500E+05
1.2873300E-07	-2.3564300E-09	1.9938200E-09	-8.7238100E+05
1.2876900E-07	-2.3731600E-09	2.0116000E-09	-8.7014900E+05
1.2880400E-07	-2.3917000E-09	2.0312200E-09	-8.6770400E+05

1.2884000E-07	-2.4098800E-09	2.0503800E-09	-8.6502200E+05
1.2887600E-07	-2.4255100E-09	2.0667800E-09	-8.6209900E+05
1.2891100E-07	-2.4365700E-09	2.0783900E-09	-8.5895200E+05
1.2894700E-07	-2.4414900E-09	2.0836300E-09	-8.5561500E+05
1.2898200E-07	-2.4392700E-09	2.0815900E-09	-8.5213900E+05
1.2901800E-07	-2.4296400E-09	2.0720900E-09	-8.4858500E+05
1.2905400E-07	-2.4130700E-09	2.0557500E-09	-8.4502300E+05
1.2908900E-07	-2.3907200E-09	2.0338400E-09	-8.4152600E+05
1.2912500E-07	-2.3643500E-09	2.0082100E-09	-8.3816100E+05
1.2916000E-07	-2.3360700E-09	1.9810200E-09	-8.3498900E+05
1.2919600E-07	-2.3081500E-09	1.9545200E-09	-8.3205000E+05
1.2923100E-07	-2.2827600E-09	1.9308100E-09	-8.2936800E+05
1.2926700E-07	-2.2617400E-09	1.9116200E-09	-8.2694700E+05
1.2930300E-07	-2.2464300E-09	1.8981100E-09	-8.2477000E+05
1.2933800E-07	-2.2375400E-09	1.8908100E-09	-8.2280100E+05
1.2937400E-07	-2.2350800E-09	1.8895700E-09	-8.2098800E+05
1.2940900E-07	-2.2384100E-09	1.8936300E-09	-8.1926800E+05
1.2944500E-07	-2.2463800E-09	1.9017300E-09	-8.1757700E+05
1.2948100E-07	-2.2574000E-09	1.9122700E-09	-8.1585000E+05
1.2951600E-07	-2.2697000E-09	1.9235400E-09	-8.1403200E+05
1.2955200E-07	-2.2814700E-09	1.9338700E-09	-8.1208000E+05
1.2958700E-07	-2.2911000E-09	1.9418400E-09	-8.0996900E+05
1.2962300E-07	-2.2972800E-09	1.9463700E-09	-8.0768500E+05
1.2965900E-07	-2.2991300E-09	1.9468200E-09	-8.0523800E+05
1.2969400E-07	-2.2962100E-09	1.9430200E-09	-8.0265200E+05
1.2973000E-07	-2.2885900E-09	1.9352200E-09	-7.9996100E+05
1.2976500E-07	-2.2767300E-09	1.9240200E-09	-7.9720900E+05
1.2980100E-07	-2.2614200E-09	1.9102800E-09	-7.9444400E+05
1.2983600E-07	-2.2436600E-09	1.8949800E-09	-7.9171400E+05
1.2987200E-07	-2.2245900E-09	1.8791500E-09	-7.8906200E+05
1.2990800E-07	-2.2053500E-09	1.8637500E-09	-7.8652700E+05
1.2994300E-07	-2.1870000E-09	1.8495900E-09	-7.8413500E+05
1.2997900E-07	-2.1704500E-09	1.8373100E-09	-7.8190100E+05
1.3001400E-07	-2.1564300E-09	1.8273300E-09	-7.7983100E+05
1.3005000E-07	-2.1454200E-09	1.8198900E-09	-7.7791900E+05
1.3008600E-07	-2.1377200E-09	1.8150400E-09	-7.7615100E+05
1.3012100E-07	-2.1334100E-09	1.8126500E-09	-7.7450500E+05
1.3015700E-07	-2.1323200E-09	1.8124900E-09	-7.7295200E+05
1.3019200E-07	-2.1341600E-09	1.8142100E-09	-7.7146400E+05
1.3022800E-07	-2.1384500E-09	1.8174100E-09	-7.7001000E+05
1.3026300E-07	-2.1445900E-09	1.8216100E-09	-7.6856900E+05
1.3029900E-07	-2.1518900E-09	1.8263000E-09	-7.6711700E+05
1.3033500E-07	-2.1595900E-09	1.8309800E-09	-7.6563900E+05
1.3037000E-07	-2.1669300E-09	1.8351400E-09	-7.6413400E+05
1.3040600E-07	-2.1731600E-09	1.8383100E-09	-7.6260400E+05
1.3044100E-07	-2.1776300E-09	1.8400900E-09	-7.6106000E+05
1.3047700E-07	-2.1797900E-09	1.8401600E-09	-7.5952000E+05
1.3051300E-07	-2.1792800E-09	1.8383400E-09	-7.5800600E+05
1.3054800E-07	-2.1759100E-09	1.8345800E-09	-7.5654300E+05
1.3058400E-07	-2.1696800E-09	1.8289700E-09	-7.5515000E+05
1.3061900E-07	-2.1608600E-09	1.8217600E-09	-7.5384500E+05

1.3065500E-07	-2.1498700E-09	1.8133300E-09	-7.5263700E+05
1.3069000E-07	-2.1372600E-09	1.8041400E-09	-7.5152300E+05
1.3072600E-07	-2.1237300E-09	1.7947400E-09	-7.5049100E+05
1.3076200E-07	-2.1100200E-09	1.7856700E-09	-7.4951500E+05
1.3079700E-07	-2.0968800E-09	1.7774600E-09	-7.4855800E+05
1.3083300E-07	-2.0849700E-09	1.7705400E-09	-7.4757800E+05
1.3086800E-07	-2.0748800E-09	1.7652700E-09	-7.4652500E+05
1.3090400E-07	-2.0670400E-09	1.7618200E-09	-7.4535000E+05
1.3094000E-07	-2.0616700E-09	1.7602400E-09	-7.4400700E+05
1.3097500E-07	-2.0588300E-09	1.7604000E-09	-7.4246100E+05
1.3101100E-07	-2.0583800E-09	1.7620300E-09	-7.4068900E+05
1.3104600E-07	-2.0599800E-09	1.7647300E-09	-7.3868700E+05
1.3108200E-07	-2.0631100E-09	1.7679600E-09	-7.3646700E+05
1.3111700E-07	-2.0671700E-09	1.7711600E-09	-7.3405900E+05
1.3115300E-07	-2.0714100E-09	1.7737000E-09	-7.3151200E+05
1.3118900E-07	-2.0750900E-09	1.7749800E-09	-7.2888900E+05
1.3122400E-07	-2.0774400E-09	1.7744600E-09	-7.2625900E+05
1.3126000E-07	-2.0778200E-09	1.7717100E-09	-7.2369300E+05
1.3129500E-07	-2.0756900E-09	1.7664500E-09	-7.2125700E+05
1.3133100E-07	-2.0707400E-09	1.7586000E-09	-7.1900800E+05
1.3136700E-07	-2.0628500E-09	1.7482800E-09	-7.1698600E+05
1.3140200E-07	-2.0522000E-09	1.7358400E-09	-7.1521100E+05
1.3143800E-07	-2.0392000E-09	1.7218300E-09	-7.1367900E+05
1.3147300E-07	-2.0245100E-09	1.7070100E-09	-7.1235900E+05
1.3150900E-07	-2.0089800E-09	1.6922400E-09	-7.1120400E+05
1.3154400E-07	-1.9935600E-09	1.6784400E-09	-7.1014600E+05
1.3158000E-07	-1.9792200E-09	1.6665000E-09	-7.0910600E+05
1.3161600E-07	-1.9668400E-09	1.6571900E-09	-7.0799800E+05
1.3165100E-07	-1.9571300E-09	1.6510500E-09	-7.0673800E+05
1.3168700E-07	-1.9505000E-09	1.6483500E-09	-7.0525000E+05
1.3172200E-07	-1.9470800E-09	1.6490000E-09	-7.0347600E+05
1.3175800E-07	-1.9465700E-09	1.6525900E-09	-7.0137800E+05
1.3179400E-07	-1.9483900E-09	1.6583400E-09	-6.9894500E+05
1.3182900E-07	-1.9516500E-09	1.6652500E-09	-6.9619000E+05
1.3186500E-07	-1.9552800E-09	1.6721100E-09	-6.9315500E+05
1.3190000E-07	-1.9581400E-09	1.6777000E-09	-6.8990700E+05
1.3193600E-07	-1.9591600E-09	1.6808700E-09	-6.8653300E+05
1.3197100E-07	-1.9574600E-09	1.6807100E-09	-6.8313400E+05
1.3200700E-07	-1.9525000E-09	1.6766300E-09	-6.7981600E+05
1.3204300E-07	-1.9441400E-09	1.6684600E-09	-6.7668000E+05
1.3207800E-07	-1.9326400E-09	1.6565100E-09	-6.7381900E+05
1.3211400E-07	-1.9186700E-09	1.6415000E-09	-6.7130800E+05
1.3214900E-07	-1.9032300E-09	1.6245100E-09	-6.6919000E+05
1.3218500E-07	-1.8875600E-09	1.6068800E-09	-6.6748200E+05
1.3222100E-07	-1.8729600E-09	1.5900700E-09	-6.6616200E+05
1.3225600E-07	-1.8606200E-09	1.5754400E-09	-6.6517800E+05
1.3229200E-07	-1.8515200E-09	1.5641600E-09	-6.6445200E+05
1.3232700E-07	-1.8462600E-09	1.5570400E-09	-6.6388200E+05
1.3236300E-07	-1.8450100E-09	1.5544200E-09	-6.6334900E+05
1.3239800E-07	-1.8474700E-09	1.5561600E-09	-6.6273600E+05
1.3243400E-07	-1.8529300E-09	1.5616300E-09	-6.6193200E+05

1.3247000E-07	-1.8603000E-09	1.5698200E-09	-6.6084200E+05
1.3250500E-07	-1.8683000E-09	1.5794200E-09	-6.5939900E+05
1.3254100E-07	-1.8755900E-09	1.5889800E-09	-6.5756900E+05
1.3257600E-07	-1.8809300E-09	1.5971100E-09	-6.5535100E+05
1.3261200E-07	-1.8833200E-09	1.6026000E-09	-6.5277900E+05
1.3264800E-07	-1.8821100E-09	1.6045600E-09	-6.4991800E+05
1.3268300E-07	-1.8771000E-09	1.6025400E-09	-6.4685700E+05
1.3271900E-07	-1.8684500E-09	1.5965100E-09	-6.4369800E+05
1.3275400E-07	-1.8567800E-09	1.5868900E-09	-6.4055100E+05
1.3279000E-07	-1.8429500E-09	1.5744300E-09	-6.3751600E+05
1.3282500E-07	-1.8279900E-09	1.5601800E-09	-6.3468400E+05
1.3286100E-07	-1.8130000E-09	1.5452600E-09	-6.3212200E+05
1.3289700E-07	-1.7989800E-09	1.5308200E-09	-6.2987200E+05
1.3293200E-07	-1.7867300E-09	1.5178400E-09	-6.2794500E+05
1.3296800E-07	-1.7767800E-09	1.5071100E-09	-6.2632200E+05
1.3300300E-07	-1.7694000E-09	1.4991100E-09	-6.2496400E+05
1.3303900E-07	-1.7645300E-09	1.4940000E-09	-6.2380700E+05
1.3307500E-07	-1.7619100E-09	1.4916700E-09	-6.2277400E+05
1.3311000E-07	-1.7610900E-09	1.4917600E-09	-6.2178100E+05
1.3314600E-07	-1.7615200E-09	1.4937100E-09	-6.2074500E+05
1.3318100E-07	-1.7626100E-09	1.4968900E-09	-6.1958900E+05
1.3321700E-07	-1.7638400E-09	1.5005900E-09	-6.1825400E+05
1.3325200E-07	-1.7647300E-09	1.5041300E-09	-6.1669700E+05
1.3328800E-07	-1.7648800E-09	1.5069000E-09	-6.1489800E+05
1.3332400E-07	-1.7640100E-09	1.5083700E-09	-6.1286000E+05
1.3335900E-07	-1.7618800E-09	1.5081200E-09	-6.1061000E+05
1.3339500E-07	-1.7583200E-09	1.5058200E-09	-6.0819600E+05
1.3343000E-07	-1.7532100E-09	1.5012700E-09	-6.0568500E+05
1.3346600E-07	-1.7464700E-09	1.4943800E-09	-6.0315100E+05
1.3350200E-07	-1.7380700E-09	1.4852200E-09	-6.0067600E+05
1.3353700E-07	-1.7280900E-09	1.4739700E-09	-5.9834000E+05
1.3357300E-07	-1.7167000E-09	1.4610000E-09	-5.9621300E+05
1.3360800E-07	-1.7042100E-09	1.4468200E-09	-5.9435200E+05
1.3364400E-07	-1.6910800E-09	1.4321100E-09	-5.9279200E+05
1.3367900E-07	-1.6779500E-09	1.4176600E-09	-5.9154500E+05
1.3371500E-07	-1.6655400E-09	1.4043400E-09	-5.9059700E+05
1.3375100E-07	-1.6545900E-09	1.3930000E-09	-5.8991200E+05
1.3378600E-07	-1.6458300E-09	1.3844300E-09	-5.8943500E+05
1.3382200E-07	-1.6398700E-09	1.3792000E-09	-5.8909600E+05
1.3385700E-07	-1.6370700E-09	1.3776400E-09	-5.8881500E+05
1.3389300E-07	-1.6375000E-09	1.3797600E-09	-5.8851200E+05
1.3392900E-07	-1.6409100E-09	1.3851800E-09	-5.8811600E+05
1.3396400E-07	-1.6467000E-09	1.3932000E-09	-5.8757000E+05
1.3400000E-07	-1.6539900E-09	1.4028400E-09	-5.8683700E+05
1.3403500E-07	-1.6616600E-09	1.4129000E-09	-5.8590200E+05
1.3407100E-07	-1.6685600E-09	1.4221200E-09	-5.8477000E+05
1.3410700E-07	-1.6735400E-09	1.4292900E-09	-5.8347300E+05
1.3414200E-07	-1.6756700E-09	1.4334100E-09	-5.8206000E+05
1.3417800E-07	-1.6743100E-09	1.4337600E-09	-5.8059000E+05
1.3421300E-07	-1.6691900E-09	1.4300200E-09	-5.7912800E+05
1.3424900E-07	-1.6604300E-09	1.4222900E-09	-5.7773600E+05

1.3428400E-07	-1.6485700E-09	1.4110500E-09	-5.7646600E+05
1.3432000E-07	-1.6344700E-09	1.3971200E-09	-5.7535600E+05
1.3435600E-07	-1.6191700E-09	1.3815500E-09	-5.7442400E+05
1.3439100E-07	-1.6038300E-09	1.3655100E-09	-5.7366800E+05
1.3442700E-07	-1.5895300E-09	1.3501600E-09	-5.7306000E+05
1.3446200E-07	-1.5772300E-09	1.3365200E-09	-5.7255700E+05
1.3449800E-07	-1.5676000E-09	1.3253800E-09	-5.7210300E+05
1.3453400E-07	-1.5610100E-09	1.3172600E-09	-5.7163200E+05
1.3456900E-07	-1.5575300E-09	1.3123800E-09	-5.7107900E+05
1.3460500E-07	-1.5569000E-09	1.3106600E-09	-5.7038100E+05
1.3464000E-07	-1.5586700E-09	1.3117500E-09	-5.6948700E+05
1.3467600E-07	-1.5622000E-09	1.3151300E-09	-5.6836200E+05
1.3471100E-07	-1.5668000E-09	1.3201600E-09	-5.6699100E+05
1.3474700E-07	-1.5717400E-09	1.3261200E-09	-5.6537900E+05
1.3478300E-07	-1.5763900E-09	1.3323100E-09	-5.6354800E+05
1.3481800E-07	-1.5801900E-09	1.3380600E-09	-5.6153400E+05
1.3485400E-07	-1.5826700E-09	1.3427700E-09	-5.5938800E+05
1.3488900E-07	-1.5834800E-09	1.3458900E-09	-5.5716900E+05
1.3492500E-07	-1.5823900E-09	1.3470100E-09	-5.5493600E+05
1.3496100E-07	-1.5792000E-09	1.3457800E-09	-5.5274600E+05
1.3499600E-07	-1.5738000E-09	1.3419500E-09	-5.5064400E+05
1.3503200E-07	-1.5661800E-09	1.3354300E-09	-5.4866600E+05
1.3506700E-07	-1.5564100E-09	1.3262400E-09	-5.4683300E+05
1.3510300E-07	-1.5446800E-09	1.3145800E-09	-5.4515300E+05
1.3513800E-07	-1.5313000E-09	1.3008700E-09	-5.4361700E+05
1.3517400E-07	-1.5167800E-09	1.2857100E-09	-5.4220200E+05
1.3521000E-07	-1.5017800E-09	1.2698800E-09	-5.4087300E+05
1.3524500E-07	-1.4871200E-09	1.2543200E-09	-5.3958900E+05
1.3528100E-07	-1.4736900E-09	1.2400300E-09	-5.3830800E+05
1.3531600E-07	-1.4624000E-09	1.2280200E-09	-5.3698500E+05
1.3535200E-07	-1.4540600E-09	1.2191500E-09	-5.3558200E+05
1.3538800E-07	-1.4492800E-09	1.2140600E-09	-5.3406600E+05
1.3542300E-07	-1.4483800E-09	1.2130300E-09	-5.3242100E+05
1.3545900E-07	-1.4512600E-09	1.2159500E-09	-5.3064600E+05
1.3549400E-07	-1.4574000E-09	1.2222700E-09	-5.2875200E+05
1.3553000E-07	-1.4658900E-09	1.2310400E-09	-5.2676400E+05
1.3556500E-07	-1.4754700E-09	1.2409900E-09	-5.2472400E+05
1.3560100E-07	-1.4846700E-09	1.2506400E-09	-5.2268100E+05
1.3563700E-07	-1.4919500E-09	1.2585000E-09	-5.2069400E+05
1.3567200E-07	-1.4959600E-09	1.2632400E-09	-5.1881900E+05
1.3570800E-07	-1.4956400E-09	1.2638600E-09	-5.1710700E+05
1.3574300E-07	-1.4904300E-09	1.2598300E-09	-5.1560200E+05
1.3577900E-07	-1.4803900E-09	1.2512100E-09	-5.1433200E+05
1.3581500E-07	-1.4661500E-09	1.2386200E-09	-5.1331100E+05
1.3585000E-07	-1.4489400E-09	1.2232100E-09	-5.1253200E+05
1.3588600E-07	-1.4304300E-09	1.2065300E-09	-5.1196900E+05
1.3592100E-07	-1.4125100E-09	1.1903000E-09	-5.1157700E+05
1.3595700E-07	-1.3970900E-09	1.1762500E-09	-5.1130400E+05
1.3599200E-07	-1.3858300E-09	1.1658400E-09	-5.1108500E+05
1.3602800E-07	-1.3799200E-09	1.1600700E-09	-5.1085600E+05
1.3606400E-07	-1.3799100E-09	1.1593700E-09	-5.1055400E+05

1.3609900E-07	-1.3856200E-09	1.1635000E-09	-5.1012500E+05
1.3613500E-07	-1.3961700E-09	1.1715900E-09	-5.0952700E+05
1.3617000E-07	-1.4100400E-09	1.1822500E-09	-5.0873800E+05
1.3620600E-07	-1.4253000E-09	1.1937500E-09	-5.0775300E+05
1.3624200E-07	-1.4398400E-09	1.2042700E-09	-5.0658300E+05
1.3627700E-07	-1.4516200E-09	1.2121300E-09	-5.0525800E+05
1.3631300E-07	-1.4589700E-09	1.2160200E-09	-5.0381800E+05
1.3634800E-07	-1.4608000E-09	1.2151900E-09	-5.0231500E+05
1.3638400E-07	-1.4567000E-09	1.2095300E-09	-5.0080500E+05
1.3641900E-07	-1.4470300E-09	1.1996300E-09	-4.9934300E+05
1.3645500E-07	-1.4328300E-09	1.1866100E-09	-4.9797800E+05
1.3649100E-07	-1.4156600E-09	1.1720200E-09	-4.9674900E+05
1.3652600E-07	-1.3974300E-09	1.1576200E-09	-4.9568000E+05
1.3656200E-07	-1.3801100E-09	1.1450900E-09	-4.9478300E+05
1.3659700E-07	-1.3654400E-09	1.1358300E-09	-4.9405300E+05
1.3663300E-07	-1.3547800E-09	1.1307800E-09	-4.9346600E+05
1.3666900E-07	-1.3489000E-09	1.1302700E-09	-4.9298600E+05
1.3670400E-07	-1.3479300E-09	1.1340000E-09	-4.9256200E+05
1.3674000E-07	-1.3513400E-09	1.1411100E-09	-4.9214000E+05
1.3677500E-07	-1.3581000E-09	1.1502900E-09	-4.9166200E+05
1.3681100E-07	-1.3668400E-09	1.1600100E-09	-4.9107100E+05
1.3684600E-07	-1.3760000E-09	1.1686800E-09	-4.9031700E+05
1.3688200E-07	-1.3841000E-09	1.1749300E-09	-4.8936000E+05
1.3691800E-07	-1.3899600E-09	1.1777500E-09	-4.8817800E+05
1.3695300E-07	-1.3927600E-09	1.1766100E-09	-4.8676700E+05
1.3698900E-07	-1.3921700E-09	1.1715400E-09	-4.8514000E+05
1.3702400E-07	-1.3883800E-09	1.1630400E-09	-4.8332500E+05
1.3706000E-07	-1.3819300E-09	1.1520400E-09	-4.8137000E+05
1.3709600E-07	-1.3736800E-09	1.1397200E-09	-4.7933300E+05
1.3713100E-07	-1.3646400E-09	1.1273300E-09	-4.7728100E+05
1.3716700E-07	-1.3557800E-09	1.1160200E-09	-4.7528400E+05
1.3720200E-07	-1.3478900E-09	1.1067300E-09	-4.7340700E+05
1.3723800E-07	-1.3415400E-09	1.1000200E-09	-4.7170700E+05
1.3727300E-07	-1.3369800E-09	1.0960900E-09	-4.7023100E+05
1.3730900E-07	-1.3341400E-09	1.0948000E-09	-4.6900900E+05
1.3734500E-07	-1.3327300E-09	1.0956900E-09	-4.6805200E+05
1.3738000E-07	-1.3323100E-09	1.0981400E-09	-4.6735500E+05
1.3741600E-07	-1.3323900E-09	1.1014800E-09	-4.6689200E+05
1.3745100E-07	-1.3325300E-09	1.1050800E-09	-4.6661900E+05
1.3748700E-07	-1.3324400E-09	1.1084800E-09	-4.6648500E+05
1.3752300E-07	-1.3320000E-09	1.1114000E-09	-4.6642900E+05
1.3755800E-07	-1.3312600E-09	1.1137800E-09	-4.6638400E+05
1.3759400E-07	-1.3304300E-09	1.1157200E-09	-4.6628600E+05
1.3762900E-07	-1.3298000E-09	1.1174400E-09	-4.6607500E+05
1.3766500E-07	-1.3296300E-09	1.1191700E-09	-4.6570700E+05
1.3770000E-07	-1.3300900E-09	1.1210500E-09	-4.6514800E+05
1.3773600E-07	-1.3312200E-09	1.1231100E-09	-4.6438200E+05
1.3777200E-07	-1.3328500E-09	1.1252000E-09	-4.6341000E+05
1.3780700E-07	-1.3346600E-09	1.1270200E-09	-4.6225100E+05
1.3784300E-07	-1.3361800E-09	1.1281300E-09	-4.6094300E+05
1.3787800E-07	-1.3369000E-09	1.1280600E-09	-4.5953300E+05

1.3791400E-07	-1.3363300E-09	1.1263600E-09	-4.5808000E+05
1.3795000E-07	-1.3341500E-09	1.1227700E-09	-4.5664300E+05
1.3798500E-07	-1.3302200E-09	1.1172000E-09	-4.5528000E+05
1.3802100E-07	-1.3246700E-09	1.1098600E-09	-4.5404400E+05
1.3805600E-07	-1.3179100E-09	1.1012300E-09	-4.5297300E+05
1.3809200E-07	-1.3105800E-09	1.0920400E-09	-4.5209100E+05
1.3812700E-07	-1.3034600E-09	1.0831500E-09	-4.5140300E+05
1.3816300E-07	-1.2973600E-09	1.0754800E-09	-4.5089200E+05
1.3819900E-07	-1.2930400E-09	1.0698700E-09	-4.5052700E+05
1.3823400E-07	-1.2910300E-09	1.0669500E-09	-4.5026300E+05
1.3827000E-07	-1.2915800E-09	1.0670500E-09	-4.5004300E+05
1.3830500E-07	-1.2946000E-09	1.0701300E-09	-4.4980400E+05
1.3834100E-07	-1.2996400E-09	1.0757800E-09	-4.4948400E+05
1.3837700E-07	-1.3059400E-09	1.0832300E-09	-4.4902700E+05
1.3841200E-07	-1.3125100E-09	1.0914700E-09	-4.4839000E+05
1.3844800E-07	-1.3182800E-09	1.0993600E-09	-4.4754600E+05
1.3848300E-07	-1.3222000E-09	1.1057700E-09	-4.4648500E+05
1.3851900E-07	-1.3234300E-09	1.1097100E-09	-4.4521600E+05
1.3855400E-07	-1.3213800E-09	1.1104500E-09	-4.4376400E+05
1.3859000E-07	-1.3158600E-09	1.1076600E-09	-4.4217100E+05
1.3862600E-07	-1.3070800E-09	1.1013800E-09	-4.4049100E+05
1.3866100E-07	-1.2956200E-09	1.0920500E-09	-4.3878000E+05
1.3869700E-07	-1.2823600E-09	1.0804500E-09	-4.3709900E+05
1.3873200E-07	-1.2684100E-09	1.0675900E-09	-4.3550000E+05
1.3876800E-07	-1.2549400E-09	1.0546100E-09	-4.3403000E+05
1.3880400E-07	-1.2430400E-09	1.0426100E-09	-4.3272100E+05
1.3883900E-07	-1.2336500E-09	1.0325800E-09	-4.3159100E+05
1.3887500E-07	-1.2274000E-09	1.0252300E-09	-4.3064300E+05
1.3891000E-07	-1.2245800E-09	1.0210000E-09	-4.2986000E+05
1.3894600E-07	-1.2251300E-09	1.0199500E-09	-4.2921800E+05
1.3898200E-07	-1.2286600E-09	1.0218500E-09	-4.2867800E+05
1.3901700E-07	-1.2345000E-09	1.0261600E-09	-4.2819600E+05
1.3905300E-07	-1.2418000E-09	1.0321600E-09	-4.2772500E+05
1.3908800E-07	-1.2496300E-09	1.0390300E-09	-4.2721900E+05
1.3912400E-07	-1.2571200E-09	1.0459100E-09	-4.2663900E+05
1.3915900E-07	-1.2634600E-09	1.0520600E-09	-4.2595500E+05
1.3919500E-07	-1.2680900E-09	1.0568500E-09	-4.2514800E+05
1.3923100E-07	-1.2706300E-09	1.0598700E-09	-4.2421300E+05
1.3926600E-07	-1.2709600E-09	1.0609100E-09	-4.2315800E+05
1.3930200E-07	-1.2692000E-09	1.0599700E-09	-4.2200500E+05
1.3933700E-07	-1.2656200E-09	1.0572400E-09	-4.2078700E+05
1.3937300E-07	-1.2606200E-09	1.0530200E-09	-4.1954900E+05
1.3940900E-07	-1.2547100E-09	1.0477200E-09	-4.1834200E+05
1.3944400E-07	-1.2483400E-09	1.0417700E-09	-4.1721500E+05
1.3948000E-07	-1.2419600E-09	1.0355800E-09	-4.1621800E+05
1.3951500E-07	-1.2359300E-09	1.0295400E-09	-4.1539600E+05
1.3955100E-07	-1.2305000E-09	1.0239200E-09	-4.1478100E+05
1.3958600E-07	-1.2258200E-09	1.0189700E-09	-4.1439300E+05
1.3962200E-07	-1.2219700E-09	1.0148400E-09	-4.1423400E+05
1.3965800E-07	-1.2189500E-09	1.0116100E-09	-4.1428900E+05
1.3969300E-07	-1.2167200E-09	1.0093300E-09	-4.1452500E+05

1.3972900E-07	-1.2152200E-09	1.0080200E-09	-4.1489500E+05
1.3976400E-07	-1.2144200E-09	1.0076900E-09	-4.1534400E+05
1.3980000E-07	-1.2142800E-09	1.0083200E-09	-4.1580500E+05
1.3983600E-07	-1.2148100E-09	1.0099000E-09	-4.1621100E+05
1.3987100E-07	-1.2160000E-09	1.0123800E-09	-4.1650300E+05
1.3990700E-07	-1.2178300E-09	1.0156800E-09	-4.1663300E+05
1.3994200E-07	-1.2202800E-09	1.0196500E-09	-4.1656900E+05
1.3997800E-07	-1.2232500E-09	1.0241100E-09	-4.1629700E+05
1.4001300E-07	-1.2265600E-09	1.0287600E-09	-4.1582700E+05
1.4004900E-07	-1.2299800E-09	1.0332700E-09	-4.1518500E+05
1.4008500E-07	-1.2331700E-09	1.0372400E-09	-4.1441700E+05
1.4012000E-07	-1.2357600E-09	1.0402200E-09	-4.1357900E+05
1.4015600E-07	-1.2373600E-09	1.0418100E-09	-4.1273200E+05
1.4019100E-07	-1.2375900E-09	1.0416400E-09	-4.1193200E+05
1.4022700E-07	-1.2361500E-09	1.0394600E-09	-4.1122900E+05
1.4026300E-07	-1.2328700E-09	1.0351400E-09	-4.1065600E+05
1.4029800E-07	-1.2277100E-09	1.0287500E-09	-4.1022600E+05
1.4033400E-07	-1.2208400E-09	1.0205600E-09	-4.0993400E+05
1.4036900E-07	-1.2126300E-09	1.0110200E-09	-4.0975300E+05
1.4040500E-07	-1.2035900E-09	1.0007600E-09	-4.0964000E+05
1.4044000E-07	-1.1943800E-09	9.9050800E-10	-4.0953800E+05
1.4047600E-07	-1.1857400E-09	9.8107800E-10	-4.0938800E+05
1.4051200E-07	-1.1783900E-09	9.7323500E-10	-4.0913400E+05
1.4054700E-07	-1.1729900E-09	9.6764100E-10	-4.0872600E+05
1.4058300E-07	-1.1699800E-09	9.6477200E-10	-4.0812700E+05
1.4061800E-07	-1.1696200E-09	9.6485100E-10	-4.0732100E+05
1.4065400E-07	-1.1718500E-09	9.6780800E-10	-4.0631100E+05
1.4069000E-07	-1.1763300E-09	9.7326600E-10	-4.0512100E+05
1.4072500E-07	-1.1824400E-09	9.8057200E-10	-4.0379100E+05
1.4076100E-07	-1.1893600E-09	9.8884400E-10	-4.0237400E+05
1.4079600E-07	-1.1961000E-09	9.9706300E-10	-4.0093400E+05
1.4083200E-07	-1.2016700E-09	1.0041800E-09	-3.9953600E+05
1.4086700E-07	-1.2051800E-09	1.0092300E-09	-3.9824400E+05
1.4090300E-07	-1.2059000E-09	1.0114400E-09	-3.9711200E+05
1.4093900E-07	-1.2034300E-09	1.0103300E-09	-3.9618200E+05
1.4097400E-07	-1.1977400E-09	1.0058100E-09	-3.9547800E+05
1.4101000E-07	-1.1891600E-09	9.9814300E-10	-3.9500800E+05
1.4104500E-07	-1.1784000E-09	9.8797200E-10	-3.9475900E+05
1.4108100E-07	-1.1664700E-09	9.7625100E-10	-3.9470500E+05
1.4111700E-07	-1.1545200E-09	9.6413000E-10	-3.9479900E+05
1.4115200E-07	-1.1437900E-09	9.5283200E-10	-3.9498300E+05
1.4118800E-07	-1.1354000E-09	9.4350900E-10	-3.9519400E+05
1.4122300E-07	-1.1302100E-09	9.3709700E-10	-3.9536500E+05
1.4125900E-07	-1.1287300E-09	9.3419300E-10	-3.9543400E+05
1.4129400E-07	-1.1310300E-09	9.3498600E-10	-3.9534800E+05
1.4133000E-07	-1.1367200E-09	9.3921700E-10	-3.9506600E+05
1.4136600E-07	-1.1449800E-09	9.4620600E-10	-3.9456400E+05
1.4140100E-07	-1.1546700E-09	9.5493000E-10	-3.9384000E+05
1.4143700E-07	-1.1644500E-09	9.6414600E-10	-3.9291200E+05
1.4147200E-07	-1.1729600E-09	9.7254600E-10	-3.9181700E+05
1.4150800E-07	-1.1789600E-09	9.7891700E-10	-3.9060800E+05

1.4154400E-07	-1.1815100E-09	9.8229600E-10	-3.8934900E+05
1.4157900E-07	-1.1800900E-09	9.8209800E-10	-3.8811100E+05
1.4161500E-07	-1.1746800E-09	9.7818200E-10	-3.8696600E+05
1.4165000E-07	-1.1657300E-09	9.7087900E-10	-3.8597600E+05
1.4168600E-07	-1.1541400E-09	9.6094400E-10	-3.8518900E+05
1.4172100E-07	-1.1411400E-09	9.4946100E-10	-3.8463600E+05
1.4175700E-07	-1.1281300E-09	9.3770400E-10	-3.8432400E+05
1.4179300E-07	-1.1165000E-09	9.2697400E-10	-3.8423500E+05
1.4182800E-07	-1.1074700E-09	9.1843400E-10	-3.8433000E+05
1.4186400E-07	-1.1019700E-09	9.1296900E-10	-3.8455300E+05
1.4189900E-07	-1.1004700E-09	9.1107400E-10	-3.8482600E+05
1.4193500E-07	-1.1029900E-09	9.1280300E-10	-3.8506400E+05
1.4197100E-07	-1.1090600E-09	9.1776400E-10	-3.8518200E+05
1.4200600E-07	-1.1178300E-09	9.2519000E-10	-3.8510100E+05
1.4204200E-07	-1.1281800E-09	9.3403100E-10	-3.8475700E+05
1.4207700E-07	-1.1388000E-09	9.4309200E-10	-3.8410200E+05
1.4211300E-07	-1.1484000E-09	9.5117900E-10	-3.8311200E+05
1.4214800E-07	-1.1558600E-09	9.5723900E-10	-3.8179200E+05
1.4218400E-07	-1.1602900E-09	9.6047600E-10	-3.8017600E+05
1.4222000E-07	-1.1611400E-09	9.6042500E-10	-3.7831900E+05
1.4225500E-07	-1.1582300E-09	9.5699300E-10	-3.7629800E+05
1.4229100E-07	-1.1517800E-09	9.5044000E-10	-3.7420200E+05
1.4232600E-07	-1.1422700E-09	9.4133400E-10	-3.7212800E+05

12.4 Summary

In this chapter, a computer program based on the numerical scheme – FDTD method has been presented for simulating an EMP environment. This will help the users of this book to develop computer code for simulating the desired EMP phenomena. For example, the users can change the EMP environment by changing the charging and discharging time of the capacitor bank, switch closure time, and the device under test, etc.

References

1 Longmire, C.L. (1978). On the electromagnetic pulse produced by nuclear explosions. *IEEE Trans. Antennas Propag.* AP-26 (1): 3–13.

2 Pandey, D.C. (1995). Five days intensive course on HPE, *organized by LRDE, Bangalore*, March 1995.

3 Ghose, R.N. (1984). *EMP Environment and System Hardness Design.* Gainesville, VA: Interference Control Technologies, Inc..

4 Pande, D.C. (1995). Nuclear electromagnetic pulse: hardness and assurance. *Intensive Course on High Power Electromagnetics (HPE-95)*, Volume 2.

5 Mar, M.H. (1995). Electromagnetic pulse (EMP) coupling codes for use with the vulnerability/lethality (V/L) taxonomy. *U.S. Army Res. Lab. ARL* DTIC 19950830064 (ARL-TR-786): 8.

6 Florig, H.K. (1998). The future battlefield: a blast of gigawatts? *IEEE Spectr.* 25 50–54.

7 Ying, W., Chemerys, V., and Dongsheng, X. (1999). The multi-function EMP simulator. *Proceedings of the 12th IEEE Pulsed Power Conference*, Monterey, USA (27–30 June 1999).

8 Baum, C.E. (1978). EMP simulators for various types of nuclear EMP environments: an interim categorization. *IEEE Trans. Antennas Propag.* AP-26 (1): 35–53.

9 Schilling, H., Schluter, J., Peters, M. et al. (1995). High voltage generator with fast risetime for EMP simulation. *Proceedings of the 10th IEEE International Pulsed Power Conference, Albuquerque, NM*, pp. 1359–1364.

10 Shen, H.M., King, R.W.P., and Wu, T.T. (1987). The exciting mechanism of the parallel-plate EMP simulator. *IEEE Trans. Electromagn. Compat.* 29 (1): 32–39.

11 Giri, D.V., Liu, T.K., and Tesche, F.M. (1980). Parallel plate transmission line type of EMP simulators: a systematic review and recommendations. *Sensor and Simulation Notes*, vol. SSN 261.

12 Bardet, C., Dafif, O., and Jecko, B. (1987). Time-domain analysis of a large EMP simulator. *IEEE Trans. Electromagn. Compat.* EMC-29 (1): 40–48.

Electromagnetic Pulse Simulations Using Finite-Difference Time-Domain Method, First Edition.
Shahid Ahmed.

13 Hoo, K.M.S. (1975). Numerical analysis of a transmission line EMP simulator. *IEEE Trans. Electromagn. Compat.* EMC-17 (2): 85–92.

14 Miller, E.K., Poggio, A.J., and Bruke, G.J. (1973). An integrodifferential equation technique for time domain analysis of thin wire structures. *J. Comput. Phys.* 12 (24): 48.

15 Shen, H.-m., King, R.P., and Wu, T. (1983). Theoretical analysis of the rhombic simulator under pulse excitation. *IEEE Trans. Electromagn. Compat.* EMC-25 (1): 47–55.

16 Klaasen, J.J.A. (1992). An efficient method for the performance analysis of bounded-wave NEMP simulator. *Sensor and Simulation Notes* vol. SSN 345.

17 King, R.W.P. and Blejer, D.J. (1979). The electromagnetic field in an EMP simulator at a high frequency. *IEEE Trans. Electromagn. Compat.* EMC-21 (3): 263–269.

18 Shlager, K.L., Smith, G.S., and Maloney, J.G. (1996). Accurate analysis of TEM horn antenna for pulse radiation. *IEEE Trans. Electromagn. Compat.* 38 (3): 414–423.

19 Utton, T.S. (1999). Ultra-wideband TEM horns, transient arrays and exponential curves: an FDTD look. Master's thesis. Air Force Institute of Technology, WPAFB, USA, *Tech. Report. AFIT/GE/ENG/99M-29.*

20 Yang, F.C. and Lee, K.S.H. (1977). Energy confinement of a bounded-wave EMP simulator. *Sensor and Simulation Notes,* vol. SSN 230.

21 Baum, C.E. (1972). General principles for the design of Atlas I and II, Part IV: additional considerations for the design of pulsar arrays. *Sensor and Simulation Notes,* vol. SSN 146.

22 Sancer, M.I. and Varvatsis, A.D. (1973). The reduction of waste energy delivered outside of a parallel-plate simulator fed by a sloped source array. *Sensor and Simulation Notes,* vol. SSN 162.

23 Robertson, R.C. and Morgan, M.A. (1995). Ultra-wideband impulse receiving antenna design and evaluation. In: *Ultra-Wideband Short-Pulse Electromagnetics 2* (ed. L. Carin), 179–186. New York: Plenum Press.

24 Kanda, M. (1982). The effects of resistive loading of "TEM" horns. *IEEE Trans. Electromagn. Compat.* EMC-24 (2): 245–255.

25 Wu, T.T. and King, R.W.P. (1965). The cylindrical antenna with non-reflecting resistive loading. *IEEE Trans. Antennas Propag.* AP-13: 369–373.

26 Baum, C.E. (1968). Admittance sheets for terminating high-frequency transmission lines. *Sensor and Simulation Notes, Air Force Institute of Technology and California Institute of Technology,* vol. SSN 53.

27 Baum, C.E. (1995). Low frequency compensated TEM horn. *Sensor and Simulation Notes,* vol. SSN-377.

28 Rushdi, A.M., Menendez, R.C., Mittra, R., and Lee, S.-W. (1978). Leaky modes in parallel-plate EMP simulators. *IEEE Trans. Electromagn. Compat.* EMC-20 (3): 444–451.

29 Giri, D.V., Baum, C.E., and Schilling, H. (1978). Electromagnetic considerations of a spatial modal filter for suppression of non-TEM modes in the transmission-line type of EMP simulators. *Sensor and simulation Notes*, vol. SSN 247.

30 Wagner, C.L. and Schneider, J.B. (1998). Divergent fields, charge and capacitance in FDTD simulations. *IEEE Trans. Microwave Theory Tech.* 46 (12): 1231–1236.

31 Labratories, B. (1975). *EMP Engineering and Design Principles*. Whippany, NJ: Electrical Protection Department, Bell Telephone Labratories.

32 Kunz, K.S. and Luebbers, R.J. (1993). *The Finite-Difference Time-Domain Method for Electromagnetics*. CRC Press.

33 Taflove, A. and Hagness, S.C. (2000). *Computational Electrodynamics: The Finite-Difference Time-Domain Method*. Artech House, Inc..

34 Yee, K.S. (1966). Numerical solution of initial boundary value problems involving Maxwell's equations in isotropic media. *IEEE Trans. Antennas Propag.* AP-14: 302–307.

35 Steich, D.J. (1995). Local outer boundary condition for the finite-difference time-domain method applied to Maxwell's equations. PhD thesis. Pennsylvania State University, USA.

36 Berenger, J.P. (1994). A perfectly matched layer (PML) for the absorption of electromagnetic waves. *J. Comput. Phys.* 114: 185–200.

37 I. The Mathworks (2019). MATLAB version 9.7.0.1261785 (R2019b).

38 GNU Octave (2019). OCTAVE version 5.1 (2019). https://www.gnu.org/software/octave/.

39 Ahmed, S., Sharma, D., and Chaturvedi, S. (2003). Modeling of an EMP simulator using a 3-D FDTD code. *IEEE Trans. Plasma Sci.* 31 (2): 207–215.

40 Rao, N.N. (1994). *Elements of Engineering Electromagnetics*. Englewood Cliffs, NJ: Prentice Hall.

41 Woan, G. (2000). *The Cambridge Handbook of Physics Formulas*. Cambridge University Press.

42 Morgan, M.A. and Robertson, R.C. (1997). Optimized TEM horn impulse receiving antenna. In: C.E. Baum, L. Carin, A.P. Stone. *Ultra-Wideband, Short-Pulse Electromagnetics 3*, 121–129. New York: Plenum Press.

43 Dick, A.R., MacGregor, S.J., McPhee, A.J. et al. (1998). An investigation into high speed gas breakdown. *23rd International Power Modulator Symposium, California, USA*, pp. 202–205.

44 Collin, R.E. (1985). *Antennas and Radiowave Propagation*. McGraw-Hill Book Company.

45 Luebbers, R., Kunz, K., Schneider, M., and Hunsberger, F. (1991). A finite-difference time-domain near zone to far zone transformation. *IEEE Trans. Antennas Propag.* 39 (4): 429–433.

46 Luebber, R., Ryan, D., and Beggs, J. (1992). A two-dimensional time domain near zone to far zone transformation. *IEEE Trans. Antennas Propag.* 40 (7): 848.

47 Taflove, A. and Umashankar, K. (1983). Radar cross-section of general three dimensional scatterer. *IEEE Electromagn. Compat.* 25 (4): 433–440.

48 Nardone, C. (1992). Multichannel fluctuation data analysis by the singular value decomposition method. Application to MHD Modes in JET. *Plasma Phys. Controlled Fusion* 34 (9): 1447–1465.

49 Huang, N.E., Shen, Z., Long, S.R. et al. (1998). The empirical mode decomposition and the Hilbert spectrum for nonlinear and non-stationary time series analysis. *Proc. R. Soc. London, Ser. A* 454: 903–995. (Printed in Great Britain).

50 Gupta, S.L. and Kumar, V. (1994). *Handbook of Electronics.* Meerut: Pragati Prakashan.

51 Press, W.H., Vetterling, W.T., Teukolsky, S.A., Flannery, B.P. (1998). *Numerical Recipes in Fortran.* Cambridge University Press.

52 Shlager, K.L. (1995). The analysis and optimization of bow-tie and TEM-horn antennas for pulse radiation using the finite-difference time-domain method. PhD dissertation. Atlanta, GA: Georgia Institute of Technology.

53 Urrutia, J.M. and Stenzel, R.L. (2000). Electron magnetohydrodynamic turbulence in a high-beta plasma. II. Single point fluctuation measurements. *Phys. Plasmas* 7 (11): 4457–4465.

54 Shoenbach, K.H., Joshi, R.P., Kolb, J.F. et al. (2004). Ultrashort electrical pulses open a new gateway into biological cells. *Proc. IEEE* 92 (7): 1122–1137.

55 Chen, J.-Y., Furse, C.M., and Gandhi, O.P. (1994). A simple convolution procedure for calculating currents induced in the human body for exposure to electromagnetic pulse. *IEEE Trans. Microwave Theory Tech.* 42 (7): 1172–1175.

56 Miklavcic, D., Pavselj, N., and Hart, F.X. (2006). Electrical properties of tissues. In: M. Akay. *Wiley Encyclopedia of Biological Engineering*, 1–12. Wiley.

Index

Printed and bound by CPI Group (UK) Ltd, Croydon, CR0 4YY

17/09/2026

14964317-0004